American Moonshot

ALSO BY DOUGLAS BRINKLEY

Rightful Heritage:
Franklin D. Roosevelt and the Land of America

The Quiet World:
Saving Alaska's Wilderness Kingdom, 1879–1960

The Wilderness Warrior:
Theodore Roosevelt and the Crusade for America

Cronkite

The Reagan Diaries

The Nixon Tapes

The Great Deluge: Hurricane Katrina,
New Orleans, and the Mississippi Gulf Coast

Wheels for the World: Henry Ford, His Company, and a Century of Progress, 1903–2003

John F. Kennedy and Europe (editor)

The Unfinished Presidency: Jimmy Carter's Journey Beyond the White House

Rosa Parks: A Life

Driven Patriot: The Life and Times of James Forrestal

FDR and the Creation of the U.N. (with Townsend Hoopes)

Dean Acheson: The Cold War Years

A new era in human history began on July 20, 1969, when Apollo 11 astronauts Neil Armstrong and Edwin "Buzz" Aldrin walked on the moon. No longer were global citizens shackled to Earth. It was John F. Kennedy's vision for an American moonshot that jump-started this epic NASA feat. Photos of moon prints became popular in newspapers around the world.

American Moonshot

John F. Kennedy and the Great Space Race

Douglas Brinkley

HarperCollins books may be purchased for educational, business, or sales promotional use. For information, please e-mail the Special Markets Department at SPsales@harpercollins.com.

FIRST HARPERLUXE EDITION

ISBN: 978-0-06-285991-4

HarperLuxe™ is a trademark of HarperCollins Publishers.

Library of Congress Cataloging-in-Publication Data is available upon request.

19 20 21 22 23 ID/LSC 10 9 8 7 6 5 4 3 2 1

To my colleagues at Rice University and
to my beloved wife, Anne

Contents

Preface:
Kennedy's New Ocean

The most difficult thing is the decision to act. The rest is merely tenacity.

—AMELIA EARHART

It will not be one man going to the moon . . . it will be an entire nation. For all of us must work to put him there.

—PRESIDENT JOHN F. KENNEDY, MAY 25, 1961

Even the White House ushers were abuzz on the morning of October 10, 1963, because President John F. Kennedy was honoring the Mercury Seven—astronauts Lieutenant Scott Carpenter (USN), Captain Leroy "Gordo" Cooper (USAF), Lieutenant Colonel

John Glenn (USMC), Captain Virgil "Gus" Grissom (USAF), Lieutenant Commander Walter "Wally" Schirra (USN), Lieutenant Alan Shepard (USN), and Captain Donald "Deke" Slayton (USAF)—with the coveted Collier Trophy that afternoon in a Rose Garden affair. (Robert J. Collier had been an editor of *Collier's Weekly* in the early twentieth century; he promoted the careers of Orville and Wilbur Wright, believing deeply that flight was going to revolutionize transportation.) The trophy had been established in 1911 to be presented annually for "the greatest achievement in aeronautics in America," with a bent toward military aviation. At the Mercury ceremony were representatives from such Project Mercury aerospace contractors as McDonnell Aircraft Corporation (designers of the capsule) and Chrysler Corporation (which fabricated the Redstone rockets for the U.S. Army's missile team in Huntsville, Alabama). Kennedy wanted to personally congratulate the "Magnificent Seven" astronauts, all household names, for their intrepid service to the country. And his remarks marked the end of the Mercury projects after six successful space missions.

At the formal ceremony, Kennedy, in a fun-loving, jaunty mood, full of gregariousness and humor, presented the flyboy legends with the prize. It was the first occasion for all seven spacemen and their wives to be together at the White House since the maiden astronaut, Alan Shepard, accepted a Distinguished Service Award for his Mercury suborbital flight of fifteen minutes to an altitude of 116.5 miles on May 5, 1961. Surrounding Kennedy as he spoke were such aviation history dignitaries as Jimmy Doolittle, Jackie Cochran, and Hugh Dryden. Instead of recounting the Mercury Seven's space exploits in rote fashion, Kennedy used the opportunity to drive home his brazen pledge of 1961, that the United States would place an astronaut on the moon by the decade's end. Scoffing at critics of Project Apollo (NASA's moonshot program) as being as thickheaded as those fools who laughed at the Wright brothers in 1903 before the Kitty Hawk flights, he turned visionary. "Some of us may dimly perceive where we are going and may not feel this is of the greatest prestige to us," Kennedy said. "I am confident that its significance, its uses and benefits will become as obvious as the *Sputnik* satellite is to us, as the airplane is to us. I hope this award, which in effect closes out the particular phase of the program, will be a stimulus to them and to the other astronauts who will carry our flag to the moon and perhaps someday, beyond."

For Kennedy, much depended on the United States going to the moon, beating the Soviet Union, being *first*, winning the Cold War in the name of democracy and freedom, and planting the American flag on the lunar surface. Just five weeks later, Kennedy was assassinated in Dallas, Texas. Writing the president's obituary in *Aviation Week & Space Technology* on December 2, 1963, editor Robert Hotz, who had been at the Collier Trophy ceremony that October, predicted that when a NASA astronaut walked on the moon in less than six years' time, Kennedy, America's thirty-fifth president, would be honored as a spacefaring seer whose eternal marching command to his fellow countrymen was "Forward!"

Even though Kennedy wasn't alive for the fulfillment of his May 25, 1961, pledge to a joint session of Congress to land a "man on the moon" and return him safely to Earth, the marvel of television made it possible for more than a half-billion people to watch the historic *Apollo 11* mission in real time, and I was one of them. On July 20, 1969, when Neil Armstrong gingerly descended from the spider-like lunar module the *Eagle* with his hefty backpack and bulky space suit, becoming the first human on the moon, I cheered like a banshee. I was only eight years old that summer, and watching all things Apollo 11—from the nearly two-hundred-hour galactic journey out of the Space Coast of Florida to splashdown in the Pacific Ocean—became my obsession. I didn't miss a moment of the long, nerve-racking chain of events that led to the *Eagle* establishing the moon base Sea of Tranquility (named in advance by Armstrong). I vividly remember our astronauts planting the American flag on the lunarscape, bouncing on the desolate moon's surface, handling instruments, and procuring moon rocks.

My family lived in Perrysburg, Ohio, and we considered Armstrong, from the nearby community of Wapakoneta, essentially a hometown boy. It was stunning that this local kid, who grew up on an Auglaize County farm with no electricity, was leading America into the new world of lunar exploration. When Armstrong said, "That's one small step for [a] man, one giant leap for mankind," every member of my family was awed at the instantaneous greatness of it all. We were hardly alone in realizing that Apollo 11 had changed all who watched it unfold or lived in its wake. I was proud of my country.

For years I longed to hear Neil Armstrong describe what it was like to contemplate Earth from 238,900 miles away, to explain, in his own words, the thermodynamics affecting motion through the atmosphere both in launching and reentry. Former Johnson Space Center director George Abbey of Houston (now a colleague of mine at Rice University) once told me that many NASA astronauts felt that looking at Earth was

John F. Kennedy was handsome, debonair, and press savvy. Often, when he visited Cape Canaveral, Florida, or Huntsville, Alabama, or Houston, Texas, to inspect NASA sites, he wore dark sunglasses, which gave the visits a touch of Hollywood glamour. Because he was six feet one in height, sitting in a cramped Mercury, Gemini, or Apollo capsule for a photo op was not an option. So he mastered the art of looking upward at rockets.

akin to a religious experience. Did Armstrong agree? What did it feel like, emotionally, spiritually, to stand on the surface of the moon? Armstrong's reticence was legendary. He was known to be media shy. But I hoped to persuade him to talk with me about his storied career. Perhaps I could get him to reflect in fresh ways on his lunar experience. In 1993, I wrote him requesting an interview (enclosing signed copies of my books *Dean Acheson: The Cold War Years* and *Driven Patriot: The Life and Times of James Forrestal*). I got a polite postcard rejection of the "Not now, but I'll keep you in mind" variety.

It wasn't until eight years later that NASA afforded me the privilege of interviewing Neil Armstrong for its official Oral History Project. I was surprised at and honored by the chance to speak in depth with the "First Man"—and thrilled when the date was set for September 19, 2001, in Clear Lake City, Texas. Then, eight days in advance of the big meeting, I saw the horrifying collapse of the World Trade Center towers on TV and listened to accounts of the two other disastrous airplane hijackings. A pervasive sense of gloom and urgency enveloped America. Like everyone else, I felt shock and repugnance at the ghastly scenes of our nation under attack, feelings that still burn to this day. I was sure my Armstrong interview would be canceled. But it didn't play out that way. To my utter astonishment, a NASA director telephoned me to say that Armstrong, no matter what, never missed a scheduled appointment. His effort to keep his word was legendary. The post-9/11 skies were largely shut to commercial aircraft, but Armstrong, whose own boyhood hero was flier Charles Lindbergh, refused to cancel his appointment at the Johnson Space Center, piloting his own plane from his adopted hometown of Cincinnati. It was a matter of honor, part of Armstrong's "onward code."

The six-hour interview went well. When I asked Armstrong why the American people seemed to be less NASA crazed in the twenty-first century than back during John F. Kennedy's White House years, he had a thoughtful response. "Oh, I think it's predominantly the responsibility of the human character," he said. "We don't have a very long attention span, and needs and pressures vary from day to day, and we have a difficult time remembering a few months ago, or we have a difficult time looking very far into the future. We're very 'now' oriented. I'm not surprised by that. I think we'll always be in space, but it will take us longer to do the new things than the advocates would like, and in some cases, it will take external factors or forces which we can't control."

Moments later, I again tried to get Armstrong to loosen up and be more expressive about his lunar accomplishment, to defuse his engineer's penchant for personal detachment. I had long pictured him in the sultry evenings at Cape Canaveral leading up to the *Apollo 11* launch, looking up at the luminous moon and knowing that he and Edwin "Buzz" Aldrin would soon be the first humans to visit a place beyond Earth. "As the clock was ticking for takeoff, would you every night or most nights, just go out and quietly look at the moon? I mean, did it become something like 'My goodness?!'" I asked.

"No," he replied. "I never did that."

That was the extent of his romantic notions about the lifeless moon. Neil Armstrong was first and foremost a Navy aviator and aerospace engineer, following military orders with his personal best. What became clear to me after interviewing him (and other Mercury, Gemini, and Apollo astronauts of 1960s fame) was that the story of the American lunar landing wasn't wrapped up in any idealized aspiration to walk on the moon surface; instead, it was all about the old-fashioned patriotic determination to fulfill the pledge made by President Kennedy on the afternoon of May 25, 1961. "I believe," our thirty-fifth president had said before Congress, "that this nation should commit itself to achieving the goal, before the decade is out, of landing a man on the moon and returning him safely to Earth."

Only one top-tier Cold War politician had the audacity to risk America's budget and international prestige on such a wild-eyed feat within such a short time frame: in John F. Kennedy, the man and the hour had met. Even Kennedy's own national security advisor, McGeorge Bundy, thought the whole moonshot gambit scientifically reckless, politically risky, and a "grandstanding play" of the most outlandish kind; and he had the temerity to voice his opinion in no uncertain terms to the president. "You don't run for President in your forties," Kennedy snapped back, "unless you have a certain moxie."

Without Kennedy's daunting vow to send astronauts to the moon and bring them back alive in the 1960s, Apollo 11 would never have happened in my childhood. The grand idea undoubtedly grew out of a series of what Armstrong called "external factors or forces"—including World War II, *Sputnik*, the Bay of Pigs, Yuri Gagarin, atomic bombs, intercontinental ballistic missiles, the inventions of the silicon transistor and microchip, and a steady stream of Soviet advances in space. Myriad new technological capabilities unfurled and coalesced with Kennedy's indomitable "Go,

40

64

the dramatic achievements in space which occurred in recent weeks should have made clear to us all the impact of this new frontier of human adventure. Since early in my term, our efforts in space have been under review. With the advice of the Vice President we have examined where we are strong and where we are not, where we may succeed and where we may not. Now it is time to take longer strides -- time for a great new American enterprise -- time for this nation to take a clearly leading role in space achievement,

Reading copy of President John F. Kennedy's special message to Congress, "Urgent National Needs," delivered May 25, 1961. In the address, Kennedy argues for increased support of the National Aeronautics and Space Administration (NASA) and the United States' landing a man on the moon by the end of the decade.

go, go!" leadership style. It's my contention that if JFK had been wired differently—if he hadn't had such a hard-driving father who raised him with the need to achieve great things or a brother who died in World War II trying to destroy a German missile facility—then the moonshot might not have happened.

For Kennedy, who himself became a World War II naval hero for his bravery in the *PT-109* incident of 1943 in the Pacific Theater, the Project Mercury astronauts were ultimately fearless public servants like him. The NASA astronauts Kennedy had fêted in the Collier Trophy ceremony weeks before his assassination volunteered for space travel duty at a pivotal moment in the Cold War. Like Kennedy, these astronauts were courageous, pragmatic, and cool; they were husbands and fathers who, as journalist James Reston noted, "talked of the heavens the way old explorers talked of the unknown sea." Kennedy's New Frontier ethos was based on adventure, curiosity, big technology, cutting-edge science, global prestige, American exceptionalism, and historian Frederick Jackson Turner's famous "frontier thesis." All six of NASA's Mercury missions occurred during Kennedy's presidency. With Madison Avenue instinct, Kennedy routinely claimed that space was the "New Ocean" or "New Sea." If so, then he was the navigator in chief ordering NASA spacecraft with noble names such as *Freedom*, *Liberty Bell*, *Friendship*, *Aurora*, *Sigma*, and *Faith* into the great star-filled unknown. His talent for converting Cold War frustration over Soviet rocketry success into a no-holds-barred competition for the moon was politically masterful. And the American public loved him for leading the effort.

There were other U.S. politicians who promoted NASA's manned space program with zeal in the late 1950s and '60s, Lyndon Johnson chief among them. But only the magnetic Kennedy knew how to sell the $25 billion moonshot (around $180 billion in today's dollars) to the general public. Due to Kennedy's leadership over 4 percent of the federal budget went to NASA in the mid-1960s. In sports terms, he built a team like a great coach, and then played to win. The faith he placed in ex-Nazi rocketeer Wernher von Braun and NASA technocrat James Webb showed that the president was a leader who instinctively knew how to tap the right talent at the right time. Although he worried about space budgets, in the end he never shrank from asking Congress for the fiscal increases that NASA's moonshot required. What makes Kennedy's leadership even more impressive was the way he wrapped both domestic and foreign policies around his New Frontier moon program in a judicious, cost-effective, and effective way. Building on Franklin Roosevelt's New Deal and Harry Truman's Fair Deal, Kennedy's New Frontier was activist federal government writ large. What the Interstate Highway System, the Saint Lawrence Seaway, and ICBM development were to Dwight Eisenhower, NASA's manned space programs were to Kennedy: America, the richest nation, doing big projects well.

It's fair to argue that NASA's Projects Mercury, Gemini, and Apollo were just a shiny distraction, that the taxpayers' revenue should've been spent fighting poverty and improving public education. But it's disingenuous to argue that Kennedy's moonshot was a waste of money. The technology that America reaped from the federal investment in space hardware (satellite reconnaissance, biomedical equipment, lightweight materials, water-purification systems, improved computing systems, and a global search-and-rescue system) has earned its worth multiple times over. Ever since, whenever we have worried about an America in decline, Kennedy's moonshot challenge has stood as the green light reminding us that together as a society we can accomplish virtually any feat.

Full of blithe optimism, Kennedy's pledge set an audacious goal, capping a three-and-a-half-year period in which the Soviet Union twice shocked the world, first by launching the first orbital satellite, on October 4, 1957, and then by sending cosmonaut Yuri Gagarin on the first manned space mission on April 12, 1961, just six weeks before Kennedy's rally cry to Congress. For a world locked in a Cold War rivalry between the Americans and the Soviets, space quickly became the new arena of battle. "Both the Soviet Union and the United States believed that technological leadership was the key to demonstrating ideological superiority," Neil Armstrong explained later. "Each invested enormous resources in evermore spectacular space achievements. Each would enjoy memorable successes. Each would suffer tragic failures. It was a competition unmatched outside the state of war." Kennedy, with depth and commitment, articulated a visionary strategy to leapfrog America's Communist rival and win that high-stakes contest in the name of the capitalistic free-market system as represented by the United States. It was just a matter of figuring out how to do it, using engineering exactitude, military know-how, taxpayer dollars, and political pragmatism.

Kennedy's moonshot plan was more than just a reaction to Soviet triumphs. Instead, it represented simultaneously a fresh articulation of national priorities, a semi-militarized reassertion of America's bold spirit and history of technological innovation, and a direct repudiation of what he saw as the tepid attitude of the previous administration. Within months of winning the presidency in November 1960, Kennedy had decided that America's dillydallying space effort was symbolic of everything that had been wrong with the Eisenhower years. According to Theodore Sorensen, Kennedy's speechwriter and closest policy advisor, "the lack of effort, the lack of initiative, the lack of imagination, vitality, and vision" annoyed Kennedy to no end. To JFK, "the more the Russians gained in space during the last few years in the fifties the more he thought it showed up the Eisenhower administration's lag in this area and damaged the prestige of the United States abroad."

Only forty-three when he entered the White House, Kennedy represented generational change. When he was born in 1917, West German chancellor Konrad Adenauer was already lord mayor of Cologne, French president Charles de Gaulle was a company commander in the French army, Soviet premier Nikita Khrushchev was chairman of a workers' council in Ukraine, and President Dwight D. Eisenhower was a newly married West Point graduate preparing to train soldiers for battle in World War I. At the dawn of the transformative 1960s, these leaders, all born in the nineteenth century, seemed part of the past, while Kennedy and his spacemen were the fresh-faced avatars of a future in which a moon-landing odyssey was a vivid possibility. "I think he became convinced that space was the symbol of the twentieth century," Kennedy's science advisor Jerome Wiesner recalled. "He thought it was good for the country. Eisenhower, in his opinion, had underestimated the propaganda windfall space provided to the Soviets."

Calculating that the American spirit needed a boost after *Sputnik*, Kennedy decided that beating the Soviets to the moon was the best way to invigorate the nation and notch a win in the Cold War. But he also understood that a vibrant NASA manned space program would involve nearly every field of scientific research and technological innovation. U.S. leadership in space required specialists who could innovate tiny transistors, devise resilient materials, produce antennae that would transmit and receive over vast distances never before imagined, decipher data about Earth's magnetic field, and analyze the extent of ionization in the upper atmosphere.

President Kennedy bet that a lavish financial investment in space, funded by American taxpayers, would pay off by uniting government, industry, and academia in a grand project to accelerate the pace of technological innovation. He doubled down on Apollo even while calling for tax cuts. Breaking up congressional logjams over NASA appropriations became a regular feature of his presidency. Though the cost of Project Apollo eventually exceeded $25 billion, the intense federal concentration on space exploration also teed up the technology-based economy the United States enjoys today, spurring the development of next-generation computer innovations, virtual reality technology, advanced satellite television, game-changing industrial and medical imaging, kidney dialysis, enhanced meteorological forecasting apparatuses, cordless power tools, bar coding, and other modern marvels. Shortsighted politicians may have carped about the cost, but in the immediate term, NASA funds went right back into the economy: to manned space research hubs such as Houston, Cambridge, Huntsville, Cape Canaveral, Pasadena, St. Louis, the Mississippi-Louisiana border, and Hampton, Virginia, to the thousands of companies and more than four hundred thousand citizens who contributed to the Apollo effort.

Because NASA worked in tandem with American industry, the agency often received bogus credit for developing popular products like Teflon (developed by DuPont in 1941), Velcro (invented by a Swiss engineer to extract burrs stuck in his dog's fur on alpine hikes in 1941), and Tang (released in 1957 as a grocery store product). The most pernicious myths were that NASA innovated miniaturized computing circuits and personal computers; it didn't. NASA, however, *did* adopt these product innovations for manned-space missions.

Even without the manifest technological and societal benefits of Apollo, Kennedy would have set a course to the moon because he believed America had an obligation to lead the world in public discovery. For though the moon seemed distant, in reality, it was only three days away from Earth. On September 12, 1962, at the Rice University football stadium, just a short walk across campus from my office at the university's history department, Kennedy offered the nation a stirring rationale for Apollo. Identifying the moon as the ultimate Cold War trophy and throwing his weight behind landing there was the most daring thing Kennedy ever did in politics. "Why, some say, the moon? Why choose this as our goal? And they may well ask why climb the highest mountain? Why, 35 years ago, fly the Atlantic? . . . We choose to go to the moon in this decade and do the other things, *not* because they are easy, but *because they are hard*, because that goal will serve to organize and measure the best of our energies and skills, because that challenge is one that we are willing to accept, one we are unwilling to postpone, and one which we intend to win."

For Kennedy, the exploration of space continued the grand tradition that began with Christopher Columbus and flowed through America's westward expansion, through the invention of the electric light, the telephone, the airplane and automobile and atomic power, all the way to the creation of NASA in 1958 and the launch of the Mercury missions that took the first Americans into space. Kennedy saw the Mercury Seven astronauts he hosted in the Rose Garden as path blazers in an American tradition that extended from Daniel Boone and Meriwether Lewis to Charles Lindbergh and Amelia Earhart. When he was a boy, Kennedy's favorite book was the chivalry-drenched *King Arthur and the Knights of the Round Table.* As president, he loved when newspapers such as the *Los Angeles Examiner* and *St. Louis Post-Dispatch* called his Mercury Seven astronauts "knights of space," with him as King Arthur. "He made a statement that he found it difficult to understand why some people couldn't see the importance of space," von Braun recalled of Kennedy's visit to the George C. Marshall Space Flight Center in Huntsville in September 1962. "He said he wasn't a technical man but to him it was so very obvious that space was something that we simply could not neglect. That we just had to be first in space if we want to survive as a nation. And that, at the same time, this was a challenge as great as that confronted by the explorers of the Renaissance."

In 1959, LA Dodgers slugger Wally Moon became known for his towering home runs over the left-field wall at Los Angeles Coliseum, hits that radio announcer Vin Scully dubbed "moon shots." The term quickly seeped its way into the culture and became synonymous with Kennedy's aspirational space vision. *Merriam Webster* still treats *moon shot* as two words. But I have chosen the singular *moonshot* through this narrative, because it is usually uttered without a pause or break. As early as Kennedy's Rice University address, in fact, the *Houston Press* called NASA's new Manned Spacecraft Center in town the "Moonshot Command Post."

A large question I try to answer is what drove Kennedy—perhaps a deep romantic strain (which his wife, Jackie, believed was his true-self)—to gamble so much political capital on his aspirational Project Apollo moonshot? Certainly, he did harbor a quixotic streak when it came to exploration, and an interest in the sea that, he once wrote, began "from my earliest boyhood" sailing the New England coast, observing the stars, and feeling the gravitational push and pull between the moon and tides. During Kennedy's Rice speech, he deemed space the ocean ready to be explored by modern galactic navigators. "We set sail on this new sea," he said, "because there is new knowledge to be gained, and new rights to be won, and they must be won and used for the progress of all people."

By the time Kennedy stepped down from the Rice dais, his memorable words had been seared into the imaginations of every rocket engineer, technician, data analyst, and astronaut at NASA. It was that rare moment when a president outperformed expectations. "The eyes of the world now look into space," he had vowed, "to the moon and to the planets beyond, and we have vowed that we shall not see it governed by a hostile flag of conquest, but by a banner of freedom and peace. We have vowed that we shall not see space filled with weapons of mass destruction, but with instruments of knowledge and understanding. Yet the vows of this Nation can only be fulfilled if we in this Nation are first, and, therefore, we intend to be first. In short, our leadership in science and in industry, our hopes for peace and security, our obligations to ourselves as well as others, all require us to make this effort, to solve these mysteries, to solve them for the good of all men, and to become the world's leading spacefaring nation."

For Kennedy, spurred onward by Alan Shepard's successful suborbital arc into space on May 5, 1961, the moonshot was many things: another weapon of the Cold War, the sine qua non of America's status as a superpower, a high-stakes strategy for technological rebirth, and an epic quest to renew the American frontier spirit, all wrapped up as his legacy to the nation. He would bend his presidential power to support the Apollo program, no matter what. How he envisioned the moonshot gambit, his day-to-day tactics and long-term protocol, and how he pulled it off are what this presidential biography is all about. It's a political epic of how Huntsville rocket genius Wernher von Braun, the Texas wheeler-dealer Lyndon Baines Johnson, and North Carolina–raised manager James Webb of NASA took up the dream that someday astronauts like Neil Armstrong and Buzz Aldrin could indeed break the shackles of Earth and walk on the moon. "I think [the lunar landing] is equal in importance," von Braun boasted, "to that moment in evolution when aquatic life came crawling up on the land."

Hundreds of U.S. policy planners and lawmakers followed the leadership directives of President Kennedy and Vice President Johnson. And then thousands of astrophysicists, computer scientists, mechanics, physicians, flight trackers, office clerks, and mechanical engineers followed the White House planners. Millions of Americans joined in the dream, too. Finally, when humans did walk on the moon, five hundred million people around the world took pride in watching the *human* accomplishment on television or listening on the radio. Even Communist countries swooned over Apollo 11. "We rejoice," the Soviet newspaper *Izvestia* editorialized, "at the success of the American astronauts." Unfortunately, Kennedy didn't live to see the *Eagle* make its lunar landing on that historic day of July 20, 1969. Everybody at NASA knew that Armstrong's "giant leap for mankind" was done to fulfill Kennedy's audacious national directive. As Kennedy dreamed, the first human footprints on the gray and powdery moon were made by mission-driven American space travelers. And given that the moon has no erosion by wind or water because it has no atmosphere, they will likely remain stamped there for time immemorial as his enduring New Frontier legacy. Someday the *Eagle* landing spot and those astronauts' footprints should be declared a National Historical site. "We needed the first man landing to be a success," Aldrin later reflected on JFK's lunar challenge, "to lift America to reaffirm that the American dream was still possible in the midst of turmoil."

Throughout the United States there is a hunger today for another "moonshot," some shared national endeavor that will transcend partisan politics. If Kennedy put men on the moon, why can't we eradicate cancer, or feed the hungry, or wipe out poverty, or halt climate change? The answer is that it takes a rare combination of leadership, luck, timing, and public will to pull off something as sensational as Kennedy's Apollo moonshot. Today there is no rousing historical context akin to the Cold War to light a fire on a bipartisan public works endeavor. Only if a future U.S. president, working closely with Congress, is able to marshal the federal government, private sector, scientific community, and academia to work in unison on a grand effort can it be done. NASA has achieved other astounding successes in the realm of space, such as exploring the solar system and cosmos with robotic craft and establishing a space station, but without presidential drive, these didn't galvanize the national spirit. Kennedy's moonshot was less about American exceptionalism, in the end, than about the forward march of human progress. For as the Apollo 11 plaque left on the moon by Armstrong and Aldrin reads, WE CAME IN PEACE FOR ALL MANKIND.

PART I
Rockets

Robert H. Goddard poses with his first liquid-fueled rocket on March 16, 1926, in Worcester, Massachusetts. Although this first rocket rose only forty-one feet, Goddard's immense body of work, covered by 214 patents, established him as one of the founders of spaceflight.

1
Dr. Robert Goddard Meets Buck Rogers

Earth is the cradle of humanity. But one cannot live in the cradle forever.

—KONSTANTIN TSIOLKOVSKY

History has taught us that artists are often decades ahead of engineers and scientists in imagining the future, and so it was with the idea of voyaging to the moon. In 1865, Jules Verne published *From the Earth to the Moon*, which detailed the story of three intrepid astronauts blasted from a gigantic cannon in Tampa, Florida, en route to a lunar landing. A devotee of hot air balloons, astronomy, and newfangled gunnery, with a mind that could easily grasp Galileo's theories on the phenomenon of lunar light, Verne consulted with French scientists about the challenges of a lunar voyage, then translated those complexities for the layman. Enormously popular, *From the Earth to the Moon* and its sequel, *Around the Moon*, inspired readers to reimagine what was possible and to beware of "certain narrow-minded people, who would inevitably shut up the human race upon this globe."

Blessed with a probing curiosity that never rested, Verne was eerily prescient. Writing around the time of the American Civil War, he accurately prophesied that the United States would beat Russia, France, Great Britain, and Germany to the moon, and that the voyage would be launched from the Florida tidal lowlands, at the approximate latitude from which *Apollo 11* blasted off in 1969. Verne's postulation that the projectile would take four days to reach the moon was likewise remarkably accurate.

Verne's novels exemplified the optimistic spirit of their times, when the potential for industrial and technological progress seemed limitless. It was a spirit that still suffused public discourse and literature into the early years of the twentieth century—especially in the United States, where, as novelist Kurt Vonnegut once noted, the enthusiastic experimenter and inventor with "the restless, erratic insight and imagination of a gadgeteer" has been an archetype since the nation's birth.

It was into this cultural milieu that John Fitzgerald "Jack" Kennedy was born on May 29, 1917, in the Boston suburb of Brookline, Massachusetts. His parents, Joe and Rose Kennedy, both grandchildren of Irish immigrants, came from families that had thrived in business and politics. Rose's father, John "Honey Fitz" Fitzgerald, was a former Massachusetts congressman and mayor of Boston. Joe's parents had worked their way into the upper middle class through the saloon business and connections in the Democratic Party. The pride of their respective families, Joe and Rose were already well established in Boston society when they started their married life. Joe first made waves as a banker with a gift for spotting opportunity in the fine print of legal documents. Eventually, he became known as a Wall Street speculator or even manipulator, amassing millions by the time Jack was a small boy. The second of what would eventually be nine children, Jack had a sheltered upbringing, wanting for nothing, though this didn't protect him from contracting scarlet fever at age two. He was quarantined with a 104-degree fever and blisters all over his body. For three weeks, his parents attended church services daily to pray for his recovery. After a month, he took a turn for the better, but he continued to suffer from various afflictions for the rest of his life, despite an outward appearance of robust good health.

Always taking life lightly, Jack could be a scamp of a boy, yet at the end of the day, he wanted only to delve into books by Mark Twain, Robert Louis Stevenson, H. G. Wells, Jules Verne, and others, which created in him a ravenous appetite for world history and high-risk adventure. An appreciation of the seashore, sailing, and maritime culture was ingrained in him at an early age. Simultaneously, his parents instilled in him a focus on politics and global affairs, with suppertime conversations typically turning on the week's news from the *New York Times* and *The Saturday Evening Post*, including the latest developments in aeronautics.

As the century progressed and real-world technological advances closed the gap on Verne's fiction, space travel moved into the realm of plausibility. In the wake of the Wright brothers' first flight, at Kitty Hawk, North Carolina, the skies were suddenly open to mankind, and even space travel seemed attainable. In 1910, a year out of the White House, Theodore Roosevelt became the first U.S. president to fly in a plane—it was the future, he declared. If Orville and

Wilbur Wright could devise aircraft controls that made fixed-wing powered flight attainable, and if TR flew, then why couldn't the moon be conquered someday? After Kitty Hawk, space was talked about as a "new frontier," to be conquered by rockets instead of Conestoga wagons and the Pony Express, and the news reported regularly on the latest advances in the burgeoning fields of aviation and rocketry.

In the early years of the 1920s, northeast of Los Angeles, astronomer Edwin Hubble was observing the solar system through the Mount Wilson Observatory's just-completed Hooker Telescope, at one hundred inches, then the world's largest. By 1924, his findings would shatter the common notion that the Milky Way encompassed the entire cosmos, proving instead that it was just one among potentially billions of galaxies in an unimaginably vast universe. CBS Radio often hosted astronomers speculating about life in other galaxies, while top-tier universities began hiring space physicists. Interest in space transcended regionalism. Every village, it seemed, had a space buff, with discerning eyes for the moon.

Wearing aviator's caps, fancy goggles, and exotic silk scarves with calf-high leather boots, aviators—like Eddie Rickenbacker—had become American heroes in the First World War. Even though airpower hadn't been a determining factor in the war per se, the U.S. government recognized its future potential in warfare. The pursuit of aeronautical innovation that would eventually take us to the moon, in fact, had come into being during World War I. A group led by Charles D. Walcott, secretary of the Smithsonian Institution, had lobbied Congress to create an advisory committee that would coordinate aeronautical innovation efforts across government, industry, and academia, with the goal of producing cutting-edge military aircraft. In 1915, with President Woodrow Wilson determined to keep the United States neutral even as war consumed Europe, congressmen had quietly slipped a rider into a naval appropriations bill calling for the creation of the National Advisory Committee for Aeronautics (NACA) to "supervise and direct the scientific study of the problems of flight to their practical solution." It was an attempt to achieve aeronautical parity with the European powers, which, in the decade since the Wright brothers' flight, had been busily studying the new technology's military applications.

Initially charged with meeting just a few times a year, the committee soon expanded its role, building America's first civilian aeronautical research laboratory. The year Jack Kennedy was born, the NACA established the Langley Memorial Aeronautical Laboratory (LMAL) in Hampton, Virginia, situated on the Little Back River off the Chesapeake Bay, at the end of a peninsula running between the James and York Rivers. The regrettable fact was that no single airplane flown by U.S. pilots in World War I had been built in an American factory. The NACA leaders wanted this lack of foresight to end. In 1920, the NACA got a public relations boost with the appointment of the legendary aviator Orville Wright to the agency's board. Soon the government-sponsored test laboratory was conducting research in aerodynamics, aircraft structures, and propulsion systems for both industrial and military flights while pioneering such innovations as wind tunnels, engine test stands, and test-flight facilities. It was at the NACA that safety solutions for flying "blind" (in fog, blizzards, and thunderstorms) were created. Although the NACA would build additional laboratories in Ohio and California during the Second World War, it would be primarily at the Hampton incubator that the idea of launching Americans to the moon would get its first serious discussion.

Six hundred miles north of Hampton, some of the most aspirational aeronautical news in the world had emanated from the working-class city of Worcester, Massachusetts, about forty miles west of where Jack Kennedy spent his toddler years. There, in 1919, Dr. Robert H. Goddard, a professor at Clark College (now Clark University), unveiled his astronautical ideas in the sixty-nine-page *A Method of Reaching Extreme Altitudes*, issued as part of the Smithsonian Miscellaneous Collections. The document, written as bait to attract grant funding, illuminated a method of constructing a two-stage, solid-propellant rocket to use for atmospheric research. But what most fired the public imagination was Goddard's assurance that a rocket fueled by a combination of gasoline and liquid oxygen would be able to dispense thrust beyond Earth's atmosphere.

Goddard had grown up in Worcester to a family with New England roots dating to the seventeenth century. Sickly as a boy, often bedridden with pleurisy and bronchitis, he engaged his mind with the telescope, microscope, and a subscription to *Scientific American* provided by his father, a mechanical inventor of the type that propelled New England textile manufacturing throughout the Industrial Revolution. Amid peers more interested in football and hockey, Goddard scoured the local library for books on the physical sciences. When he was sixteen, he read H. G. Wells's *The War of the Worlds* (1897), became fixated on the cosmos, and attempted to construct a balloon out of aluminum, crafting the metal in his home laboratory, dreaming of the moon and Mars.

Imbued with a visionary imagination, mapping out a future career as a physicist, Goddard was valedictorian of his high school class, delivering a speech that included the optimistic observation, "It has often proved true that the dream of yesterday is the hope of today, and the reality of tomorrow." He attended Worcester Polytechnic Institute and then Clark College, from which he received an MA and PhD in physics in 1910 and 1911, but Goddard was more than a blackboard genius and armchair theorist: he was his father's son, and he wanted to build and launch his own rockets to prove his bold theories. Always self-motivated, he had already conducted several rocketry experiments while scouring for funding to conduct more.

The crucial theoretical turning point for Goddard occurred when he realized that Isaac Newton's Third Law of Motion applied to motion in space, too. That opened up a universe of possibilities in his mind. By 1912, he became the first American to credibly explore mathematically the practicality of adopting rocket propulsion to reach high altitudes and even the moon. Following in the time-honored tradition of Alexander Graham Bell and Thomas Edison, and wanting to make money on his innovations, Goddard applied for patents steadily, eventually receiving 214 of them from the federal government, including the first for a multi-stage rocket.

During World War I, Goddard lent his mechanical talents to the U.S. Army, developing the prototype of a tube-based rocket launcher that would later become the bazooka, a light infantry weapon ubiquitous in World War II. The main impetus for Goddard's work was his unwavering quest to prove that a rocket could navigate space. After the Smithsonian published *A Method of Reaching Extreme Altitudes*, which included the idea of launching a rocket loaded with flash powder at the moon, such that the impact would be visible from Earth, newspapers across America reprinted his eye-popping pronouncement. "Communication with Moon Is Made Possible," trumpeted the *Fort Worth Star-Telegram*. "Rocket for Moon, Plan of Professor" was the front-page headline in the Colorado Springs *Gazette*. Even while noting that shooting a flash powder rocket at the moon would not be "of obvious scientific importance," Goddard believed that for all the inherent logistical and engineering difficulties, such launches depended "on nothing that is really impossible."

Though the paper laid the theoretical foundation

for U.S. rocket development in the twentieth century, the contemporary media and public remained uncertain. On January 12, 1920, the *New York Times* ran a front-page story about Dr. Goddard's wizardry, titled "Believes Rocket Can Reach Moon." The next day, however, the *Times* editors admitted to "uneasy wonder" over the idea, maintaining that Goddard's proposed rocket would "need to have something better than a vacuum against which to react." Goddard, the *Times* editors derided, "lacked the knowledge ladled out daily in high schools," comparing him uncomplimentarily to Albert Einstein and equating some of his ideas to "deliberate step[s] aside from scientific accuracy" as observed in Verne's *From the Earth to the Moon.* Even this criticism wasn't as bad as that of the *Philadelphia Inquirer,* which compared the Worcester rocketeer's idea to a Mother Goose nursery rhyme.

The snarky reaction in the *Times* and elsewhere made it open season on Goddard. Insipid jokes about his pie-in-the-sky moon trip abounded, even as respected aeronautical engineers stepped forward to support his principal contention: that a rocket could indeed function in a vacuum, needing no atmospheric pressure to push against. Goddard, an introverted man, balding, with a close-trimmed little mustache, and always impeccably dressed in a tailor-made suit, cringed at the skepticism and sophomoric humor that greeted his ideas, and he was determined to prove his detractors wrong. Though averse to showmanship, he conducted public demonstrations before an assembly of undergraduates, rigging a .22-caliber pistol loaded with a blank cartridge to the top of a spindle, inserting it into a bell jar, and then pumping out the air to mimic the vacuum of outer space. When fired remotely, the gun kicked back and made four full revolutions on its spindle, dramatically demonstrating thrust and velocity. As he watched the pistol spin, Goddard remarked dryly, "So much for *The New York Times*." More than fifty subsequent simulation tests using vacuum chambers proved beyond question that rocket propulsion could indeed operate in a void. Eventually, both the times and the *Times* caught up with his ideas: to the newspaper's credit, it issued a public retraction of its 1920 commentary forty-nine years later, after *Apollo 11* was launched to the moon.

By 1921, still adhering to the hard, everyday, hermetic work of experimentation, Goddard was convinced that he would achieve greater thrust by switching from solid- to liquid-fueled rockets, utilizing cylindrical combustion chambers with impinging jets to atomize and mix liquid oxygen and gasoline. To test his ideas, he needed money. As the decade unfolded, Goddard realized that firing a rocket into space would require funding far in excess of the five-thousand-dollar grant he'd received from the Smithsonian. His own estimate was one hundred thousand dollars. In an era when few federal dollars were flowing into aviation technology research, even for military purposes—the NACA was still in its infancy—Goddard knew he had to think outside the box. Though secretive by nature, he determined that the garish publicity swirling around his high-minded plans for space rocketry held a wow factor that could potentially compel interest from private-sector investors. After all, he was that rare scientist of stature in America: one who'd proposed launching a rocket to the moon.

Publicity can be a double-edged sword, and as a colleague later observed, Dr. Goddard had "early discovered what most rocket experimenters find out sooner or later—that next to an injurious explosion, publicity is the worst possible disaster." Yet Goddard himself stoked the disaster, or at least the publicity, announcing in April 1920 that he had already received nine applications from brave men who wanted to soar to the moon in the first rocket. Of course, there was no manned rocket planned—Goddard had never written anything serious about that possibility—but it didn't matter. In homes like that of the Kennedys, in tree-lined Brookline, a new debate replaced the old one, as people stopped wondering whether a moon rocket was really possible and turned to the even more adventurously romantic questions: Would you go? Do you have the mettle?

Goddard's public relations instincts had proved well grounded, and as his work progressed, he fed the public just enough information to tantalize, always mentioning an eventual moon launch to stoke their imaginations, even though privately his focus remained on "the more practical objects of my experiments." Public consensus held that a mission to the moon could easily happen in their lifetime, but public consensus did not equate to public funding. Though Clark University and the Smithsonian continued their limited financial support, Goddard's solicitations for Americans to support his "Rocket to the Moon" research-and-development fund came up short, yet again.

If Robert Goddard's early 1920s experiments kicked up a storm of excitement in the United States, they had just as sharp an effect in Germany, where two other rocketry pioneers were toiling in similar isolation but even greater obscurity. One of them was Hermann Oberth, a German national born in Romania's Transylvania region. After moving to Germany, the gangly Oberth became a university student who had, like Goddard, been fascinated with the idea of space travel since reading Verne's *From the Earth to the Moon*, even constructing a small replica of the rocket engine described in the novel. Oberth wrote his doctoral dissertation at Heidelberg University on the futuristic prospect of launching a rocket into space, but while his hypothesis was cited as a brilliant examination of possible modes and methods, it was peevishly rejected on the grounds that it covered both physics and astronomy rather than focusing on a single discipline.

As the bitterly disappointed Oberth prepared a revised version of his monograph for the printer, he heard about Goddard's *A Method of Reaching Extreme Altitudes* and wrote to the American inventor, requesting a complimentary copy. This audacious letter made Goddard uneasy because he disdained the German penchant, as demonstrated in World War I, for turning scientific inventions into tools of war; nevertheless, he acceded. The two rocket engineers engaged in a friendly correspondence, and a grateful twenty-nine-year-old Oberth quickly added a description of Goddard's work to the end of his book *Die Rakete zu den Planetenräumen* (*The Rocket into Planetary Space*), published in 1923. Anchored in the claim that liquid-propelled rockets could indeed escape Earth's atmosphere, thereby making interplanetary travel possible, Oberth's book was an instant sensation in German scientific circles. Within months, word of *Die Rakete zu den Planetenräumen* reached the Soviet Union, where it straightaway elicited an unexpectedly passionate response from university-trained scientists. Following the Russian tsar's 1917 abdication and the civil war that followed, the new Communist government kept the nation largely closed off from the Western world through the early 1920s, as the Leninist revolution celebrated its most intellectually creative period. An account in Moscow's leading newspaper, *Izvestia*, praised Oberth's work while also describing it as providing support to "the American professor Dr. Goddard, who has recently presented a sensational plan to send a rocket to the moon." As a result, Goddard became respected, even idolized, in the Soviet Union.

As word of Goddard's and Oberth's separate but parallel research spread, a blind schoolteacher in a small town south of Moscow was outraged. Konstantin Tsiolkovsky had written similarly about rockets in

1903, in a book called *Explorations of the Space of the Universe by Jet-Propelled Instruments.* In meticulously constructed calculations and equations, Tsiolkovsky's volume had proposed many of the same principal features of space rockets described by Goddard and Oberth. More recently, Tsiolkovsky had written a novel, *Beyond the Planet Earth* (1920), about an international cabal of space travelers who reach the moon: an American named Franklin, an Englishman named Newton, an Italian named Galileo, and the calm and controlled Russian hero of the story, Ivanov. Though far-fetched and kitschy, Tsiolkovsky's novel was ingenious in envisioning future moonshots and space stations in the twentieth and twenty-first centuries.

With Tsiolkovsky as the homegrown hero and Goddard as the acknowledged leader in the global rocketry revolution, a "mass fascination with space travel . . . exploded in Soviet Russia in the 1920s," according to historian Asif Siddiqi. "Students, workers, writers, journalists, artists and even filmmakers," he wrote, "explored various dimensions of the possibility of cosmic travel." Meanwhile, on the ground in Moscow, the new Communist state was busily retooling the largely agrarian nation for a future of technology and industry.

During this nascent era of rocketry, an aristocratic boy named Wernher von Braun was growing up in the city of Wirsitz, in the province of Posen, in Prussian Germany. Born on March 23, 1912, five years before John Kennedy, von Braun was raised to be an aristocrat. When he was five, his family moved to Berlin. Wernher was a talented young pianist who dreamed initially of becoming a composer in the vein of Beethoven or Bach. Then, at twelve years old, he heard the siren call of space. "For my confirmation," von Braun recalled, "I didn't get a watch and my first pair of long pants like most Lutheran boys. I got a telescope."

Having read in a popular magazine about automobiles being powered by rockets, young Wernher built his own contraption using a coaster wagon with six large firework skyrockets fastened on the back bed. Rolling the strange vehicle onto Berlin's swank Tiergartenstrasse, he lit the fuse. "I was ecstatic," he later recalled. "The wagon was wholly out of control and trailing a comet's tail of fire, but my rockets were performing beyond my wildest dreams. . . . The police took me into custody very quickly. Fortunately, no one had been injured, so I was released in charge of the Minister of Agriculture—who was my father."

Although he was grounded by his father for his dangerous public experiment, young von Braun's obsessive pursuit of rocketry continued unabated. His jet wagon opened up the cosmos to him. The moon became his fixation. At fifteen, he read a magazine article describing an imaginary lunar voyage. "It filled me with nomadic urge," he later wrote. "Interplanetary travel. Here was a task worth dedicating one's life to! Not just to stare through a telescope at the moon and the planets, but to soar through the heavens and actually explore the mysterious universe! I knew how Columbus felt." As he matured, young von Braun read Oberth's *The Rocket into Planetary Space*, and though stymied by the profusion of mathematical equations and scientific terminology in the text, he was ignited with a burning desire to learn more.

Realizing that to be a rocket scientist meant learning calculus, von Braun bore down on his academic studies. Wherever he went, he carried a slide rule and compass. At sixteen he transferred to a special college preparatory school on Spiekeroog Island, in the North Sea, where, much like Jack Kennedy, he learned to sail and swim in rough seas and began to associate the vastness of the ocean with the vastness of space. Obsessed with the stars, von Braun persuaded the school's headmaster to purchase a five-inch refracting telescope and led the effort to construct an astronomical observatory on campus. "As soon as the art of orbital flight is developed," he wrote in his diary, "mankind will quickly proceed to utilize this technical ability for proactive application." His love of astronomy spilled out in "Lunetta" (Little Moon), a five-page short story about living on an Earth-orbiting space station, written for the school magazine.

Beginning in 1930, von Braun attended the Technische Hochschule Berlin (Technical University of Berlin), where he apprenticed under his idol Dr. Hermann Oberth and conducted liquid-fueled rocket tests as part of the embryonic German rocket team. The military took note of their work. The provisions of the Treaty of Versailles had neutered German aspirations in traditional weaponry, but it hadn't banned rocket development. Von Braun believed that what Oberth was trying to prove to the world was fourfold: that a machine could soar beyond Earth's atmosphere; that humans could leave the gravity of Earth; that humans could survive flight in a space vehicle; and that space exploration could be financially profitable. The last goal was, at the time, elusive. Among von Braun's duties for Oberth was fund-raising for rocket science research at a Berlin retail department store, where he would stand for eight hours a day soliciting money beside a display on interplanetary exploration. From that experience with sales, he learned that the cash barrier was one of the hardest obstacles for a rocketeer to surmount. As part of his 1930 pitch, von Braun would bark, "I bet you that the first man to walk on the moon is alive today somewhere on this Earth!"

That very year, future moon walker Neil Armstrong was born on a farm near the small town of Wapakoneta, Ohio.

In the United States, Goddard's promised launch of the world's first liquid-fueled rocket was continually being postponed. Other scientists had theorized about the use of liquid fuels to replace the explosive gunpowder that for centuries had been used to propel primitive rockets, but it was Goddard who had translated theoretical blackboard mathematics to the actual physical world. Liquid fuel (once jetted into a combustion chamber in the presence of liquid oxygen and flame) indeed brought forth expanding gases against the top of the engine, a force called thrust (which was more substantial than the heaviness of the rocket), and it lifted the rocket upward.

Finally, on March 16, 1926, Goddard and two assistants gathered on a snowy cabbage field in Auburn, Massachusetts, five miles from Worcester. The experiment was a family affair: the field itself was owned by Goddard's aunt, Effie Ward, and Goddard's wife, Esther Kisk Goddard, would serve as the day's photo documentarian. Goddard mixed liquid oxygen with gasoline in the rocket's carefully positioned propellant tanks, and his friend Henry Sachs used a blowtorch to touch off the black-powder igniter. On cue, the contraption, built out of slender pipes, rose gently into the sky "as if," Goddard wrote afterward, "it said, 'I've been here long enough; I think I'll be going somewhere else, if you don't mind.'"

Dr. Goddard's ten-foot-tall rocket reached an altitude of forty-one feet before crashing downward like spent fireworks debris. Yet that short distance put humans a thousand times closer to the moon. No one could now deny the viability of liquid-fueled propulsion, which provides greater thrust than gunpowder while allowing greater control over how long the rocket burns. Goddard's achievement, however, was not widely reported—a consequence, perhaps, of the hype he himself had generated with his talk of a moon rocket. Anything less seemed a disappointment, if not an abject failure, to an impatient Roaring Twenties public. This throttle ability factor, however, had important future space applications, such as landing on the moon, when astronauts had to control the deceleration beyond what the kick of a solid motor could provide.

In the Soviet Union, Goddard's lunar ambitions

continued to generate excitement. In the months after his first successful liquid-fueled launch, an erroneous rumor spread in Moscow that Goddard would visit the USSR, prompting Russian scientists to prepare a litany of questions for the wunderkind of Worcester. Though in reality Goddard had no plans to travel to the Soviet Union, he did correspond with the leading space organization there, the Society for the Study of Interplanetary Communications. He also wrote a letter to the All-Inventors' Vegetarian Club of Interplanetary Cosmopolitans. Despite its frothy name, the group had thousands of members and played an integral part in organizing the first Soviet exhibition dedicated to space travel. The Russians who thronged to the Moscow show were, like Americans, interested mainly in moon travel, and they crowded around a mock-up of a moon rocket with intense curiosity.

While his rocketry experiments generally received little attention, Goddard couldn't duck the wave of publicity that followed his ambitious test of the world's first multistage rocket, on July 17, 1929.

It was not the good kind of publicity.

Over the previous year, Goddard had scraped together just enough money to conduct his field test, in which he would send aloft an enlarged version of his earlier designs, accommodating a series of liquid-fuel compartments designed to detonate in succession and carry aloft a scientific payload that included a barometer, a thermometer, and a camera. Gathering on a hot and humid summer afternoon in the Auburn cabbage field, Goddard and his team lit the fuse, setting off a series of sonic roars that could be heard two miles away as the rocket ascended to ninety feet.

When neighbors looked skyward to discern what the fracas was, they saw something tall and thin falling to Earth. The weird phenomenon was the rocket, but reports that a pilot seemed to have fallen out of an airplane soon sent ambulances, police cars, and even a search plane rushing to the cabbage field. After the clamor dissipated, residents of Auburn forbade Goddard from ever launching another rocket within city limits, forcing the professor to find another launch site. Refusing to be diverted, he embraced Edison's maxim that, in the scientific realm, an experimenter did not fail but "found ten thousand ways that won't work." And Goddard seized on an offer to experiment at an army base in nearby Devens, but soon found the controlled bureaucratic setting drearier than the cabbage field. Somehow, even the weather seemed more overcast, gray and dismal.

In Germany, Professor Hermann Oberth wasn't faring much better. In 1928 and 1929, Fritz Lang, a risk-taking motion picture director, was in production on *Die Frau im Mond* (*The Woman in the Moon*), a feature film about a rocket trip to the moon. Lang had contracted Oberth to be a science consultant on the film, with his principal task being the design and construction of a working rocket that would be launched as publicity for the film's premiere. For the project, he was assisted by rocketry enthusiasts from the Verein für Raumschiffahrt (Society for Space Navigation, or VfR), which produced a monthly magazine covering news of high-altitude rockets. The Oberth project was widely covered in the United States, where it elicited an envious response from Goddard. He resented competitors under the best of circumstances, but especially when his own fortunes were low. By the late 1920s, the status of rocket science in America was deficient. No level of government was interested in lending support. Universities were ungenerous, and the nascent aeronautical industry was openly skeptical. Under the tightfisted circumstances, an aeronautical engineer holding a contract with someone willing to underwrite a rocket, even one that was a glorified movie prop, could only inspire envy.

Oberth's happiness was short-lived. Overwrought from the strained collaboration with Lang, and in the dumps, he was rumored to have suffered a nervous breakdown. He later denied this, but the stress of scientific innovation indeed drove the gifted rocketeer to the brink of exhaustion. As *Die Frau im Mond*'s October 15, 1929, premiere approached, it became clear that Oberth's rocket would not be ready. Despite being dismissed by critics as claptrap for fourteen-year-old boys, the film was an enormous box-office hit. Nevertheless, Oberth's failure to meet his obligations soured the producers on lending him any further support. Burned out and depleted, he was forced to scour elsewhere for financial backing, but had limited success.

In the United States, Goddard was in something of the same situation, trying to maintain his dignity while hunting desperately for funding for his visionary aerodynamics and rocketry projects. Then the famous aviator Charles Lindbergh came knocking. Two years after his epochal 1927 solo flight across the Atlantic in the *Spirit of St. Louis*, the Lone Eagle had become American royalty, moving in the most elite social and business circles and able to secure private meetings with experts and engineers in whatever field intrigued him. By 1929, erroneously believing that piston-engine aircraft were nearing their top capacity for speed, Lindbergh had become keenly interested in rockets as a potential new airpower source. Having read an article in *Popular Science* about Goddard's liquid-fueled test launch, Lindbergh visited Clark University specifically to meet the distinguished professor. Wandering around the green-leafed campus and chatting for hours, the two men were impressed with each other. Although Goddard was normally suspicious of technology thieves and spies, he immediately trusted Lindbergh and described his rocketry work to him with unprecedented candor. Likewise, Lindbergh respected Goddard's up-from-the-tinker's-bench perseverance, his bedrock belief that there was no prayer more powerful than desire.

Lindbergh made himself Goddard's chief fund-raiser and advocate. No scientist could have asked for a better backer, especially given that their discussion took place in November 1929, just weeks after the October 29 stock market crash that ushered in the Great Depression. As the economy cratered, failing banks and panicked investors weren't focused on giving loans to mad scientists, and even Lucky Lindy, with all his flyboy cachet, made little headway. After a string of discouraging calls, he notched only one small grant to support Goddard's experiments. Nonetheless, an undaunted Lindbergh kept trying, rattling the tin cup to the millionaire crowd as he proclaimed that Goddard was a prodigy destined to open up space for human exploration someday.

In the new year, Lindbergh finally landed a big donor in Daniel Guggenheim, a philanthropic New Yorker who had made more than $250 million for his family in mining. Through his Daniel Guggenheim Fund for the Promotion of Aeronautics, Guggenheim offered $100,000 to underwrite Goddard's theoretical research, in payments spread over four years. Armed with this grant, Goddard immediately moved both his rocket experiments and his household to Roswell, New Mexico, a tumbleweed-strewn Eden Valley ranching town that offered Goddard three things his work required: flat and arid terrain, mild weather, and a sparse population. Secluded in this new desert environment, away from the public glare, Goddard made incremental advances over the following years. In the spring of 1937, he launched a rocket that reached an altitude of nearly nine thousand feet in just 22.3 seconds, but nobody paid much attention to isolated feats conducted under the ceramic-blue skies of New Mexico. "Morning in the desert," Goddard wrote in his diary from Roswell, New Mexico, in June 1937, "when the impossible not only seems possible, but easy."

Even as Goddard labored as America's only serious rocket experimenter, pushing toward more powerful multistage rockets with the thrust to escape Earth's gravitational pull, five thousand miles away the field of rocketry was moving in a more ominous direction as Europe prepared for another world war.

John F. Kennedy and his friend Lem Billings with the dog Dunker in The Hague, Netherlands, on August 24, 1937.

2

Kennedy, von Braun, and the Crucible of World War II

Only a pure determinist could designate the V-2 a sine qua non of the origins of the Space Age in our time. What the German engineers did, with their clever fabrication of what seemed even in World War II a "baroque arsenal," was to prod their enemies to the East and West into premature fear and rivalry and to make themselves and their blueprints the most prized spoil of the war.

—WALTER A. McDOUGALL,
THE HEAVENS AND THE EARTH (1985)

As 1930 began, twelve-year-old John Kennedy was living at his parents' home on Pondfield Road in the suburb of Bronxville, in Westchester County, New York. Jack was a good-natured boy with green-gray eyes and a Huck Finn cowlick, admired for his quick wit and beatific smile. With both his father and mother traveling extensively, sometimes away from the household for months at a time, it was Jack's older brother, Joseph Jr., who served as the day-to-day model of everything Jack both did and did not want to become. Blessed with a deft intellect and preternatural drive, Joe was self-contained, slender, vigorous, a bit humorless, and a natty dresser with a distinctive aura of future greatness. On the downside, he also had a notoriously quick temper and could be brusque when under pressure.

As a youth, Jack held none of these attributes.

From a distance, Jack was a fortunate son, living in the lap of luxury even as most American children were beginning to suffer the grips of the Great Depression. Money was tight as banks foreclosed at an alarming rate. Privileges for the Kennedy kids came in droves. But Jack's affluence came with its own complications and challenges. Determined not to raise carefree children, his parents were hard on him, and the constant obligation to be as tough and resilient as Joe Jr. also bore down. Once, when Jack was a boy, he and Joe engaged in a game of chicken on their bicycles, pedaling into each other. Joe was barely hurt, while Jack had to get twenty-eight stitches. After that incident, Jack resisted any expectation that he be like his older brother, developing a distinct personality of his own rather than emulating Joe's. Confronting the mores of the high and the rich, he adopted an air of amused, slightly sardonic detachment, as though he were a social scientist clinically observing an esoteric subspecies of man from afar. Yet he did so with such confidence and goodwill that most were charmed rather than offended. Burdened with high expectations, he met them in a charismatic and exemplary way that would one day enter the language: Kennedyesque.

As Jack was learning to negotiate his formative years, a new, wildly popular cartoon and radio character appeared to help those of his generation escape to the stars, if only in their imaginations. *Buck Rogers*, originally a short story by Philip Francis Nowlan, debuted as a comic strip in January 1929 and achieved enormous popularity through syndication in the early 1930s. In each strip, the eponymous hero, an earthling of the twenty-fifth century, roamed the stars in search of adventure. By 1932, a *Buck Rogers* radio show was airing on CBS, bringing the concept of space exploration to millions. To capitalize on the craze, futuristic Buck Rogers toys, such as the ZX-31 Rocket Pistol and the XZ-44 Liquid Helium Water Pistol, were rushed to the market, where they pushed aside cowboys-and-Indians playthings and sold like hotcakes to children of all ages.

Capitalizing on the popularity of *Buck Rogers*, other newspaper syndicates unveiled their own original science-fiction comics. *Flash Gordon* debuted in 1934 and was centered entirely on interplanetary travel in rockets. Becoming even more famous and successful than *Buck Rogers*, *Flash Gordon* took place contemporaneously, in the early 1930s, starting when Flash, a handsome polo player and Yale graduate, meets a scientist working in isolation on a space rocket. If that smacked a bit of what Americans already knew of their real-life rocket scientist, Dr. Robert Goddard, the story continued from there with everything that the wizard of Roswell couldn't yet supply: a trip into outer space and the cosmic, gravity-defying feat that Americans craved. Although Kennedy, in his teen years, was now more interested in novel-length adventures such as *Treasure Island* and *Robinson Crusoe*, *Buck Rogers* and *Flash Gordon* were making his generation believe in space as the next frontier.

After earning a mixed report card at his private day school in the Riverdale section of the Bronx, Jack Kennedy, having relocated to New York with his family, spent the summer of 1930 preparing to leave for boarding school. He expected to follow his brother to the elite boarding school Choate, in Wallingford, Connecticut, but at the very last moment, at his mother's insistence, he enrolled instead at a different Connecticut school: Canterbury, in New Milford, which reflected his family's Roman Catholic values. Attending Canterbury in 1930–31, he felt isolated and was undoubtedly homesick, but he didn't complain. "Please send me the *Literary Digest*," he implored his father in a letter. Although Jack suffered an attack of appendicitis at Canterbury, he kept a stiff upper lip. As Rose Kennedy put it, the family was "accustomed to the idea that every now and then he would be laid up by some disease or accident." While different from his father, by late adolescence Jack exhibited the same dauntlessness, the absolute belief that complaints were a bore and a nuisance to those within earshot.

In 1931, Jack transferred to Choate, in part because it functioned as a direct conduit into Harvard. However, the teenager didn't fit into the old-money mold that dominated the school. A Catholic in a WASP milieu, restless and unfocused, he was inattentive and couldn't master his schoolwork, plan his days, keep his possessions in order, stand out in sports, or manage his spending money, much less gain the respect of his teachers. Jack was more of an undisciplined big boy, popular but unfocused. During his Choate years, he veered toward flirting and frittering around. "Jack has rather superior mental ability without the deep interest in his studies or the mature viewpoint that demands of him his best effort all the time," the Choate headmaster wrote of Kennedy at eighteen. "We have been and are working our hardest to develop Jack's own self-interest, great enough in social life, to the point that will assure him a record in college more worthy of his natural gifts of intelligence, likeability, and popularity."

In a more relaxed, less competitive family and school, Jack might have been judged an above-average teen with good manners. But as a Kennedy and Choate student, he relied on whimsical irony and unrivaled charm to excel. Blessed with a fine winning smile and coolheaded demeanor, he was liked for his insouciance and good cheer. At the same time, he was constantly plagued with bouts of illnesses, was thin as a rake, and in 1934 was struck by a digestive disorder that caused fatigue, weight loss, and spells of pain. Physicians at the Mayo Clinic in Rochester, Minnesota, determined he had colitis. He spent the spring semester of that year in a hospital room filled with books, magazines, newspapers, and, somewhere underneath it all, a phonograph to play Bing Crosby records.

As he approached commencement in 1935, graduating 65th in a class of 110, the mere fact of his attendance at Riverdale, Canterbury, and Choate practically guaranteed admission to the Ivy League, whose schools were weathering the Great Depression by accepting nearly every applicant with an elite prep school on his résumé. In the fall he began studies at Princeton, but had to withdraw during his first semester due to illness. Upon his recovery, he set his sights on Harvard, his father's alma mater.

In the mid-1930s, the president of Harvard described his method of "fishing" for students from families with an annual income above five thousand dollars—this at a time when Jack himself would soon be receiving as much as *fifty* thousand dollars per year from a trust fund set up by his parents. But even if his prep school background and income hadn't made Jack a shoo-in, his father's new position as chairman of the Securities and Exchange Commission under President Franklin Roosevelt probably would have. It couldn't have hurt, too, that one of the reference letters submitted with Jack's Harvard application was written by Harry Hopkins, one of FDR's closest advisors. "I feel that Harvard can give me a better background and a better liberal education than any other university," Jack wrote in his application. And in fact, Harvard accomplished both those goals.

Kennedy started at Harvard in the fall of 1936, and by his junior year his innate interest in history had led him to studies in government, political science, and foreign relations. Yet still he rebelled against leading too proscriptive a life. Life in Cambridge was a steadying influence, but the prankster playboy ways Kennedy had developed at Choate still occupied his weekends, when he would typically journey to New York City for parties and nightclubbing. By this point, his family had created a new home base, in the town of Hyannis Port, a village on Cape Cod, in Massachusetts. Jack thrived on that peninsula during the summer months, swimming and sailing, or driving along the Atlantic coast in a convertible with other hellcats from the wealthy towns nearby.

During Kennedy's school years, rocketry was a popular fad in Germany—in comics, in novels, and in cinema. In the German public's imagination, it seemed that it would be just a matter of time before space travel became an empirical reality. Yet for all the optimism about rockets, there was a creeping concern within Weimar Germany that the nation was leaning increasingly toward fascism, particularly after the 1930 election.

In 1930, eighteen-year-old Wernher von Braun arrived at the Technische Hochschule Berlin, enrolling in the college's engineering school but focused entirely on studying mechanical engineering, with an eye on rocketry. His hero was Dr. Hermann Oberth, a high school science teacher in Romania and rocket theoretician who had made a splash in the Central European media. "Oberth was the first, who when thinking about the possibilities of a spaceship, grabbed a slide-rule and presented mathematically analyzed concepts and designs," von Braun recalled. "I, myself, owe to him not only the guiding star of my life, but also my first contact with the theoretical and practical aspects of rocketry and space travel."

Von Braun met Oberth a couple of times in 1930. The rocketry concepts developed by Oberth would remain imprinted on von Braun's spongelike mind even as he diverged from his hero over one of the most important issues facing their scientific community: the militarization of rocketry by the German government.

In January 1933, Adolf Hitler, an Aryan racist genocidal provocateur against Jews, Slavs, and countless others, was appointed chancellor of Germany and quickly consolidated power. In this charged atmosphere, the rocketry community was divided. Some distrusted Hitler's Nazi Party outright. But the young von Braun led a contingent that saw full-bore cooperation with the military as the path to space rocketry success. With Oberth having returned to his home in Romania, von Braun entered enthusiastically into the German Army's new rocket program. On December 18, 1934, von Braun launched two advanced A-2 rockets, named *Max* and *Moritz* (after German comic characters), to altitudes of 6,500 feet.

Both Kennedy and von Braun were primed with an inevitable sense that they would be called to military duty if World War II erupted. Each was patriotic and wanted to serve his country. By the late 1930s, global events were setting the stage for a cataclysmic conflict between fascism and democracy. The difficult, somber early years of the Great Depression were giving way to a gathering crisis in Europe, as the world careened toward the measureless suffering of another Great War. Recognizing that the United States might soon be drawn into the European conflict, and also yearning for government funding, Robert Goddard tried to interest President Franklin Roosevelt's top generals and admirals in the development of long-range military rockets, but most of America's military leaders considered his ideas marginal or even crackpot. Even the engineers at the NACA, in Hampton, Virginia, had to abandon more visionary aerodynamic experiments in favor of practical military aviation advancements. However, when a NACA delegation toured Germany in 1936, they were flabbergasted by the sophistication of the Third Reich's aeronautical technology and enthusiasm for rocketry.

Major Jimmy Doolittle was also aware of Germany's daunting aviation hardware. Born in Alameda, California, in 1896, Doolittle became interested in space science as a boy. By the time of World War I, Doolittle was an accomplished U.S. Army Air Corps pilot, serving as a flight instructor in Ohio. After the war, he flew a de Havilland DH-4 in the first cross-country flight, from Jacksonville Beach, Florida, to San Diego. Besides being a brave pilot, Doolittle was a master of aviation technology, having earned a doctorate in aeronautical engineering from MIT, and helped develop new flight instruments that let pilots navigate through fog, cloud banks, darkness, and rainstorms. He was also the first U.S. pilot to recognize the psychological effects of flight, especially how a pilot's hearing and vision were affected by high altitudes. As a civilian in the 1930s, he routinely shattered aviation speed records.

In 1937, Doolittle went to Germany on Shell Oil

business and was stunned to learn that Hitler was mass-producing modern fighter planes and bombers at a frightening rate. Touring airplane factories such as Junkers, Heinkel, Dornier, Messerschmitt, and Focke-Wulf, Doolittle realized that the United States was way behind the Germans. Once back in America, he met with Army Air Corps leaders, sounding the Paul Revere–like alarm. In October 1938, Doolittle traveled to the New Mexico desert to discuss rocket propulsion with Goddard. While other military men held a low opinion of rocketry, Doolittle wrote a memo praising Goddard, detailing how a rocket could be used in warfare. The memo ended with a reference to Goddard's spaceman reputation, admitting that while "interplanetary transportation" was a dream of the very distant future, "with the moon only a quarter of a million miles away—who knows!"

For the most part, Goddard was protective of his work during the 1930s. Frank Malina from Caltech tried to convince Goddard to join a rocketry project at his university, but he refused. In a September 1936 letter, Goddard wrote disparagingly of Malina to Caltech's Robert Milikin. Goddard commented that he had tried to help Malina with some of his questions, but "I naturally cannot turn over the results of many years of investigation, still incomplete, for a student's thesis."

In the summer of 1937, just before his sophomore year at Harvard, Jack Kennedy traveled through England, France, Italy, Austria, Germany, and the Netherlands in an automobile with a school friend named Lem Billings. Avoiding first-class travel to accommodate Billings's modest budget, the two met and mingled with people from outside Kennedy's typically upper-crust milieu. To speak with people beyond the tourist areas, Kennedy insisted that they pick up hitchhikers, especially in countries where publicly conversing with an American was deemed a bit risky. Seeking libertine pleasures was also built into the itinerary.

In Germany, the rise of fascism intrigued both Jack and Lem.

Despite newspaper reports on the human rights abuses of Hitler's government and the overt militarism that fueled German economic growth, admiration for the Nazis was common among the U.S. and British upper classes in the mid- to late 1930s. Many American intellectual-corporate elites believed Hitler's rule had led to efficiencies in business and factories and to social stability. Both Joe Kennedy Sr. and Jr. shared these views. Joe Sr. was astonished that the hyperinflation of the early 1920s, which left the German market shattered in economic depression, had been replaced in the 1930s by robust economic growth. An equally impressed Joe Jr. thought that Germany had bested the United States in railways, aviation, medicine, forestry, and, quite ominously, social engineering.

One reason for Jack's European trip that summer was to judge for himself the situation in Germany since Hitler's rise—an admirable independent-mindedness, considering his father's and Joe Jr.'s strong opinions. Arriving in cities such as Berlin, he immersed himself in local culture, observing closely and jotting down impressions in his diary. By the time he and Lem arrived in Munich in August, trade unions and all political parties other than the Nazis had long since been dissolved. By rebuilding the military, including the Luftwaffe (air corps), and moving German troops into the disputed Rhineland, Hitler had abrogated the Treaty of Versailles. Meanwhile, military pacts were being formalized with Italy and Japan. The Dachau concentration camp was already in operation. Jews were being segregated and stripped of their most basic rights, and Romanies (Gypsies), Slavs, Jehovah's Witnesses, and homosexuals faced daily persecution. Hitler was actualizing the Aryan supremacy theories first spewed in his book *Mein Kampf*, which he wrote while in jail for treason.

During the trip, Jack was amazed by the quality of

Germany's new roads, train depots, and dams but he scorned the outward trappings of Nazi fanaticism. Fueled by German beer, he and Lem mocked their way through the Nazi "Heil Hitler" as they made their way across the country. On one occasion, traveling near Nuremberg, Jack "had the added attraction of being spitted on" for his antics. Nevertheless, his overall impression was positive. "The Germans really are too good," he wrote in his journal. "It makes people gang up against them." Lem, in his own diary of the trip, recounted the conclusion drawn by two Ivy League travelers while speeding down the autobahns in their Ford Cabriolet: that the broad new highways had a military purpose first and foremost. They knew increasingly, in their guts, that another world war was likely.

Back in the United States, fierce debates were erupting between internationalists and isolationists. President Roosevelt had no inclination to think about rockets as ballistic missiles—the whole concept seemed remote and ridiculous. The isolationists, including Charles Lindbergh and carmaker Henry Ford, were a diverse group, loud and well organized, and encompassing those who acutely remembered the horrors of World War I and those who hoped for German domination of Europe—which, by 1938 and early 1939, seemed more than possible. In March 1938, Roosevelt appointed

Joseph Kennedy Sr. as ambassador to the Court of St. James's, hoping that America's most famous Irish-Catholic businessman could influence millions of Irish Americans to drop their ethnic enmity and support Britain in the European conflict to come. More selfishly, FDR thought Kennedy could be an asset for his own reelection bid in 1940 (for an unprecedented third term).

As Jack Kennedy's interest in foreign affairs deepened, the timing of his father's appointment couldn't have been better. On July 4, 1938, the Harvard undergraduate sailed to England with brother Joe Jr., staying and working at the American embassy in London. There, Jack learned in a visceral way how European nations responded to Hitler's aggression against smaller countries such as Czechoslovakia.

After spending the fall semester of 1938 back at Harvard, Jack left for an extended overseas tour early in 1939. For the next few months, his experiences traveling in western Europe, the Balkans, and the Middle East, as well as working at the American embassy in London, would substitute for his Harvard coursework. By the time he arrived in London, his father's ambassadorship was tenuous. During his early tenure, Joseph Sr. had often expressed isolationist sentiments under the guise of arguing for peace. In at least one case, the State

Department had felt compelled to intervene, quashing his remarks before they went public. The Roosevelt administration was aghast. Joe Sr. apparently failed to understand that any opinions he expressed as ambassador had to remain within the bounds of the Roosevelt administration's foreign policy—a policy that was then walking a fine line. Given the strength of isolationist sentiment in America, official policy on developments in Europe remained guardedly neutral. But this was not the position of the president, who privately believed that the United States could not hope to remain aloof from a European conflict and who was working to subtly shift public opinion. Nazi-appeasing rhetoric from the American ambassador to Great Britain could do real damage, convincing fascist leaders that the United States would remain on the sidelines. To the surprise of some in the administration, FDR chose to retain Kennedy in his post. Among other things, he was reluctant to overreact: European diplomatic matters were delicate enough in the late 1930s without the turmoil that would ensue from the removal of a high-level appointee. In 1940, however, Joe Sr. went too far yet again, and FDR fired him.

While working unofficially on minor assignments at the London embassy, Jack made two visits to the Continent, including to the Soviet Union, Germany, and

Poland—traveling to the last just a few weeks before the Nazis invaded it. He also attended the coronation of Pope Pius XII in Rome, where his family was granted a private papal audience. These experiences, aided by Joe Sr.'s State Department connections, proved a vitally important part of Jack's education, but it was clear he was already in his element. Friends of the family were impressed by how much he knew about Hannibal, Caesar, Napoleon, and the rise and fall of great powers. U.S. ambassador Charles Bohlen, Jack's mentor while in Russia, noted, "We were all struck by Kennedy's charm and quick mind, but especially by his open-mindedness about the Soviet Union, a rare quality in those pre-war days. . . . He made a favorable impression." That September, Kennedy was in the Visitors' Gallery in the British House of Commons as Prime Minister Neville Chamberlain announced war on Hitler's Germany for attacking Poland with planes, tanks, and infantry. "For a twenty-two-year-old American," historian Richard Whalen wrote, "it was a unique opportunity to look behind the scenes as the stage was set for the Second World War."

Back at Harvard that same September 1939, Jack was fully engaged intellectually with regard to a Europe on the brink. Initially, he struck an isolationist posture like that of his father, authoring an unsigned

Harvard Crimson editorial, "Peace in Our Time," which implored the United States not to overreact to the Polish defeat by entering the hostilities. Especially problematic was his naïve belief that Hitler would disarm if allowed to run a puppet Poland and have a free hand in Eastern Europe. But as he worked on his honors thesis, "Appeasement in Munich," which detailed Neville Chamberlain's failed policy toward Hitler, his thinking shifted decisively.

In June 1940, Jack graduated cum laude with a bachelor of science degree. Later that year, his thesis was published in book form under the title *Why England Slept*, with a glowing introduction by Henry Luce, the *Time* magazine publisher, who was a close friend of Joseph Sr. With help from other family friends, it became a bestseller. For Jack, the book marked the start of an unending process of edging away from his father politically while continuing to benefit from the massive help Joe Sr. could provide. In this political realignment, Jack, the new interventionist, was not alone: with disdain for Hitler's aggression on the rise, American isolationism was ebbing, and it was Jack, not his statesman father, who had correctly read the tides of history.

What Jack had learned during his European trips in the 1930s was that military technology was a prior-

Future American president John F. Kennedy sits at a typewriter in 1940, holding open his published Harvard thesis, *Why England Slept.*

ity for the Nazi regime, and so was secrecy about the many projects undertaken. The disarmament provisions of the Treaty of Versailles meant that the German military had mastered the art of both concealment and opportunism, seizing on the treaty's catastrophic omission of rocketry to leapfrog their enemies technologically.

After World War I, several nations were vying to see who could develop and test the first jet airplane. Italy won with its successful prototype in 1940—or so it thought. Unknown to the Italians, a German Heinkel jet had already been flying for a year. Equal secrecy was accorded to the German Army Ordnance's rocket development group, headed by Dr. Walter Dornberger, a hardworking artillery officer who had been captured by U.S. Marines during World War I and spent two years imprisoned in France. After the war, he studied mechanical engineering in Berlin as a junior officer. In April 1930, Dornberger was appointed to Weapons Testing of the German Army (*Reichswehr*) specializing in ballistics and munitions, tasked with clandestinely designing a solid-propellant rocket that could be mass-produced and transcend the range of traditional artillery.

One of Dornberger's primary assistants was the young Wernher von Braun. Years later, Dornberger recalled how impressed he was by von Braun's opportunistic zeal, calculated shrewdness, and "astonishing theoretical knowledge." Von Braun was soon leading Dornberger's research team at the German military's rocket artillery unit at Kummersdorf, about thirty miles south of Berlin. It was here that von Braun began developing and testing the first of his Aggregat series of rockets, the A-1 and A-2.

By the mid-1930s, Dornberger and von Braun's rocketry work had attracted the support of the Luftwaffe, Nazi Germany's newly formed air force. And the project enjoyed support from the top ranks of the army. In 1936, Colonel Professor Dr. Karl Becker, head of the Research and Development Division of the German Army's Ordnance Department and a longtime rocketry supporter, gave Dornberger some advice: "If you want more money, you have to prove that your rocket is of military value."

Dornberger and von Braun drew up the specifications for a game-changing ballistic missile. Their description was based on an extra-long howitzer, like the monumental German weapon used against Paris near the end of World War I. Then the longest-range artillery weapon known, it was capable of lobbing massive shells to a target eighty miles away. Looking for even more destructive capability, Dornberger and von Braun planned a single-stage rocket-propelled missile that would be launched vertically and then programmed in flight to an elevation angle of forty-five degrees. The rocket they envisioned would carry nearly one hundred times the weight of the explosives in one of the advanced howitzer shells and have a range twice that of the big gun. Designed for "arrow stability," the rocket would be limited in length to not more than forty-two feet, which would allow it to be transported in a single piece, either by truck on normal roads or on a single railroad car; also, its over-the-fins diameter needed to be below nine feet, so it could fit through all European railroad tunnels. It was quickly calculated that a burnout speed of 3,600 miles per hour would be required for the rocket to achieve its military objectives.

Colonel Becker authorized the rocket's development, but the Kummersdorf base was deemed both too small and too close to population centers to guarantee secrecy. In the fall of 1937, following the advice of Dornberger and von Braun, the Third Reich founded an enormous top-secret war rocket facility at Usedom, an island off the coast of the Baltic Sea. Christened Peenemünde after a small fishing village nearby, the Heeresversuchsanstalt (Army Research Center) officially separated from the Luftwaffe in 1938, around the time of Jack Kennedy's happy-go-lucky summer tour of Germany, and soon became the most modern rocket research-and-development station in the world. With von Braun and a staff of about three hundred drawn mainly from military ranks, Peenemünde brought the full weight of German rocketry expertise under one roof, with the manpower and facilities to build and test the weapons that would give Hitler control of the skies. When the facility was at peak production, thousands of soldiers, employees, foreign laborers, and prisoners toiled to build German rockets there.

Secrecy reigned at Peenemünde. Even within the German Army, very few people were aware of the effort to turn rockets into long-range military weapons. In the project's first years, regular rocket launches accelerated the research, but required additional staff. The army began to siphon top talent from other army programs and universities and send them to the remote Baltic base (which, in addition to rocket development, also hosted units developing a winged cruise missile eventually called the Vergeltungswaffe 1, or V-1). Between 1937 and 1941, von Braun launched more than seventy next-generation Aggregat rockets there, including the A-5, a scaled-down test model of the proposed A-4, which would soon be known to the world as the V-2, an abbreviation of Vergeltungswaffe 2, or "Vengeance Weapon 2."

Rocketry's destructive potential was the overriding goal, but its greater potential was never far from von Braun's mind. On one occasion he interrupted a presentation for Colonel Becker on the progress of armed rockets to talk enthusiastically about breaking gravity's grasp on mankind and "going to the moon." And it was true: the same V-2 rockets that were being developed as long-range artillery could conceivably be adapted to spaceflight. But Becker and Dornberger's willingness to indulge von Braun's enthusiasms went only so far. Afterward, von Braun's superiors, with faces that looked carved out of stone, sternly forbade him from talking about space travel in front of other officers—and especially in front of Hitler, with whom he occasionally interacted. Von Braun was ordered to channel his moon enthusiasm into helping the Third Reich control Europe by developing military rockets.

There is currently a debate over how enthusiastic the Nazi rocketeers were in supporting Hitler's war machine. Some were supportive, it seems, while some were enthusiastic, others were opportunistic, and a few were opposed. Biographer Michael Neufeld says that von Braun fell into the opportunistic camp. The rocket engineers who truly opposed the Third Reich (such as Willy Ley) left Germany. Hitler, preoccupied with Germany's hegemony in Europe, wasn't much interested in space research on any level. His initial indifference toward von Braun's project stemmed from disappointments with such wizard weaponry during World War I. One blustery spring day in 1939, Hitler arrived in Kummersdorf accompanied by Field Marshal Walther von Brauchitsch, General Becker from Army Ordnance, Deputy Führer Rudolf Hess, and Martin Bormann, Hitler's personal secretary. This Nazi leadership squad had come to tour the facility and witness test engines. Smiling happily, full of bravura, and eager to show off his wares, von Braun tried to sell the Führer on the prototypes of his liquid-fuel rockets, touting their potential as weapons that could, in a year, be unleashed on France and Britain. On one level, Hitler was dismayed by von Braun's fantasy weapons. ("Even now I still don't know how a liquid-propellant rocket can fly," he said. "Why do you need *two* tanks and *two* different propellants?") On another, he was frustrated that even though Peenemünde and Kummersdorf were the leading rocket research centers in the world, progress was coming only in fits and starts.

Von Braun was disappointed that Hitler didn't realize that at Kummersdorf and Peenemünde the Germans were inventing the ballistic missile future, and that budgetary restrictions were their largest obstacle. "There were thousands of major problems for which there was no answer at that time," Dornberger later recalled of the work at Peenemünde and Kummersdorf. Over the first two years at Peenemünde, the team made steady but slow advances toward its goal of having ballistic missiles ready by 1943, but considering the state and pace of their research-and-development efforts, that deadline did not seem possible. Nevertheless, on September 5, 1939—four days after the Nazi war machine invaded ill-prepared Poland, and Jack Kennedy heard Neville Chamberlain declare war on Germany in the House of Commons—Dornberger promised a finished rocket for 1941.

It was a risky gamble, but the payoff was worth it: an order from Field Marshal von Brauchitsch preserved the Peenemünde program's funding and protected key personnel from transfer to military units. At the end of October 1939, von Braun successfully launched his A-5 rocket. The time for mass production was near. As for any true engineer, one test result for him was worth one thousand opinions. "It was an unforgettable sight," von Braun recalled. "The slim missile rose slowly from its platform, climbing vertically with ever-increasing speed and without the slightest oscillation until it vanished in the overcast."

Starting in September 1940, Jack Kennedy took classes as a Stanford University graduate student in business, economics, and political science, but his mind was on the war in Europe. As of May 1940, when Chamberlain resigned as Britain's prime minister, Kennedy's new all-seasons hero was Winston Churchill, who had formed a coalition government. By the time Hitler invaded the Netherlands and assaulted France, Kennedy was firmly in favor of U.S. intervention. By that summer, almost all of western Europe had fallen to the Nazis, and only Great Britain remained. From August through October, the Luftwaffe unleashed wave after wave of Junkers and Heinkel bombers, guarded by Messerschmitt fighters, on a systematic campaign against British towns, cities, fleets, ports, airfields, and radar bases. "The shocking German success and the dire threat to England forced Jack," biographer Michael O'Brien wrote, "like many Americans, to revise his thinking about the war and America's role in it."

Hitler had put his faith in the Luftwaffe and its proven aircraft to destroy Britain's Royal Air Force

(RAF), paving the way for British surrender or, as a last resort, an invasion by German troops. The theory was sound, except for one thing: in the face of relentless attacks, the British and their airmen grew tougher, not weaker. As the campaign ground on, Hitler became impatient, directing the Luftwaffe to switch gears from engaging the RAF in aerial showdowns to bombing civilian targets in London, Coventry, and other large cities. But the Blitz, as it became known, only strengthened Britain's stoic perseverance.

Frustrated, Hitler shocked the world in June 1941 by turning his attention away from the British campaign and toward the east, attacking Germany's supposed ally the Soviet Union with a lightning-fast blitzkrieg offensive involving more than 3.5 million German, Finnish, and Romanian troops and 3,500 tanks—the largest invasion force in history. Virtually all German bombers were redirected from Britain to the USSR. Initial Soviet losses were so devastating that Hitler and his high command began confidently planning their next move, prioritizing conventional air and naval resources for a renewed push against the British. In a war economy riven with infighting, conflicting priorities, and raw materials shortages, this meant fewer resources for rocket development.

One of the problems Dornberger and von Braun

faced in this competition was that Hitler's support for their rocket program was erratic. Germans faced steel shortages and other economic demands that stunted the fast growth of the rocket program. According to most sources, Hitler was skeptical of Germany's being able to launch full-blown rocket attacks on other European countries from the safety of the homeland. Faith in missile rockets meant no pilots, no soldiers, no ships, and no sailors—a strange new reality that was hard for him to fathom.

In this make-or-break moment, Dornberger and von Braun tried a new tactic to persuade Hitler of rocketry's military efficacy, arguing for its effect as a psychological weapon. In a meeting with Hitler on August 20, 1941, they made the case that their planned rocket could succeed where hordes of Luftwaffe aircraft had failed, and finally break the morale of the British people. Traveling at 3,500 miles per hour, the rocket would appear seemingly from out of nowhere, they argued. By the time anyone on the ground saw it coming, only seconds would be left before it slammed thousands of pounds of explosives into a crowded neighborhood, destroying blocks. Dornberger and von Braun sweetened their pitch by promising cooperative projects with the Luftwaffe and, significantly, rockets for even more distant targets. The stabilizing wings and two-stage propulsion they envisioned would enable a rocket to reach the big eastern cities of America. Hitler was adequately convinced, by both the A-4's destructive capabilities and its potential to instill terror in the British populace, and gave the project his lukewarm backing, subject to final testing. Despite the cost (far higher in both manpower and resources than more conventional weapons), the A-4 appealed to Hitler's sense of his own mythic power and Germany's technological superiority. "The Führer," wrote Dornberger after the meeting, "emphasized that this development is of revolutionary importance for the conduct of warfare in the whole world."

Over the course of a half century, American, German, and Soviet dreams of putting a human on the moon threaded past numerous vital turning points. World events, both large and small, accelerated rocket engineering at an astonishing rate. As the complexity, scale, and malevolence of Hitler's drive for world domination became manifest, his August 1941 turn on the subject of rocketry was a critical pivot point. No other nation at the time had Germany's momentum in the field, paired with a dedication to leveraging new science for secret weaponry. Although the connection wasn't fully understood at the time, Hitler's commitment to the V-2 advanced the pursuit of a moonshot by perhaps decades. Though Hitler had no expressed interest in reaching the moon, the uncomfortable fact is that the darkest shafts and foulest backwater of human savagery helped bring this loftiest of human dreams to reality. Indeed, the engineers at Peenemünde were solving essential questions about celestial navigation and mechanics, about how to innovate easily applicable ways of determining position and velocity when away from Earth's surface, which would prove all-important in future U.S. lunar voyages. German ballistic missile technology—built to kill people—laid a foundation for spaceflight.

Had Hitler demurred in the August meeting and continued only a halfhearted accommodation of the strange new technology, the course of World War II might have changed. According to many senior Nazi officers at the time (and some military historians since), Hitler's commitment to the V-2 actually *decreased* Germany's chances of winning the war. The voracious appetite of the Peenemünde project drew off resources when an accelerated program for conventional weapons and Luftwaffe aircraft might have made for a stronger German military machine. But to Hitler, the V-2's perceived value went beyond dollars and cents. Over the following years, as the tides of war shifted, he came to see it as a superweapon that could finally deliver German victory over the Allied nations.

Portrait of Ensign Joseph P. Kennedy Jr. and Lieutenant (jg) John F. Kennedy in their Navy uniforms. The photograph was taken in May 1942 at Turgeon Studios.

3
Surviving a Savage War

He is really home—the boy for whom you prayed so hard . . . what a sense of gratitude to God to have spared him.

—ROSE KENNEDY, DIARY ENTRY, JANUARY 1944

Jack Kennedy knew all the rules in sailing, but the rules didn't know him. The fiercer the Atlantic wind, the choppier the whitecaps, the more exhilarated Kennedy became. He longed to test his mettle against the raw elements, to feel part of the barrel and roar of the sea, to infuse his days with wind-whipped adventure. Some called his extreme sailing reckless, others a death wish, but the fact was that Kennedy simply possessed a "blue mind," experiencing his greatest contentment at or beside the sea. Whether speed-racing in the Edgartown regattas, beachcombing around Hyannis Port, collecting shells on Monomoy Island, challenging the inward-pressing tide on Nantucket Sound, or suntanning in Palm Beach, Kennedy was his most authentic self, his freest in the old transcendentalist sense, near the ocean.

That love of a maritime environment guaranteed that Jack Kennedy would join the U.S. Navy as war engulfed the world. On September 25, 1941, at age twenty-four, he was sworn in as an ensign. Because of his grim history of physical ailments, his commission had been anything but automatic, requiring that his father help arrange the appointment via a former colleague. That Joseph Kennedy Sr. could facilitate such things wasn't news to his two oldest sons, but increasingly that overbearing influence had to be contained. Jack's older brother, Joe Jr., was very cognizant of not getting smothered by his father. Taking a break from Harvard Law School and politics—he had been a 1940 Democratic Convention delegate—Joe Jr. had joined the Naval Reserve and, during the second half of 1941, was training to become an aviator. Although he had dabbled in anti-interventionist thinking, he was now fully committed to helping Great Britain defeat Germany.

Joe Jr. earned his naval wings in March 1942 and was already thinking of his postwar political career. Strikingly handsome with a touch of arrogance in his smile, he had a lofty conception of obligation and devotion to America, and believed that with ardor, focus, and drive, there was nothing he couldn't achieve. Though he never explicitly expressed the desire to be a senator or governor, a sense of destiny swirled around him. Occasionally he hinted that being president was a noble pursuit. To his credit, he intuited the danger of being tarred by his father's reputation for isolationism and appeasement, which had fallen out of favor in the Democratic Party. Indeed, some scholars believe he joined the navy with "a private mission" to prove that the Kennedys "were not cowards or defeatists." The Kennedy brothers were fitting themselves into navy service, determined to test their patriotic grit along with millions of other Americans. But more than most families, the Kennedy children coveted their status within the confines of their own clan.

When the Japanese bombed Pearl Harbor on December 7, 1941—"a date which will live in infamy," as President Roosevelt termed it—Jack and Joe Kennedy were already in uniform and ready for combat duty. They would never have to scrub off the taint of using their father's influence to avoid military service. Four days after Pearl Harbor, Nazi Germany and its Axis partners declared war on the United States. "Industrial mobilization" became an urgent catchphrase across the nation. Military bases were built on empty land. Factories were adapted to switch from manufacturing consumer goods to producing war matériel. Between 1940 and 1943, enlistment in the U.S. Armed Forces expanded from fewer than five hundred thousand to more than nine million. Mobilization efforts swelled in every direction, from far-flung Honolulu to the shipyards of Norfolk.

Jack Kennedy, it seemed, was headed for a wartime career at the Office of Naval Intelligence, in Washington. With his precarious health, vaguely journalistic ambitions, and rarefied background as a best-selling author, a stateside desk job that involved researching and writing reports seemed the ideal post. However, Kennedy had no intention of staying put in Washington, and he almost immediately began jockeying for a naval combat role in the Pacific. Trying to prove his valor, he hopscotched around the country from one military base to another, from Rhode Island to South Carolina to Illinois, and he stayed on course, despite back surgery that required a two-month respite in the hospital.

While Jack Kennedy was exchanging his rich, footloose lifestyle for the disciplinary navy, Wernher von Braun had made what amounted to a Faustian bargain for the advance of rocketry. He joined the Nazi Party and became an SS officer. He later argued that he'd had no choice, claiming he was watched carefully by the Gestapo for any sign of disloyalty. It must be observed, however, that some of von Braun's contemporaries found ways to resist the fascist regime, often perpetrating small, undramatic acts of slowdown, sabotage, or resistance. But not the coddled team at Peenemünde. Many of these engineers were Hitler loyalists who looked forward to the day when Aryan Germans would have the promised Lebensraum (room to live) granted by conquest of Europe. Desperate in later years for an excuse to vindicate himself from his close association with Hitler, von Braun would contend that he was merely an earnest engineer who put his craft above all else. This excuse didn't prevent him from enjoying his wartime position or being fêted by the Third Reich's rich and powerful.

What wasn't disputed was the intense demands von Braun faced at Peenemünde. The pace was grueling, and he worked long hours. Fellow SS officers found him fast-minded, articulate, brusque—only his proclivity for inserting literary and historical allusions in everyday conversation made him unusual. Hitler had set a goal of making five thousand V-2 ballistic missiles annually; frustrated, von Braun knew that was an improbability. The unfortunate reality was that a practical version of the V-2 had yet to fly, despite the many test launches he and an ever-larger corps of engineers fast-tracked.

As a perfectionist, von Braun, in later years, was fond of saying that "crash programs fail because they are based on the theory that, with nine women pregnant, you can get a baby a month." But time was of the essence for the Third Reich. The first official V-2 test rocket lifted off from Peenemünde in July 1942, reaching an altitude of one mile before blowing up over the Baltic. With the Eastern Front in a stalemate following Germany's monumental failure to defeat Russia, armed rockets were needed for a renewed campaign against Britain. Another V-2 was tested in August 1942 and reached an altitude of seven miles before disintegration.

Then von Braun and Dornberger, the military commander of the rocket team, had a success. On October 3, 1942, a V-2 launched from Peenemünde broke the sound barrier and traveled to an altitude of 52.5 miles at a range of 120 miles, nosing into the ionosphere, that wide band of extremely thin air that separates earthly atmospheric bands from outer space. The test marked the first time a man-made projectile had technically ever flown beyond the bulk of Earth's atmosphere. Their first inclination was to think about how much the new weapon would help the Third Reich military. Deep down, they also knew that the test had longer-term implications for future space travel. This "was a true ancestor of practically every rocket flown in the world today," says historian Paul Dickson. "It was a true spaceship in that it carried both its own fuel and oxygen and could, if needed, work in a vacuum."

Grasping the significance of the moment, von Braun and Dornberger kept their eyes glued to binoculars that game-changing afternoon until their forty-six-foot-long liquid-propellant missile disappeared. Some historians have erroneously claimed that October 3, 1942, ushered in the epochal Space Age. Such statements about the V-2 are questionable because it never reached the one-hundred-kilometer altitude until after the war. But the V-2 did mean that all that would be needed for futuristic moonbound rockets was increased thrust for larger payloads. "It was an unforgettable sight," Dornberger recalled. "In the full glare of the sunlight the rocket rose higher and higher. The flame darting from the stern was almost as long as the rocket itself. The air was filled with a sound like rolling thunder." While von Braun might have preferred to build space exploration vehicles, during the war years they found themselves working on ballistic missiles, rocket engines for aircraft, and other lethal machines.

In early 1943, setbacks in the Mediterranean theater convinced Hitler, previously a skeptic, that the V-2 was now essential to victory, putting Peenemünde at the nerve center of Nazi war plans. When von Braun and Dornberger showed Hitler film footage of their V-2 success that July and promised him that the rocket could deliver 2,200-pound warheads across the English Channel into London and other cities, Hitler grew visibly intrigued. Inflated with revenge, he hoped to use the terror weapon on civilian targets in England as payback for the Allied bombings of German cities. "Europe and the world will be too small from now on to contain a war," Hitler said. "With such weapons humanity will be unable to endure it."

As von Braun and Dornberger inaugurated the V-2 Age, Jack Kennedy was completing the Naval Reserve Officers Training School at Northwestern University in Chicago. The previous June, at Midway, the United States had halted the Japanese naval advance in the Pacific. Two months later, at Guadalcanal, in the Solomon Islands, U.S. troops halted Japan's island-hopping advance toward Australia. During those tense times, Kennedy attained the rank of lieutenant, junior grade, on October 10, 1942. He immediately volunteered for the patrol torpedo (PT) boat service, a fairly new command whose development had been accelerated following the attack on Pearl Harbor. PT boats were long and fast, with a relatively low profile and a top speed of forty knots (about forty-six miles per hour). Modestly armed, they were intended mainly to interrupt enemy supply lines by sinking barges, freighters, and sometimes the warships that accompanied them. The most intimidating of the enemy ships that might be encountered on the supply routes in the Pacific Theater was a Japanese destroyer, which averaged a speed of thirty-four knots and lacked the PT's maneuverability. Attacking in groups and buzzing around their Japanese targets, PTs were known to American sailors as "mosquito boats."

Looking something like an enlarged, fortified racing speedboat, each PT boat carried two officers and a crew of eleven. Not wanting to waste experienced career officers and other key personnel on such small-scale commands, the navy sought graduates of first-rate colleges, preferably men who had played team sports. An even higher priority was given to graduates who sailed or motored their own boats in private life. Because recruiters favored college-educated yachtsmen most of all, leadership of a PT boat became something of a snob assignment. But there were two catches: PT skippering was extremely dangerous, and the navy wanted men who weren't married. Jack Kennedy was suited in all respects except one: he hadn't been a leader in any organization, with the possible exception of his gang at Choate—and they'd all come close to expulsion for his offbeat pranks. And while he was a fine sailor on Nantucket Sound and Buzzards Bay, and had won sailing races, there was a huge difference between taking day outings in Atlantic waters and commanding a boat under the stress of war.

Despite their yachting-set cachet, PT boats weren't frivolous or experimental; they were choice weapons in the Pacific Theater. And unlike rockets, they could be mass-produced easily by companies such as Elco Motor Yachts, in Bayonne, New Jersey, and Higgins Industries, in New Orleans, Louisiana. William Liebenow of Virginia, who joined the PT boat service a few months before Kennedy, perfectly summarized the navy's need: "Our big-ship navy had just about been destroyed at Pearl Harbor," he explained. "So PT boats were kind of something that they could manufacture quickly and get them out to the war zones to harass the Jap fleet as much as possible."

In March 1943, Kennedy was given command of *PT-109* as a lieutenant. He was ordered to the Solomon Islands and took command on April 23. At the helm, he initially participated in preparations for troop movements or invasion. That summer, nighttime patrols for enemy supply barges were also ongoing. All the enlisted men admired Kennedy from the get-go. Not only was the lieutenant a gallant leader, but he was also one of the boys and experienced the same homesickness as the rest. One afternoon, he tore the *PT-109* patch off his uniform and mailed it in a letter to a cousin struggling in boarding school. "I'm not so crazy about where I'm at either, kiddo," he wrote. "Be brave, wear the patch, and we'll get through this."

Naval lieutenant John F. Kennedy on board the torpedo boat *PT-109* he commanded in the southwest Pacific.

Early on August 2, 1943, *PT-109* was moving as quietly as possible in Blackett, North Solomon, a strait used for a so-called Japanese Express of supply ships and escorts fortifying troops on nearby islands. At about 2:00

a.m., the *109* was mostly powered down. With Lieutenant Kennedy and the crew stealthily looking for a floating target in the channel, their boat was rammed by a Japanese destroyer. Ten seconds later, Kennedy's vessel had broken apart and was sinking fast. Two crew members had died. The *PT-109* went up in flames. Stranded in the middle of the Pacific, an injured Kennedy later said he didn't know exactly what had caused the disaster, whether human error or the bad luck of their boat being effectively unmaneuverable under low power. All he knew was that death was knocking. "People that haven't been there certainly can't, they just can't understand how Jack Kennedy got hit by a destroyer," observed Liebenow, "but I can see where it could happen."

Clinging to debris, Kennedy remained coolheaded in the shipwreck's aftermath. He saved the badly injured Patrick McMahon by pulling him in a three-hour swim to a small island a few miles away. Over the course of the next twelve hours, all the survivors miraculously made it there. Desperate to stay alive, Kennedy ordered his crew to swim to another island, where food and water were available. Thinking fast, Kennedy carved an SOS message in a coconut and handed it to two natives willing to row in a primitive boat seeking help for the marooned Americans.

On August 8, six days after *PT-109* was rammed, the

badly sunburned surviving crew and their commander were rescued. Kennedy was welcomed back to the PT base as a full-fledged naval hero, earning the Navy and Marine Corps medals for leadership. Although he found it hard to explain how his boat had drifted into the middle of the sea alone, he certainly wasn't blamed for the disaster; instead, he was almost immediately assigned to a different PT boat. "Most of the courage shown in the war came from men's understanding of their interdependence on each other," Kennedy would later reflect. "Men were saving other men's lives at the risk of their own simply because they realized that perhaps the next day their lives would be saved in turn. And so there was built up a great feeling of comradeship and fellowship, and loyalty."

By the beginning of the summer of 1943, Heinrich Himmler, leader of the German SS, with assistance from Walter Dornberger and Arthur Rudolph, brought hundreds of slave laborers to Peenemünde to work on rocket assembly lines as part of an effort to fast-track V-2 production. They were following orders given by Albert Speer's Armament Ministry. The Baltic facility was now operating at a frenetic pace.

On the night of August 17–18, 1943, a fleet of 596 British bombers, one of the largest air raid forces ever

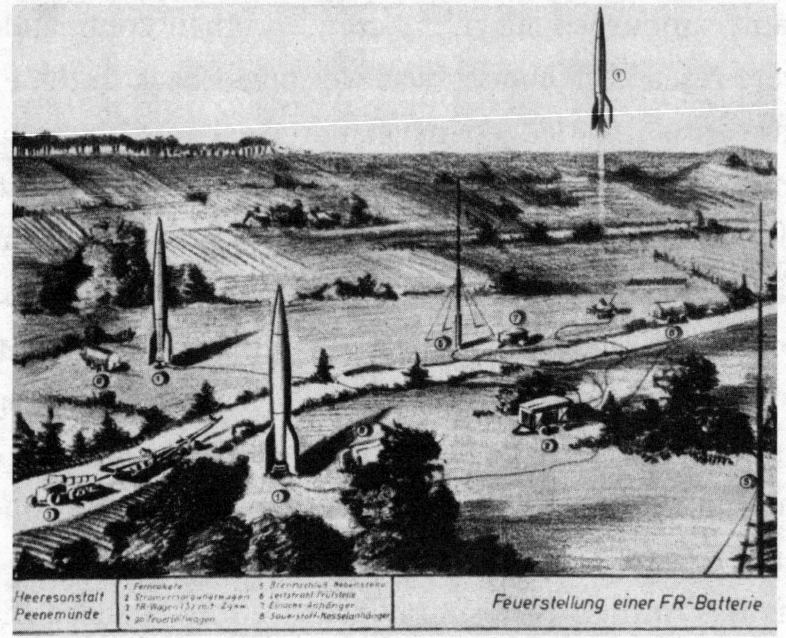

Heeresversuchsanstalt (Army Research Center), Peenemünde, 1942–1945, draft for a V-2 rocket launching position. The drawing was done on June 3, 1942.

assembled by the RAF, filled the skies over northeast coastal Germany on a mission to find Peenemünde, which it did despite clever camouflage that included artificial fog. For two hours, British bombs rained down on the Baltic Sea facility, killing 735 people, including more than 500 "foreign workers," some of whom were POWs. At first light, Dornberger's test facilities seemed devastated, but it soon became clear that the rocket development facility was still operable. "After four weeks of cleanup work," Dornberger later bragged, "Peenemünde worked full-time again. . . . In the case of the [overall] V-2 offensive, the bombing neither delayed it nor reduced it to any extent."

The British RAF attack on Peenemünde forced the German Army and the Armaments Ministry to establish new manufacturing sites at hard-to-find locations. Seeking a more secure spot for production of their ballistic missiles, the Nazis developed a secret underground facility carved inside Kohnstein Mountain, near Nordhausen, in the Harz Mountains. Using concentration camp slave labor, they transformed a tunnel complex into a facility that could produce thousands of V-2s. The location in the Harz Mountains was already a top-secret storage facility for the Nazis. Now it would be known as Mittelwerk (or "Central Works"), and it would employ slave labor from Dora, a subcamp of the Buchenwald concentration camp erected inside the tunnels themselves. The German aim was to avoid RAF and U.S. Army Air Corps bombing strikes, which had been destroying cities and industrial plants.

The burden of preparing the Mittelwerk factory under the topographical restraints of a mountain area should have taken years, but von Braun, Dornberger, and their powerful Nazi Party colleagues in charge of the top-priority program didn't have years. They barely had months. With Hitler's strongest armies having been decimated by the Soviet Union on the Eastern Front and Allied troops seizing control of southern Italy, the V-2 could buy time to secure the Third Reich's survival, but only if it arrived quickly.

Conditions at Mittelwerk and Dora as 1944 began were nothing short of a living hell: there was no fresh air, little water or food, vermin and lice, plumbing that consisted of open barrels, and grueling work without end—slave laborers were tortured and beaten if caught working at less than a double-time clip. These physical strains were combined with the oppressive knowledge that illness or injury might mean instant execution by their Nazi overseers. The daily piles of corpses awaiting cremation—upon their arrival, the inmates were greeted with a somber Nazi induction speech of the "endurance or death" variety. Only some Italians were granted POW status.

More than twenty thousand slave laborers perished from disease, exhaustion, malnutrition, torture, and beatings at Dora while building V-1 and V-2 rockets throughout 1944. Words can never aptly describe how difficult life at Dora-Mittelbau was for its workers. Von Braun, a colonel in the SS, was deeply complicit in these war crimes. He was a regular visitor to Mittelwerk.

In later years, a disingenuous von Braun would profess to have been just an apolitical space lover from the ex-Weimar amateur rocket group tasked by the Nazis to engage in military missile development. He repeatedly pointed out that Himmler had the Gestapo arrest him in March 1944, supposedly for sabotaging war rocket projects. Incarcerated by the SS for the crime of articulating "frivolous dreams" of rockets orbiting Earth and the moon and expressing doubts about the war, he was tossed into a Stettin prison cell, and feared for his life. However, Dornberger and Reich Minister of Armaments Albert Speer vouched for him, and he was released under strict orders never to utter a word or even think about space exploration. After the war, trying to escape culpability for the Mittelwerk crimes, von Braun flaunted this prison story as inoculation against charges of Nazi collaboration.

A back injury John Kennedy suffered in the sinking of *PT-109* led to his medical discharge. Soon after his return to the United States in January 1944 (as Allied troops landed near Anzio to begin the six-month campaign to capture Rome), he was sitting in a Manhattan nightclub with John Hersey, who the previous year had published a *Life* magazine story about a PT boat squadron in the Solomon Islands. Upon hearing Kennedy's saga, Hersey suggested another article in *Life*, about the demise of the boat and the fate of its crew. After checking with his father, Kennedy agreed to cooperate. By the time he and Hersey rendezvoused for in-depth interviews about the Solomon campaign, they had to meet in a hospital room in Boston, where Kennedy had undergone surgery for his back injury. He spent most of late winter and spring in various hospitals and resorts, recuperating and being treated for malaria, a common ailment among service members returning from the Pacific Theater.

In May 1944, Jack reported to the Submarine Chaser Training Center in Miami. His father had purchased the six-bedroom La Querida (roughly translated as "my dear one"), in Palm Beach, for $120,000 in 1933. The Mediterranean Revival home, set on two acres, was expanded to eleven bedrooms; multiple second-floor balconies offered sweeping views of the Atlantic. The sprawling estate, designed by architect Addison Mizner, showcased two hundred feet of pristine Atlantic beachfront. Jack retreated there to enjoy the swimming pool, soak up the Caribbean trade winds, and assess his future while still serving in American wartime.

Part of Jack's job in Miami entailed piloting PT boats around South Florida waters in anti-German-attack defensive exercises. Getting to watch Floridians and tourists frolic on beaches was far better than bloody combat. "They all wait anxiously for D-day," he wrote a friend, "and you can find beaches crowded every day with people—all looking seaward and towards the invasion coast."

Jack always loved being in Florida. Early in the war, his brother Joe Jr. had been stationed at Jacksonville, learning to fly PBY Catalina twin-engine patrol planes ("flying boats") off St. John's River. In May 1942 Joe Jr. received operational training at Banana River, Florida. His naval air station was next door to Cape Canaveral, from which NASA's rockets would be launched during the 1960s, when Jack served as president.

In May 1944, Jack decided to learn how to fly, following in the footsteps of his brother Joe Jr., who had earned his naval aviator wings two years before. He enrolled at the Embry-Riddle Seaplane Base, in Miami, and spent ten days piloting Piper J-3 Cub floatplanes around southern Florida, joining his new interest in aviation with his long-standing connection to sea and shore. (Kennedy's flight log confirming these first attempts at piloting aircraft was not discovered until 2018, during research for this book. It now resides in the historical collection of the Shapell Manuscript Foundation.)

After two years of planning, on June 6, 1944, the Allies launched Operation Overlord, sending 156,000 American, British, and Canadian troops (most under twenty years old) across the English Channel to the beaches of Normandy. It was the largest amphibious military assault in history, intended to liberate Western Europe while forcing the Nazis to divide their efforts between their Russia campaign and this new front. On D-day, Allied troops landed on five beaches along a fifty-mile stretch of the heavily fortified Normandy coast. Overwhelming or evading the Nazis' supposedly impenetrable defenses, the Allies gained a crucial foothold in Europe and decisively changed the direction of the war. Supreme Allied Commander Dwight Eisenhower later speculated that if German V-2 missile technology had been on a more "accelerated schedule," it might have caused him to scrub Operation Overlord, thereby dramatically changing the endgame of the war. For Jack Kennedy, that historic June 6 set in motion unanticipated events that would connect the very highest and lowest points of his life.

Just two days after the D-day landings began, Kennedy took his last flying lesson. For forty-five minutes he piloted over the Port of Miami, realizing that the war in Europe would soon be over and that his brother Joe Jr. would be coming home. For the first time in months, he worried that he was frittering away his time in sunny Florida. On June 11, one week after D-day, Kennedy was honored for his bravery in the South Pacific. Meanwhile, in Europe, Allied troops were spending the summer liberating Rome and pushing slowly inland from the coast of France. On the Eastern Front, the Soviets were launching a massive offensive in eastern Belarus, destroying the German Army Group Center and driving westward. By late July, the Anglo-American forces would break out of the Normandy beaches and begin racing southeastward toward Paris. By August, the Russians were pushing toward Warsaw, in central Poland.

On June 13, desperate for a counteroffensive but still awaiting deployment of von Braun's V-2 (which needed one more round of testing), the German High Command ordered V-1 flying bombs launched against Britain from sites near Pas-de-Calais, on France's north-central coast, east of the Normandy landings. The V-1—a relative of the V-2 in name and provenance only—was tested by a separate group at Peenemünde. The V-1 was a far less complex weapon, essentially a pilotless 1940s jet with a simple autopilot to regulate altitude and airspeed. Between June and September 1944, more than eight thousand of these "buzz bombs," or "Doodlebugs" (so called for the sound of their pulse-jet engines), were launched against London, flying at around 450 miles per hour and packed with a ton of high explosives. Though dubbed one of the Reich's *Vergeltungswaffen* (vengeance weapons), the V-1 sometimes failed, suffering from frequent malfunctions and guidance errors and being fairly vulnerable to Allied defensive measures, including fighter planes and barrage balloons. Even V-1s that hit their targets sometimes failed to detonate, leaving giant unexploded bombs to be unearthed in Greater London years and even decades later. Overall, only about a quarter of the V-1s launched against England between June and October 1944 hit their targets, but with more than one hundred coming in daily during the peak of Germany's offensive, those that did get through created massive destruction. And the V-1 diverted a lot of Allied resources into the defense of London. For months, Allied planes were kept busy intercepting them in midair while military aviators and the U.S. Army Air Corps tried desperately to locate and destroy V-1 launch sites on the Continent.

With as many as a quarter of Britain's air assets dedicated to fighting the V-1s, it was the Allies' turn to rush untried technology into battle. Aware that the Germans had another secret weapon on the docket, General Eisenhower was eager to knock out the underground Mittelwerk V-2 manufacturing facility in the Harz Mountains in central Germany. The Allied planners, however, erroneously believed that launching the V-2 would require some kind of large, stationary pipe or cannon along the French coast. The most daunting such location was a heavily fortified hill compound in northeastern France called Mimoyecques. American bombers had attacked that complex several times, killing hundreds and inflicting damage, yet the shrouded work there continued unabated. U.S. intelligence reports identified the Nazi complex as a massive bunker protecting a series of tunnels and cavelike workshops. One agent reported that he'd heard from a source that "a concrete chamber was to be built near one of the tunnels for the installation of a tube, 40 to 50 meters long [131 to 164 feet], which he referred to as a 'rocket launching cannon.'" Presuming that Mimoyecques would be the V-2's major launch site, American intelligence officers in London searched for an innovative way to obliterate the facility. (In truth, the Mimoyecques hill hid yet another secret German weapon: the V-3, a long-range gun that would, according to its designers, be capable of firing large-scale explosives on London at a rate of five shells per minute, around the clock. The Nazis were building fifty of these V-3s at Mimoyecques.)

Allied intelligence also learned that von Braun, following his V-2 success on October 3, 1942, was designing the world's first transatlantic ballistic missile, which Hitler wanted to use to obliterate New York, Boston, and Washington. The SS called the effort Projekt Amerika. U.S. Army general Henry "Hap" Arnold was extremely worried that this cruise missile, if properly developed, meant the Atlantic Ocean was no longer an effective barrier protecting America's Eastern Seaboard from Europe. "Someday, not too far distant," Arnold wrote, "there can come streaking out of somewhere (we won't be able to hear it, it will come so fast) some kind of a gadget with an explosive so powerful that one projectile will be able to wipe out completely this city of Washington."

Soon after V-1s began to rain down on Great Britain early that summer, Mimoyecques became a prime target for Allied bomber pilots. American commanders operating out of England, believing that they were attacking the V-1 compound and unfinished big-gun complex before it could be fully operational, approved a daring plan, Operation Aphrodite, to destroy the facility and other related rocketry sites in France. To realize the secret bombing raid, Eisenhower needed experienced U.S. pilots and B-17 and B-24 bombers that had reached the end of their military service but wanted one last dangerous mission.

On the home front that June, Jack Kennedy became the talk of the town when "Survival," John Hersey's piece about *PT-109*, appeared in *The New Yorker* (rather than *Life*, as originally planned). The article came to play a seminal role in Kennedy's life. After an arm-twisting campaign by Joseph Kennedy Sr., *Read-*

er's Digest agreed to reprint the heroic tale, delivering it to what was then the country's largest magazine readership. Later, Joseph Sr. would privately publish the story in pamphlet form, for free distribution. The historian John Hellmann observed that the article began the "construction of John F. Kennedy as a public image of fiction-like, even mythic, resonance."

Kennedy's PT service was emblematic of two heroic archetypes of the times: young Ivy League men of distinction, and harrowing military action in the Pacific Theater. Jack fit the bill, but so did most, if not all, of the other men who commanded the mosquito boats. William White wrote a wartime best seller, *They Were Expendable*, about the heroic 1942 exploits of Medal of Honor winner Lieutenant John D. Bulkeley on *PT-41* in the Pacific. Even before the war was over, *They Were Expendable* was produced as a movie starring the sought-after actor Robert Montgomery, who had himself been a decorated PT boat commander. In the popular imagination, the men of the PT boats were like the flying aces of World War I, whose stories were always told with dash and glamour, never mind the grim realities. In a naval war that was essentially fought between hulking battleships carrying thousands of sailors, the mythos of the PT boat sailors appealed to the American ideal of the independent, maverick hero.

Kennedy's steel-eyed courage in leading the survivors of the *PT-109* to safety elevated him above even other PT officers. Exuding grit, quick thinking, and endurance, Jack had valued the lives of others more than his own. Virtually overnight following the publication of "Survival," Kennedy went from being a published author and spoiled child of privilege to a naval version of U.S. Army hero Audie Murphy. "I firmly believe," Kennedy wrote years later, "that as much as I was shaped by anything, so I was shaped by the hand of fate moving in World War II."

Just as the *Reader's Digest* article was about to appear that August of 1944, twenty-nine-year-old Lieutenant Joe Kennedy Jr. was in a highly enviable position as a U.S. pilot based in England. Having completed his twenty-fifth mission, he was entitled to return home. Instead, he bravely volunteered as a pilot in Operation Aphrodite, the effort to target supposed German V-1 missile facilities and other fortified Nazi compounds in France. Some have suggested that it was brother Jack's emergence as a Pacific war hero that prompted Joe Jr.'s determination to make his own heroic name. The timing supports that theory, and certainly the brothers were fiercely competitive, but the idea doesn't hold up. In the first place, Aphrodite (named after the butterfly, not the Greek goddess) was a secret of the first magnitude, utilizing risky technology that couldn't be discussed for the duration of the war. The pilots who volunteered didn't do so for fame and glory. Second, as a pilot who had flown twenty-five missions, Joe Jr. would already have been regarded as a military hero wherever he went in the United States.

Joe Jr. was a multifaceted character, impossible to simplify in any aspect, but his thinking was probably similar to that of the other experienced B-17 Liberator and B-24 Flying Fortress pilots who volunteered for Aphrodite: if they could destroy targets such as the secret Nazi base at Mimoyecques before the Germans mass-produced weapons developed there, thus helping to hasten Allied victory in the war, then it was worth the personal risk. But Joe Jr. soon learned that there was a novel, frightening new aspect to Operation Aphrodite: his specially configured Liberator would need pilots only for takeoff. Once aloft, he and his copilot would throw switches, surrendering their Liberator to remote control from a nearby mother ship, and activate a detonator wired to twenty-five thousand pounds of Torpex explosives, 1.7 times more powerful than TNT. If all went well, they'd parachute out, leaving the Liberator, now a kind of drone aircraft, to fly onward under remote control to crash into the mouth of the caves and bunkers at Mimoyecques.

Naval Air Corps pilots recruited for Operation Aphrodite realized the extreme danger they'd face. In addition to requiring that they fly planes packed with nearly twice the explosive charge they normally carried, Aphrodite was hastily organized. The planes' interiors, for example, were so completely stripped and remodeled that pilots were confused by their surroundings, even in the cockpit. Expertise with a B-17 or B-24 was almost a liability on the high-risk mission, not an asset, because nothing in the Liberator was quite where it was supposed to be.

Lieutenant Joe Kennedy Jr. and his copilot, Lieutenant Wilford "Bud" Willy—a handsome athletic type from Fort Worth, Texas, who was the executive officer of Special Air Unit One—took off on August 12 in a converted B-24 Liberator. This was the U.S. Navy's first Aphrodite mission. With the weather cooperating and hearts pounding, they reached the designated altitude of two thousand feet, flying over the English countryside and headed toward the French coast. The bull's-eye target of Mimoyecques wasn't far away. According to plan, Kennedy and Willy switched their dronelike Liberator to remote-control mode and began arming the fuses, but they never got to bail out. Death pounced on them: an explosion shredded the plane ten minutes before the planned bailout, killing Kennedy and Willy instantly. A Naval Air Corps plane following three hundred feet behind the Liberator, with Franklin Roosevelt Jr., the president's son, aboard, was seriously hobbled by the midair explosion and forced into an emergency landing. The Kennedy-Willy aerial blast was so intense that more than fifty buildings in the town of New Delight Wood were damaged. A fragment of a radio was the only piece of the U.S. plane ever recovered.

What had gone so horrifically wrong? One possibility is that when Kennedy and Willy moved through the narrow passage between the cockpit and the bomb bay for their parachute jump, one of them inadvertently kicked a detonation wire. But that was just speculation. More likely, a circuit malfunctioned or a fire erupted in the aircraft. A review board later concluded that there had been no pilot error.

Beyond the simple tragedy of Joe Kennedy Jr.'s death is the fact that unknown to the Allies, the German V-1 launch sites near Calais were already obsolete when Operation Aphrodite began. Following earlier bombing raids and anticipating an Allied invasion by sea, Hitler had begun incrementally pulling out of Mimoyecques in the spring of 1944. The abandonment of the gargantuan project was one sign among many that the tide of the war had shifted, and particularly after the D-day landings, both sides were busy analyzing the Nazis' remaining strategies. Most of the possibilities revolved around the German weapons that had yet to be unveiled, specifically von Braun's super-secret V-2 and intercontinental missile, which Hitler hoped would leave New York City in flames. On September 5, the Canadian Third Infantry Division captured the Mimoyecques complex and finally learned its purpose, reporting to Supreme Allied Commander Eisenhower that it contained enormous V-3 guns in various states of completion, but none of the infrastructure to suggest that it was intended as a base for von Braun's savage V-1s and V-2s.

Two priests arrived at the Kennedy home at Hyannis Port to inform the family of Joe Jr.'s death. Their reaction was private. Over the following days, they proudly received Joe Jr.'s Navy Cross and Air Medal. And when a new destroyer was named the USS *Joseph P. Kennedy Jr.* in 1945, the family attended the christening. But even Joe Kennedy Sr., with all his clout, was kept in the dark about the mission that had killed his son. Only in 1963, when Jack was president, was he able to learn the exact details of his brother's death and properly reach out to the copilot's family in shared grief.

Jack did not let the melancholy he felt over his brother's death that August become emotionally debilitating. On Labor Day weekend of 1944, Lenny Thom, his *PT-109* executive officer, visited the Kennedys at Hyannis Port along with his wife, Kate. Ted, the youngest of the Kennedy boys, recalled in his memoir, *True Compass*, that the clapboard house became "an oasis of love." Kate Thom was keenly aware that Joe Jr. had been killed only three weeks before. She noticed, though, that Jack didn't speak about his brother's death over the fun-filled weekend. As far as Kate could ascertain, the household was functioning normally. Jack, feeling far better physically than he had months earlier, led the way in swimming, tennis, and a boat race. One evening, he and his friends were sitting on the front porch with some of his sisters. Amid the banter, a friend of Jack's started singing "Hooray for Hollywood." Another friend, named Barney, goof-danced to the peppy song. Everyone watched with amusement. Kate idly thought that the two boys were a couple of hams. "While he's singing," she said later, "Barney's dancing and upstairs, somebody yelled out, 'Jack! Have some respect for your dead brother.' (It was his father.)

"And we just froze," she continued. "And within minutes we were all gone. But that was the only thing that happened. It was a happy time, you know."

Rocket engineer Wernher von Braun spent the Second World War in the service of Hitler's Third Reich, building V-2s to be used against France, Belgium, the Netherlands, and Great Britain. Here he is at the Army Rocket Center at Peenemünde, in 1944, meeting officers of the Wehrmacht, during a demonstration launch of the V-2 rocket. At the time he was technical director at the Army Rocket Center.

4
Who's Afraid of the V-2?

Starting from unlikely, even utopian origins in the Weimar spaceflight movement, and ending even more strangely with ineffective weapons and emaciated slaves, the German Army rocket program and its Peenemünde center without a doubt changed the face of the twentieth century.

—MICHAEL J. NEUFELD,
THE ROCKET AND THE REICH (1995)

Even as England was suffering Germany's V-1 barrage, rumors swirled about the next super-weapon in Hitler's arsenal. "Britons pondered the possibility of worse to come," the *New York Times* reported that summer of 1944, "perhaps huge explosive rockets. The horrifying prospect that Germany's threatened new secret weapon might be a ten-ton explosive rocket—a robot bomb ten times the size of the V-1, or 'buzz bomb,' now being used against England—'may not be sheer propaganda' a commentator at an Allied advance command post said." British intelligence, having secured photos of a test V-2, understood its engineering structural and destructive capacity and surmised that the rocket was to be used against London. Another key (and essentially correct) assumption on which the Allies based their strategy was that the V-2 had a range of 210 miles.

Allied troops pushed to find and destroy every potential V-2 launch site within a 210-mile radius of London, a task that took them across the northern coast of France and into Belgium and Holland. In his usual forthright fashion, British prime minister Winston Churchill pronounced a German V-2 rocket attack and another Luftwaffe blitz like the one of 1940–41 the direst threats to his nation's survival.

As the Allies raced to suppress the German V-2 campaign even before it began, the war-ravaged British capital was on constant high alert. Rushed by the devastating German loss on D-day, von Braun conducted a missile test on June 22, during which his V-2 crossed the Kármán line, the boundary between Earth's atmosphere and outer space, reaching an altitude of 109 miles—and became the first man-made object to reach outer space.

Despite von Braun's stunning success, the German High Command was becoming impatient with progress on the V-2, specifically its intricate engineering and the unending need for technical refinements. Too many V-2s were breaking up on entry. On August 6, 1944, Heinrich Himmler, consumed with frenzied urgency, installed SS general Hans Kammler as director of the V-2 project, with marching orders to prepare for utilization of the rocket "as quickly as possible." Dictatorial, brusque, and a fanatical adherent of Nazi ideology, the hard-nosed Kammler clashed with many at Peenemünde. Ironically, his insistence that he be granted military command of what he declared would be Germany's most glorious campaign delayed the V-2 offensive even more. Von Braun loathed Kammler, but continued his trial-and-error experiments on the V-2. Even though von Braun thought Kammler dangerous, he nevertheless advised his superior officer on how to speed up mass production. Fearful of the Armaments Ministry, von Braun had learned how to obey orders without challenging assumptions. As the German Army became desperate for men of fighting age, von Braun found a noncombat management position for his younger brother, Magnus, at the Mittelwerk compound, where meaningful jobs came with one of the Third Reich's highest-priority security clearances.

On August 29, Hitler ordered V-2 attacks to begin as soon as possible. The Allies continued to methodically target suspected V-2 launch sites within a 210-mile launch radius of London, but what they didn't know was that Dornberger had specified early on that the V-2 launch apparatus be mobile, able to be quickly put in position, utilized, and then moved elsewhere for another round of launches, frustrating Allied efforts to track the weapons.

On September 8, 1944, the Nazis fired a V-2 rocket at Paris, which had been liberated by the Allies just two weeks before. This rocket landed near the Porte d'Italie in the French capital, causing no casualties. Within four minutes of takeoff, a second V-2 came screaming back to Earth. This ballistic missile killed six people and changed the world forever. The moment the officer in charge shouted, "Zundung!" (ignition) represented the culmination of more than twenty-five years of German engineering brilliance, technological innovation, military desperation, and rank brutality. At just under forty-six feet in length, the V-2 weighed over fourteen tons, including about one ton of explosives. Following designs innovated by Oberth and von Braun, it lifted off when liquid fuel and oxidizer were pumped into the combustion chamber at the rate of thirty-three gallons per second. After about one minute aloft, with the fuel depleted, the rocket traveled through space at an altitude of just under fifty miles, at which point it arced ballistically toward its target, reaching speeds of up to thirty-six hundred miles per hour. Midcourse corrections had been made at Peenemünde with three components: sophisticated gyroscopes that measured the movements of the rocket; an onboard analog computer that calculated necessary changes; and rudders that could be adjusted by the computer.

The following day, a detail of German soldiers surprised the residents of a quiet suburban neighborhood in The Hague, knocking on doors and ordering the occupants to vacate immediately. The next day, military vehicles hauling V-2s aboard Meillerwagen (mobile launch platforms) pulled into the now-deserted Dutch streets with orders to launch two V-2s at London. Well trained by Dornberger, the mobile units had the first combat V-2 rocket in position within an hour. Within another half hour, their job would be done.

Of the London-bound rockets, one did no serious damage, but the other wreaked havoc in the western suburb of Chiswick. John Clarke was six years old and playing in the bathroom of his family's home when the rocket struck, causing his building to crumble and sending a shard of metal from the V-2 through his hand. His three-year-old sister, sleeping in the next room, was killed. "There wasn't a mark on Rosemary," Clarke, who was permanently deafened in the attack, remembered sixty years later. "The blast goes up and comes down in a mushroom or umbrella shape, but in the process of that my sister's lungs collapsed. She was deprived of air." An off-duty serviceman walking nearby and a woman sitting in her house were also killed, making a total of three deaths from the first V-2 strike on England.

That evening, the true dawn of the missile rocket age, von Braun was informed of the successful launches on Paris and London. Witnesses at Peenemünde provided different versions of his initial reaction to the fulfillment of his dream. One of his loyal secretaries, Dorette Kersten Schlidt, recalled the mood in the office. "Von Braun was completely devastated," she said. "In fact never before or afterward have I seen him so sad, so thoroughly disturbed. 'This should never have happened,' he said. 'I always hoped the war would be over before they launched an A-4 [the scientists' name for the V-2] against a live target. We built our rocket to pave the way to other worlds, not to raise havoc on earth.'"

Another colleague at Peenemünde remembered the night much differently. "When the first V-2 hit London, we had champagne," he noted. "And why not? We were at war . . . we still had a Fatherland to fight for." Perhaps von Braun was truly remorseful at the stark reality of his invention killing civilians; after all, his interest in rocketry had begun in dreams of spaceflight. But once his V-2s began raining down on Britain, France, and later Belgium and the Netherlands, he crossed a damning threshold: the man who dreamed of the moon was now Hitler's agent of mass destruction, working in a Europe imbued with sorrow and death.

At the time of the V-2 launches, about five million Germans had already been killed in the war. Nearly one in three Nazi servicemen would die in the bloody conflict. Allied bomber attacks sometimes killed thousands of people at a time. Nor was von Braun always a mere distant observer of the carnage that became commonplace during the Nazi reign of terror. At the wretched Mittelwerk factory, more than 10,000 slave laborers from the Dora-Mittelbau prison camp perished producing V-2 missiles, V-1 flying bombs, and other weapons. If von Braun wasn't directly responsible for their

A V-2 rocket in launching position at Peenemünde (German Army Research Center) during World War II.

deaths, he was certainly complicit. From September 1944 until the end of the war the following May, more than 4,300 V-2s were launched by the German Wehrmacht against allied targets, mostly in London, Antwerp, and Liege. These missiles killed an estimated 9,000 civilians and military personnel, caused serious injury to an additional 25,000, and damaged or destroyed over a million homes.

So driven was von Braun by his quest for V-2 glory, and for Germany to win the war, that he grew inured to the chaos and loss his advanced work was causing. Like Lieutenant Colonel Nicholson in Pierre Boulle's 1952 novel *The Bridge over the River Kwai*, von Braun was task driven, obsessed with completing a job no matter the cost. "With a satisfied eye he witnessed this gradual materialization," Boulle wrote of Nicholson, "without connecting it in any way to humble human activity. Consequently, he saw it only as something abstract and complete in itself: a living symbol of the fierce struggles and countless experiments by which a nation gradually raises itself in the course of centuries to a state of civilization."

Von Braun was convinced that rocketry was just that kind of symbol, elevating the technological excellence of a civilization. Much later, his convivial comments about the technical achievement of the V-2 betrayed no genuine anguish, only the arch observation that scientific breakthroughs since the fifteenth century had often begun in the sphere of weapons development. But the truth was that rather than elevating civilizations, his V-2 work for the Nazis degraded it, becoming nothing more than a tool by which a three-year-old British girl had been killed willy-nilly, and which would claim thousands more innocents to come. In that respect, Germany's *Wunderwaffe* (miracle weapon) was no different from a simple club, hatchet, sword, or bayonet. Or, for that matter, the incendiary bombing raids the USAAF and RAF conducted on Germany.

Both versions of von Braun's initial reaction to the launch of the V-2 on France, Belgium, and Great Britain might be entirely accurate. The enigmatic engineer often told people what they wanted to hear and showed the colors they hoped to see. If he was a hypocrite and accomplice, he believed, then so be it—that was how his rocket program had survived in the Third Reich. If he had an intense concern about the Holocaust happening around him, he never expressed it. Everywhere von Braun went in high Nazi circles after September 8, 1944, he was congratulated. With his broad shoulders, groomed hair, and splendid physique, gazing up in the air at parties with thoughtful self-importance, he was treated as the proud exemplar of German rocketry genius, and he possessed an exalted opinion of himself. Von Braun's amoral hunger to construct rockets governed his embrace of an evil regime. Both during World War II and after, he accepted accolades as an engineering visionary who foresaw the potential of human spaceflight, never admitting that he was essentially a fast-track Nazi arms merchant who developed brutal weapons of mass destruction. In the late 1960s, as the United States was involved in a race with the Soviet Union to land a human on the moon first, humorist Tom Lehrer wrote a song about von Braun's opportunistic approach to serving whoever would let him build rockets regardless of their purpose:

Don't say that he's hypocritical, say rather that he's apolitical
"Once the rockets are up, who cares where they come down?
That's not my department," says Wernher von Braun.

Lehrer's biting satire captured well the ambivalence of von Braun's indifference on moral questions associated with the use of the V-2 and other rocket technology.

After that first September launch, German V-2s began striking targets in Great Britain at a rate of about two hundred per month. The hastily built rockets, however, were far from perfect. Some V-2s hit their targets but failed to explode. Others burst prematurely in the air, sending burning debris showering down for miles around, like spent fireworks. In October, the Third Reich launched V-2s on Antwerp, Belgium, determined to prevent that all-important European port from becoming an Allied stronghold. At the government level, British officials initially promulgated the fiction that ruptured gas pipes had caused the early V-2 damage, but the fact was that Whitehall didn't want to release any information about the destruction that might be helpful to the Nazis. "The enemy," Churchill recalled, "made no mention of his new missiles until November 8, and I did not feel the need for a public statement until November."

The German propaganda announcement of V-2 attacks made headlines in the United States. Reporters searched for anyone knowledgeable to discuss rocket engineering. What did the V-2 mean? Did the United States have a similar program? One of the few Americans in a position to know was keeping a very low profile. Dr. Robert Goddard, then working on developing jet-assisted takeoff units for navy aircraft in Maryland, received intelligence on the V-2 that November, writing matter-of-factly in his diary entry from Annapolis, "V-2 type rocket appears to be of interest."

Not surprisingly, Goddard felt threatened by von Braun's engineering breakthroughs on the V-2, which made all previous rocketry trials seem quaint. Refusing to be crushed by the knowledge, though, within weeks he was echoing the Roosevelt administration's efforts to promote a patriotic fallacy: that the V-2 had been copied directly from Goddard's hundreds of static tests and earlier flight tests conducted at a ranch near Roswell, New Mexico. Starting early in 1945, articles in publications such as *National Geographic Newsletter* claimed that Goddard was the father not only of America's infinitesimal rocket capability, but of Germany's burgeoning efforts as well. This exaggeration took hold in the United States, despite the many basic technical points that separated Goddard's last, lonely prewar rocket tests from the grand-scale work being done by the hundreds of scientists at Peenemünde. Even if the U.S. propaganda effort did a disservice to engineering history, it did spark a public demand for an increased American role in liquid-fueled rocketry and military missiles.

At the NACA headquarters in Washington, DC, located in a corner of the Navy building, leadership kept a close eye on the V-2. Described as the "Force Behind Our Air Supremacy," the NACA research-and-development team tested highly sophisticated superchargers for high-altitude bombers such as the B-17 and B-24, as well as airfoils that are still being used in twenty-first-century aircraft manufacturing. While unable to develop anything as sophisticated as a V-2, the NACA engineers were responsible for a number of innovations that were vital to the war effort. Month by month, engineers at the NACA's Aircraft Engine Research Laboratory, in Cleveland, Ohio, made vast improvements to fighter plane engines—for example, improving engine cooling capabilities in the B-29 Superfortress, a bomber essential to Pacific war strategy. The NACA's first jet-engine test was performed in Cleveland's Altitude Wind Tunnel. And when its Icing Research Tunnel opened in 1944, the notion of future manned space travel became more likely.

There is no record that Jack Kennedy, honorably discharged from the navy with the full rank of lieutenant on March 1, 1945, ever publicly commented on the V-2 rockets raining down on London. Like all Americans, he was happy that the V-1 and V-2 seemed to have appeared too late in the war to save the Third Reich, but having experienced the horror of war, he surely understood the weapon's ghastly potential for mass destruction. In early 1945, he submitted an article to the *Atlantic Monthly* about the need for global peace, but the editors weren't enthusiastic, and despite the intervention of his father, the piece was never published.

Like most returning veterans, Kennedy needed time to readjust. Seeing so much death had left the soldiers of his generation shell-shocked. He took his time, philosophically deliberating on his future and convalescing in the Arizona desert after his back operation. Eager to pay homage to his late brother, he edited a book of memorial essays about Joe's short life, which would be privately published. But he also needed to contend with the stressful issue of claiming his new place as the eldest son in his family. With Joe Jr. gone, the expectation of a life in politics fell squarely on Jack's shoulders. He told friends he felt obligated to fulfill his father's longtime ambition to see one of his sons in high office, though at that point he scorned electoral politics as involving too much handshaking, baby kissing, and general pandering. Faced with paternal pressure and the prospect of having to abandon the more bookish and journalistic careers he'd contemplated before the war, Jack described plans for a future in public service, a vague goal with a range of possible options, including the diplomatic corps.

As a young intellectual, Kennedy showed promise, reserving special excitement for the field of international relations. His book *Why England Slept* had been a best seller and defined him as a thoughtful strategist with a natural affinity for global affairs. In person, Jack was known for his sense of humor, loving to laugh while downing a comradely glass of beer. Never humble but eager for self-improvement, he unquestionably had the kind of bright curiosity and original, analytical mind that made a good basis for a political career. But a certain indefiniteness was also a part of his personality. He could occasionally be high-strung, he was frequently in poor health, and he was always an unrepentant partier and womanizer—not, perhaps, the perfect recipe for someone contemplating a high-profile public life.

From his parents' Palm Beach home, Kennedy was enthralled that the Third Reich was being squeezed on all sides. As the Americans and their allies pushed into the heart of western Europe in the months after D-day, the Soviets were closing in on Germany from the east. The war had cost the lives of an astounding twenty-four million Soviet civilians and soldiers, but Joseph Stalin's government was still standing. On January 12, 1945, the Soviets launched a new offensive that liberated Warsaw and Krakow, captured Budapest on February 13 after a two-month siege, drove the Germans and their Hungarian collaborators out of Hungary in early April, forced the surrender of Slovakia with the capture of Bratislava on April 4, and captured Vienna on April 13. This crystallized the reality to Kennedy that World War II in Europe was almost at an end.

Even though the V-2 never struck a death knell to British morale, Hitler continued to authorize rocket attacks on Britain and Belgium, his two major target nations. Civilian deaths numbered around five thousand but were far short of the massive levels the German High Command had predicted. Many Allied four-engine bomber air raids killed more people in one night than a month of V-2 attacks. Winston Churchill claimed that two people were killed for each V-2 launched against his country—numbers that, while still tragic, were nowhere near enough to change the trajectory of the war, which had tilted decidedly toward Allied victory.

Production of V-2 rockets at Mittelwerk accelerated through January 1945, keeping thousands of slave laborers struggling to stay alive, including Jews diverted from concentration camps to work at the underground factory. The facility was manufacturing almost 700 V-2s per month when production ceased in March. The death rate at Mittelwerk accelerated in the last months of the war due to the collapse of the food supply and increased vicious repression in the Dora camp. And then, on March 27, the last V-2 was launched, killing thirty-four-year-old Ivy Millichamp in the English town of Orpington. All told, the 2,500 V-2s launched by Germany had killed 2,742 people in England and seriously wounded a further 6,467. Casualties in Belgium and France were fairly low. "The V-2's role in the war was at the end," historian Christopher Potter wrote in *The Earth Gazers*, "but its role in history was about to begin a new chapter."

When U.S. troops crossed the Rhine River at Remagen in March 1945, Hitler issued his *Nerobefehl* (Nero Decree), ordering the complete destruction of German infrastructure to prevent its use by the invading Allied forces. Human assets, too, were in a precarious position. As German defeat had become inevitable, an emboldened U.S. Army Ordnance Corps used every means at its disposal to identify technical assets within Germany. One of the Army's aims was to secure the country's top scientists before they were captured by the Soviets or escaped to the Middle East or South America. Aeronautical engineers, synthetic-fuel experts, naval weaponeers, physicists, chemists, arms manufacturers, and thousands of others were on the list, but none was prioritized higher than the rocket scientist Wernher von Braun.

With Peenemünde bound to be overrun by the

Soviet Union, von Braun followed orders by General Kammler to evacuate the facility on February 17. Von Braun helped transport thousands of personnel, equipment, blueprints, and research documents toward the Mittelwerk facility aboard a fleet of trucks, cars, and trains. All were marked with the insignia of the VZBV (Vorhaben zur besonderen Verwendung, or Project for Special Disposition), a nonexistent, allegedly top secret agency they'd invented as a ruse to get the items through SS checkpoints. Reaching Nordhausen, the site of the Mittelwerk factory, the convoy occupied abandoned buildings and began secreting V-2 documents underground. Then, in mid-March, von Braun received orders from General Kammler to evacuate again, with five hundred of his key personnel, to Oberammergau, in the Bavarian Alps.

Monitored by the SS at Oberammergau, von Braun and his rocket team itself became bargaining chips for Kammler, who, using lawyer's logic, hoped to trade them to the Allies for leniency. If no such bargain seemed possible, he would perhaps look for a deal with the Soviet Union.

Von Braun, however, was the kind of man who, while walking, exuded the feeling of always knowing his destination and how to get there. With the Soviets encircling Berlin on their final offensive and Germany collapsing, he decided that the best option for his team was to surrender to the U.S. Army. "We despise the French," one member of von Braun's rocket team later explained. "We are mortally afraid of the Soviets; we do not believe the British can afford us; so that leaves the Americans." Taking advantage of the chaos and conflicting orders of those final days, von Braun bluffed his way out of confinement. After suffering a broken arm in a car wreck and demanding that it be quickly set in a cast so that he could continue on, he and his team settled at Haus Ingeborg, in Oberjoch, a resort town near the Austrian border. There they rendezvoused with General Dornberger and von Braun's brother, Magnus, and waited for news. Within days, they learned that Adolf Hitler had committed suicide on April 30 in Berlin, beginning a process that would end a week later with Germany's unconditional surrender.

With elements of the advancing U.S. Seventh Army only a few miles away, it was decided that Magnus von Braun (who was near fluent in English) would depart to make contact. Climbing on a bicycle, he set off down a country road toward Allgäu. On May 2, in one of those fortunate moments in history, he soon encountered a surprised private, Fred Schneikert of Wisconsin, who aimed an M-1 rifle at him while ordering, "Hands up." Magnus told Schneikert that the inventor of the V-2 was nearby, ready to surrender. "I think you're nuts," Schneikert told him. Nevertheless, Schneikert relayed the message to his U.S. Army superiors.

Enter into the unfolding drama Holger Toftoy, from Illinois, a graduate of the West Point class of 1926 and chief of the U.S. Army technical intelligence teams assigned to Europe to search for, examine, and appraise captured German weapons and equipment. The snatching of von Braun and other Peenemünders was the biggest bonanza imaginable to Toftoy, who was earning a reputation in army circles as "Mr. Missile." If the army could take custody of V-2 research and parts, then in one fell swoop the United States would soon be the premier rocket-builder in the world. Von Braun's surrender was the kind of gift horse that Toftoy could only dream about.

Arrangements were made for the Peenemünders to be escorted through the lines that night and held in a safe haven. Following information provided by von Braun and Dornberger, General Toftoy had U.S. Army troops race toward Nordhausen (in what was soon to be the Soviet Occupation Zone) for the spoils of the V-2 program and to Mittlewerk (where the corpses of slave laborers were piled up like cordwood). Having captured the area around Nordhausen and Mittelwerk, Toftoy set up a special V-2-related mission. In one of the great technology grabs in history, U.S. forces collected fourteen tons of blueprints and design drawings from these Nazi facilities and von Braun's secret mine shaft, and enough parts to fabricate one hundred V-2s. Just days after the U.S. grab, Peenemünde was seized by the Soviets. They confiscated missile hardware and production facilities and the remnants of von Braun's left-behind production team—but most of the vital personnel, documents, and equipment were already in the American occupation zone. The V-2 materials apprehended in Germany by Toftoy's team at Nordhausen were shipped by the U.S. Army to Antwerp, and then onward to the United States.

Realizing he was in a technology race with America and was already behind the eight ball, Stalin fumed over being a step behind in the apprehension of Peenemünde technicians, and von Braun in particular. "This is absolutely intolerable," Stalin pronounced. "We defeated Nazi armies; we occupied Berlin and Peenemünde, but the Americans got the rocket engineers. What could be more revolting and more inexcusable." This agonized realization came far too late, for the United States had already nabbed the most valuable technology. The Red Army was, however, able to seize research facilities on the island of Usedom, as well as hundreds of executive-level German technicians and engineers. As a further consolation prize, the USSR discovered plenty of machine and rocket components at Mittelwerk. In 1946, the Red Army relocated the captured German assets to the USSR, where the V-2 began a second career as the Soviet R-1.

Safely ensconced with the American forces, von Braun was confident that his team's priceless ballistic missile know-how would buy them immunity, even though their work had led directly to the deaths of thousands of Allied civilians and ten to twenty thousand prisoners. Undoubtedly the German officers who'd ordered the V-2's use on civilian populations would be charged with war crimes. Certainly, the soldiers who fired the rockets would be captured and treated as enemy combatants. But the Nazi missilers and mechanical engineers responsible for refining the ballistic rocket technology to kill as many people as possible? For their willingness to surrender and put their knowledge to work for their new American hosts, they would be respected and treasured by the U.S. government (although von Braun got only about 25 percent of the engineers and foreman-level craftsmen he had requested). The double standard was in play. In a dramatic reversal of fate, instead of being treated as war criminals in the months that followed Hitler's death, von Braun and his team were housed with their families in Landshut,

Bavaria, in southern Germany, in a comfortable dormitory complex built for the 1936 Olympics. Von Braun's team's alibis were that the tentacles of circumstance had forced them to work for their German Fatherland in wartime. Now, with the Allied victory, they would use their hard-learned rocket engineers' expertise to help the great United States become a leader in ballistic missile production and, perhaps someday, the only Space Age superpower.

After German surrender in 1945, Charles Lindbergh, working for United Aircraft as a test pilot, was recruited by the U.S. Navy to conduct a study of German rocketry and jet propulsion accomplishments. With a .38 automatic in a shoulder holster and dressed otherwise like a regular GI, Lindbergh explored Germany, stunned at the level of annihilation and decay. Following an intelligence lead, he tracked down Willy Messerschmitt, the designer of the German Messerschmitt warplanes. Messerschmitt saved his postwar hide, as von Braun had with the U.S. Army, by telling Lindbergh all the secrets of Nazi jet propulsion science. What interested Lindbergh the most was the ballistic missile technology of the Third Reich. Working for the navy, he then made his way to Nordhausen to investigate the catacombs where V-2 rockets were constructed in the highest elevation of north Germany, in a rugged mountain terrain. What he saw there, the stunning technology the Germans had developed in the ballistic missile realm, left him flabbergasted. "Imagine," he recalled, "finding the demon of sheer space hiding in a mountain like a giant grub?"

Within weeks of Germany's capitulation, the U.S. Army was shipping V-2 rocket parts back home, most eventually making their way via New Orleans to the White Sands Proving Ground in New Mexico. A complete V-2 was confiscated and shipped to the Annapolis Experiment Station for examination, where U.S. government propogandists deemed it quite similar to Goddard's Roswell rockets. Under Toftoy's leadership, the U.S. Army was preparing to guide missiles into its postwar weapons program at a rapid pace. In July, at the Potsdam Conference, the Big Three leaders Winston Churchill, Joseph Stalin, and the new American president, Harry Truman—Roosevelt had died of a stroke while in Warm Springs, Georgia, on April 12—disingenuously promised to share all German scientific assets discovered. As historian Walter A. McDougall aptly put it, this so-called agreement "was a sham."

With the Pacific war still raging, the United States wanted to accelerate America's missile program, mostly by copying the V-1. Plus, there was no faster route than the wholesale acceptance of the V-2 scientists, parts, components, and complete systems from Peenemünde and Mittelwerk to upgrade America's long-term missile capability. That June, U.S. troops had conquered Okinawa, the last stop before the Japanese islands. But military leaders knew that sending U.S. troops into Japan itself could easily result in over one hundred thousand casualties. The hope was to bomb Japan into submission.

Von Braun was held at Kransberg Castle for a couple of days, which under the code name "Dustbin" served as an Anglo-American detention center for German scientists, doctors, and industrialists. The interrogation that was integral to his intake process went well. On July 20, 1945, U.S. secretary of state Cordell Hull green-lighted the transfer of all German technology experts as part of a secret recruitment program that had been named Operation Overcast. Initially, von Braun fretted that he and his team would be squeezed for information, perhaps bullied, and then shipped back to West Germany to stand trial for war crimes, but he needn't have worried. With the war in the Pacific still raging and competition with the Soviets already heating up, the rocket engineers were given protected status by Toftoy. Working clandestinely, the Joint Intelligence Objectives Agency (JIOA) classified the records of

German "Peenemünders," as they were dubbed, expunging any evidence of Nazi Party membership that might pose a security threat. With this whitewashing complete, the German rocket engineers, physicists, chemists, and others were provided security clearances to work in America, inoculated against prosecution for war crimes.

In full cooperation with Toftoy, von Braun's team exhaustively explained the captured V-2s to their new American colleagues, rebuilt the rockets, and commented on current missile-related projects not yet actualized. With the same sureness of touch, von Braun even began campaigning for the financial resources needed to make more advancements. His sharp and level pitch was convincing: in just a few years, the V-2 would be considered last year's military hardware, replaced by warheads one hundred times more lethal. What surprised the U.S. interrogators most was von Braun's determination to fire V-2s beyond the "top layer of the atmosphere"—that is, into space. To his interrogators, von Braun praised the pioneering work of Dr. Robert Goddard, who had died on August 10, 1945, in Baltimore, after a battle with throat cancer. Over the years, von Braun gave several accounts of Goddard's influence on the V-2, indicating that it had been minimal. His abundantly generous comments in

1945, however, marked an almost symbolic turning point for American rocketry. Once Goddard was buried in Worcester, Massachusetts, leadership in the field of rocket engineering fell to the newcomers who had surpassed him. In a calculated ploy, von Braun, facile at scheming and maneuvering to promote his work, had nothing to lose and everything to gain by citing Goddard so extravagantly.

In November 1945, Operation Overcast was renamed Operation Paperclip by JIOA for security breach reasons. The name was chosen by officers who would fasten a paper clip to the folders of Nazi rocket experts they chose to hire. In its first year, 119 German rocket engineers were brought to the United States, cleared of war crimes, and put to work under Operation Paperclip. While von Braun and the others were assigned to work at forlorn Fort Bliss, near El Paso, Texas, Toftoy himself directed the army's new guided missile program from an office in Washington, DC.

In the entire twentieth century, the splitting of the atom during World War II would prove to be the only human event on a par with the American moonshot. The Atomic Age was born on July 16, 1945, when the Trinity Test was successful in New Mexico. What this meant to human civilization became abundantly clear on August 6, 1945, when the *Enola Gay* dropped the first atomic bomb, "Little Boy," on the city of Hiroshima, Japan. Within four months, the acute effects of the atomic bombing had killed somewhere between 90,000 and 146,000 people in Hiroshima, and 39,000 and 80,000 people in Nagasaki, where a second bomb, nicknamed "Fat Man," was dropped three days later; about half the deaths in each city occurred during the first twenty-four hours. There was nothing iffy about the total annihilation of these cities; each was turned into a smoking wasteland of rubble. Confronted with the possibility of total destruction of their nation, the Japanese surrendered on August 14, 1945. (The formal surrender ceremony was held aboard the USS *Missouri* in Tokyo Bay on September 2.) While most Americans cheered, relieved that U.S. armed forces had been spared a bloody invasion of the Japanese homeland, the reality began to sink in that humanity had entered a dangerous new epoch. "Dropping the bombs ended the war," President Truman later boasted, "saved lives and gave the free nations a chance to face the facts."

Created in Los Alamos, New Mexico, hours northwest of Goddard's Roswell ranch, the atomic bomb ushered in an age in which humans held the godlike power to end life on Earth. The potential combination of the bomb with the new forms of ballistic missile technology conceived by von Braun meant the end could come without warning, at any time. Ironically, Germany's V-2 program had been partially spurred by unreasonable fears of American rocket development—U.S. rocketry was then effectively nonexistent—while America's conviction that Germany would soon invent an atomic bomb had spawned the Manhattan Project, which produced the bombs used on Japan. (In fact, Germany canceled its atomic program long before conceiving a viable weapon.) Significant though the Soviet breakthroughs in military aviation had been, after World War II, the United States held a virtual monopoly on both V-2 rocket and atomic bomb technologies.

The atomic bomb instantly transformed the nature of warfare and rendered all previous strategy moot. Armies didn't have to roll down the street with tanks to conquer. War no longer had to mean the rumble of diesel engines or the thunder of big guns in the distance. Because it could not be defeated by conventional forces, the American superweapon became a new source of fear and dread to the Soviets. Some thought America's monopoly over the atom gave it the leverage to obtain a postwar settlement largely on its own terms. Others knew that the Soviets, whose decisive victory over the Nazis had just won them the first real security they'd enjoyed since 1917, would never accept U.S. hegemony.

The atom bomb also made individuals think differently about their own lives. Global citizens could no longer feel fully safe in their own homes. Rockets didn't offer parents time to run and save their children. Any given second could be the last. Ominous portents were self-evident. "A screaming comes across the sky," novelist Thomas Pynchon wrote decades later in *Gravity's Rainbow.* "It has happened before, but there is nothing to compare it to now."

According to the U.S. federal government's own estimate, at the close of the war, America had been eight years behind the Germans in rocket capability. With the arrival on American soil of von Braun and the other Peenemünde engineers, that gap vanished all at once.

Meanwhile, in late 1945, von Braun and his rocket team were assigned work contracts with the U.S. Army Ordnance Corps, which was interested in developing rockets and artificial satellites out of the public glare. They were officially called "War Department Special Employees." For the next five years, these German scientists, stationed at Fort Bliss, Texas, near El Paso, were engaged mainly in rebuilding captured V-2 rockets, which would then be sent to the army's White Sands Proving Ground in nearby New Mexico, where army and navy personnel and technicians from

General Electric would conduct tests designed to improve the rockets' destructive capacity. New Mexico had already had a world-altering role as the site of the first-ever atomic bomb test, on July 16, 1945; now this isolated White Sands tract of land the size of the state of Connecticut was populated by scientists, pilots, and engineers, its beautiful arid landscape dotted with spider-like antennas, cargo trucks, watchtowers, and optical telescope laboratories. Federal purse strings had been loosened for rocket construction in the region.

At first the status of the German rocket engineers at Fort Bliss was murky. Considered wards of the army, the men in the von Braun team had no passports or visas, their mail was censored, and they weren't allowed off the Fort Bliss base without an escort. But their families were allowed to join them from Germany, which was a real boon. These engineers weren't exactly prisoners: they could return to Germany, if they insisted, though they couldn't move anywhere else in the United States. Von Braun himself was well paid for his work by the federal government, earning an annual salary of six thousand dollars—more than twice the average American income of the era.

Although exasperated by their limbo, few on von Braun's team opted to leave El Paso during those first years. Located across from Juarez, Mexico, on the northern side of the Rio Grande, the city offered a better, more secure lifestyle than a now-divided Germany, especially for a valuable scientist. In El Paso, the expatriate scientists occupied their nonworking hours hiking the Franklin Mountains (with escorts), watching local football games, going to movies, and learning English. Von Braun dubbed himself a POP (prisoner of peace) rather than a POW, and he nurtured his genuine interest in local history by talking with cowboys and drifters about the curve of the Big Dipper and the nonvisible four points of the Southern Cross at local gatherings. At Fort Bliss, he said, "The GIs sized me up with uncomfortable accuracy," referring to his German accent. "But they also invited me to join their black jack and poker games."

Though isolated and confined, the Germans at Fort Bliss knew they'd been lucky to escape what could have been much worse fates after the war's end. While they were comfortably under contract in Texas, rocket engineers in Russian-occupied East Germany faced the real prospect of being kidnapped and sent to the USSR under house arrest. Ultimately, the Soviets would appropriate four thousand German rocket engineers, along with several complete V-2 assembly lines. At the same time, many old colleagues from the Third Reich's scientific and industrial hierarchy were standing trial in Nuremberg for war crimes. Thousands of other Germans were under scrupulous investigation as part of the Allies' denazification process.

Von Braun and his scientists had also been denazified, but in a more clandestine way, having had their security ratings changed by the U.S. Army intelligence to make them acceptable emigrants. Any nostalgia for their German homeland—the dark forests, beer halls, and autobahns—had been subordinated by a newfound love of American democratic institutions by most. But some of the imported Germans were sent back home, labeled as "ardent Nazis," unfit for U.S. residency. Others left voluntarily. Those that stayed, deemed not "ardent Nazis," were the ones the Army wanted to keep.

Arthur Rudolph, a close colleague of von Braun and chief operations officer at the Mittelwerk V-2 production facility, where he oversaw the slave labor, was an ardent Nazi and anti-Semite. Nevertheless, he was quietly accepted into the United States and granted citizenship in the mid-1950s. Keeping a low profile at Fort Bliss, working in secrecy, he immediately began making contributions to the American rocket program, as did ex-Peenemünder Kurt Debus. Hubertus Strughold, another beneficiary of Operation Paperclip, was a physician who had been the director of the Luftwaffe

Institute of Aviation Medicine during the Nazi era. In the United States, he was given a similar job, heading the new Air Force School of Aviation Medicine, in West Texas. Even as Strughold was settling in America, quite at liberty to enjoy the privileges of democracy, his former colleagues were facing trial at Nuremberg for war crimes he'd known about. In one experiment, Jewish inmates from concentration camps had been forced to squat in a chamber as the pressure inside the chamber was altered in a matter of seconds, simulating a depressurized airplane dropping from high altitude. The doctors watched as the prisoners died or permanently lost their minds.

The first successful V-2 launch in New Mexico occurred in May 1946, the rocket soaring to an altitude of sixty-seven miles; it was deemed a failure. Over the next five years, about seventy V-2 tests were conducted at White Sands, two-thirds of them successfully. Barely a day went by when von Braun wasn't daydreaming about Mars and the moon. During the early Cold War, he managed to design his first trajectories for a potential flight to the moon and started hunting for funding allies in Washington. Work associates knew that when von Braun got a faraway gleam in his eyes, he was daydreaming about a lunar voyage, bringing his imagination to bear on the thousands of necessary steps.

Nevertheless, the conundrum that had existed since the 1920s still haunted rocket engineers: long-range rockets were ideally suited to be used as weapons. With World War II over and geopolitical competition with the USSR already under way, interest and investment in rocketry as a military asset surged, while interest in and funding for peaceful applications such as manned space exploration and satellite telecommunications continued to lag. Attempting to alter this vast imbalance, Lieutenant General Ira Eaker, deputy commander of the Army Air Force, wrote the War Department in May 1946 requesting budget support for dozens of top secret test launches. The plan was to use modified and unmanned V-2s in what he described specifically as a "scientific endeavor" to explore space for peace. In truth, these V-2 tests were aimed at providing the United States with guided missile technology, deemed essential to deter the Soviet Union from overplaying its postwar hand in Europe. But massive U.S. budget cuts of 1946–48 ended many potential missile projects.

Eaker, the son of tenant farmers in Texas, had become a single-engine pilot during World War I and in 1930 made history by piloting the first transcontinental flight using in-flight refueling. A serious writer on military aviation, he coauthored the highly respected *This Flying Game* (1936), *Winged Warfare* (1939), and

Army Flyer (1942). While Eaker was commanding the American air effort in Europe during World War II, he improved the strategy of precision daylight bombing, which allowed for round-the-clock attacks against the enemy. Even though he was known mainly as a hardened, no-nonsense commander of bomber groups, his 1946 proposal to the War Department saw a moon voyage as a practical measure concomitant with the United States leading the world in space technology, although he admitted it was a largely unknown field. "If we may assume that the future of air conquest will bring with it a conquering of outer space," he wrote, "then clearly this experience and the enthusiasm which this project will generate will be very beneficial in the long run." Apparently the feeling in Washington was that the proposal was a highly fungible waste of money; space wasn't going to be the next domain of airpower.

Around the time von Braun's group surrendered to the U.S. Army, Jack Kennedy was in San Francisco, reporting for the Hearst syndicate on a conference called to negotiate the charter of the new United Nations, an organization designed to foster global peace and cooperation. When he returned east, his father arranged for him to spend the summer touring war-ravaged Germany with fellow Irish American James Forrestal,

President Truman's navy secretary. It was apparent to Kennedy that London was pummeled, Paris disgraced, Rome tarnished—the old European capitals he had visited with Lem Billings before World War II were dysfunctional compared with cosmopolitan San Francisco and New York City. Somewhat paternally, Forrestal tutored Kennedy on the postwar national security imperatives facing the United States and how the decisions of the United Nations Monetary and Financial Conference held in Bretton Woods, New Hampshire, in 1944 guaranteed that the U.S. dollar would be the reserve currency of the postwar world.

Meanwhile, Joseph Sr. commissioned a lecture bureau to schedule speaking engagements for Jack, met with local Massachusetts politicians to lay the groundwork for his son's future, and hired a team of political veterans to organize a congressional campaign in Massachusetts's Eleventh District (which included Cambridge, the home of Harvard University). In early 1946, Jack publicly announced his run for the seat. He wasn't experienced in retail politics and he wasn't local, either, but he quickly took up residence in the heavily Republican district to begin his run as an anticommunist liberal. According to Joseph Kennedy biographer David Nasaw, the patriarch acted as Jack's behind-the-scenes campaign manager, marketing him as a "fresh-faced, charming young war hero, with a bit of glamour and a wholesome down-to-earth quality, a Harvard man and a man of the people, a book-writing intellectual who was everyone's friend."

The Kennedy family formed a devoted juggernaut on the front lines of the 1946 campaign, and Jack reciprocated by giving it his all. He didn't want to let his siblings down, and he knew the campaign was his audition for a run for higher office down the line. Some people were already touting him for the Massachusetts governorship, the position that had long eluded his maternal grandfather, John "Honey Fitz" Fitzgerald, the popular two-term Boston mayor.

And yet, Jack knew that he wasn't the typical Bay State politician. On paper, he had little in common with a predominantly working-class electorate in factory cities of the Eleventh District, people who were eking out a low-paid livelihood. He was also a practicing Roman Catholic (which was not necessarily an asset), he wasn't married, and he didn't belong to any local clubs or civic groups in his district. Hair tousled and clothes casual, Kennedy was also a private man, preferring to keep to himself or pass the time with his best friends. Pandering to voters wasn't really in his makeup. With smiling ease, he could be full of good conversation while never completely connecting with anyone or revealing a thing. Nevertheless, during the course of the 1946 campaign, a persona was forged at the overlap between JFK's authentic self and the fresh-faced leader for whom voters hungered. Having rushed headlong into politics, he milked his sardonic charm, terrific drive, unforgettable smile, and unique sense of irony that connected him to a wide range of idiosyncratic audiences through laughter.

Self-assured and pleasant, Kennedy clung to his abiding sense of intellectual aspiration, resisting the politician's instinct to dumb down his rhetoric for a broader audience. Amid the well-worn traditions of Boston's Democratic circles, Jack was in danger of being deemed professorial. On this first campaign, he honed his meet-and-greet skills, but avoided pandering. "It was said," marveled Tip O'Neill, who later represented the Eleventh District, "that Jack Kennedy was the only pol in Boston who never went to a wake unless he had known the deceased. He played by his own rules."

On the campaign trail, Jack was full of intoxicating kick-and-go. Recognizing that politics was a learning process, he spent long days canvassing for votes throughout the district, leaving himself only six hours a night for sleep. While his father lavished the campaign with more money than all the other candidates combined, and used his influence, high and low, in every conceivable way, Jack was proving his mettle in face-to-face interactions. The spoiled scion, as it turned out, was a sincere candidate with something fresh and cogent to say about postwar America. What Kennedy understood was that the economy was the most overwhelming concern for voters. On a theoretical level, citizens wondered whether the Great Depression would grimly pick up again where it had left off before wartime industrialization. On the more practical level of daily life, people struggled with shortages of housing, food, and other basic necessities, and worried that events such as the massive labor union strikes occurring that year could unravel the fragile postwar economy.

Two other issues, lying just under the surface of the race in 1946, were the performance of President Truman and the direction of the Democratic Party itself. A year after the death of Franklin Roosevelt elevated him to the presidency, Truman seemed a dreary placeholder compared with his dynamic predecessor. The man who had made the epochal decision to drop atomic bombs on Japan the previous August was now largely perceived as an ineffective leader, with an approval rating of only 37 percent. Beyond the presidency, the entire Democratic Party was having difficulty defining itself after the nearly thirteen-year Roosevelt presidency. Taking a dim view of Truman's inability to excite the public and his uneven defense of labor, many Democrats glumly accepted in 1946 that the spirit of the New Deal was over, while others clung to its promise. After going so far as to tell the president to remain on the sidelines in these midterm elections, Democratic Party leaders distributed radio spots to congressional candidates featuring voice clips of the late FDR, tacitly admitting that their past still overshadowed their present. While Democrats scrambled for some kind of new governing vision, the Republicans grasped the upper hand by embracing a staunchly anti-Soviet, anticommunist posture that became the defining issue of the times.

It was the beginning of the decades-long geopolitical Cold War between the United States and the Soviet Union. After their joint victory in World War II, festering distrust coalesced quickly into opposition. In March 1946, with the Soviets consolidating their control over Eastern Europe, Winston Churchill delivered a speech in which he declared, "From Stettin in the Baltic to Trieste in the Adriatic, an iron curtain has descended across the continent." Stalin was the new Hitler, in this view, willing to purge his own citizens, bent on expansionism and the creation of a Communist world order. Holding back the tide would require a unified West and the continuing presence of American military might in Western Europe.

Among many Americans, including Kennedy, suspicion of the Kremlin metamorphosed into a creeping paranoia, igniting a Red Scare against alleged Communist infiltrators at many levels of American society. The furor swept up both voters who were rightly concerned about the Soviet threat and those who were easily led by their fears. The murkiness of the line between reason and hysteria and the whisking away of serious deliberation bothered many old New Deal Democrats. But any liberal candidates who attempted a nuanced discussion were attacked as being soft on communism. It seems incredible in retrospect that in 1943, Stalin had been *Time*'s "Man of the Year," an ally against Hitler, but a few years later he was rightfully vilified in the United States and elsewhere.

As for the new Democratic congressional candidates of 1946, Jack Kennedy was uniquely prepared for the Red Scare, having lived under the same roof as one of its earliest anticommunist mouthpieces, Joseph Kennedy Sr. A rock-ribbed political pragmatist, Jack unabashedly supported containment of the Soviet Union and a strong arm against domestic Communists. He was willing to stand up to big labor and steer clear of alliances with groups that might define (or, worse, constrain) his ambition. Unleashing his inner Churchill, Kennedy lashed out at everything from the gulag concentration camps in Siberia to the Kremlin's repression of journalists. In October, he gave a speech to the Young Democrats of New York, scolding Henry Wallace, the former vice president and recently dismissed secretary of commerce, for espousing pro-Soviet views. Later, in a Boston radio broadcast, he recounted what he had told the group about Soviet totalitarianism. "I told them that Soviet Russia today is a slave state of the worst sort," he recounted. "I told them that Soviet Russia is embarked upon a program of world aggression. I told them that the freedom-loving countries of the world must stop Soviet Russia now, or be destroyed. I told them that the iron curtain policy and complete suppression of news with respect to Russia, has left the world with a totally false impression of what was going on inside Soviet Russia today."

Choosing to distance himself from Truman, Kennedy epitomized the new liberalism that historian Arthur Schlesinger Jr. yearned to see in the postwar atmosphere. Schlesinger, who had been in Joe Jr.'s class at Harvard, was within six months of Jack's age. Having won literary praise for his 1945 history *The Age of*

Jackson, he was back at Harvard as a teacher when he began thinking about his next book, a study of liberalism in the modern era. Schlesinger wasn't pondering Kennedy when he wrote *The Vital Center*, but the iconoclastic young Democrat fit the bill. In the first place, he was not an obsessive New Dealer. Schlesinger had no patience with those hanging on to the old era. Having come to reject the idea that humans could be perfected if they received enough government help and kindness, he realized that they could, through an equal and opposite reaction, be cowed into subservience by dictators. To form a bulwark against extremism, Schlesinger called for a tough new liberalism that would extend civil rights but give no truck to totalitarians, whether fascists on the right or Communists on the left.

Yet there was a pivotal difference between Kennedy and others (notably Richard Nixon, then a candidate for Congress in California) who were tough on communism, and an even bigger difference with the political philosophy Schlesinger was then formulating. Even as he consolidated his staunch anti-Soviet views, Kennedy retained his bright-eyed idealism. A romantic at heart, he believed, like Thomas Jefferson, in human beings' inherent perfectibility, and he brought the priority of peace to every discussion of foreign affairs.

As one of the youngest candidates running for federal office in 1946, Jack Kennedy offered something other than life experience. Instead, the war hero presented himself as a reflection of his times. The fight against fascism had shaped Kennedy and his generation, forging in them a fortitude and resilience no parental wisdom, no college education, no career experience ever could have. "The war made us get serious for the first time in our lives," he said. "We've been serious ever since, and we show no signs of stopping." He also knew that the dawn of the Atomic Age had amplified that seriousness in a way no previous generation had ever had to face. In coming years, the United States would conduct more than a thousand nuclear tests, in the Pacific Ocean on Amchitka Island, in Alaska, in Colorado, Mississippi, Nevada, and New Mexico. "What we do now will shape the history of civilization for many years to come," he said in his first major speech. "We have a weary world trying to bind up the wounds of a fierce struggle. That is dire enough. What is infinitely far worse is that we have a world which has unleashed the terrible powers of atomic energy. We have a world capable of destroying itself."

That November 5, 1946, at age twenty-nine, Jack Kennedy was elected as the U.S. representative for the Eleventh Congressional District near Boston. The

Democrats that year mustered only 188 seats. Out in California, navy veteran Richard Nixon was swept into Congress on a Republican wave, along with 245 other members of the GOP.

Jack Kennedy's public career was born amid the postwar reality of atomic bombs, ballistic missiles, and the still-unrealized threat of intercontinental long-range rockets. He favored an international body to oversee atomic weapons, largely as a way of maintaining America's monopoly by dissuading other nations from building their own nuclear arsenals. But as far as Kennedy and most other politicians were concerned, rockets were simultaneously the V-2 past and the *Flash Gordon* future. In a country trying to regain a sense of normalcy, it seemed that the only group really focused on rocketry was the military, including the U.S. Air Force, a new division of the armed forces that had been established in the fall of 1947 via the National Security Act. Like the Army, this branch had benefited from having sponsored hundreds of Paperclip scientists and engineers. The Air Force was keenly interested in the Luftwaffe's technology pertaining to transsonic and supersonic aerodynamic research.

The previous autumn, Wernher von Braun had made his first public speech in America, to the El Paso

Rotary Club. "It seems to be a law of nature that all novel technical inventions that have a future for civilian use start out as weapons," he said, before going on to predict a future where rocketry took its proper role of propelling satellites and space stations into orbit and enabling missions to the moon and beyond. Von Braun got a thundering ovation and was cheered by the support, but as usual, his ideas were ahead of their time. Before his space dreams could take flight, rocketry would enter new and even more dangerous territory with the postwar development of the first intercontinental ballistic missiles.

The only sensible thing for von Braun to do was, once again, to lie low and develop his ballistic missiles in Fort Bliss–White Sands for U.S. Army purposes while keeping a moon and Mars voyage as a long-term interior motive. In 1946, the Army Signal Corps succeeded in bouncing radio waves off the moon and received the reflected signals back on Earth. This was a stunning achievement, for it established that radio transmissions through space and back to Earth were possible. This public discovery didn't mean anything to Kennedy, running for Congress. But to von Braun, it was proof that such signals could in the very near future be adapted to control manned and unmanned spacecraft alike.

Congressional candidate John F. Kennedy leans back in a chair under a shelf that holds up a 1946 campaign poster (The New Generation Offers a Leader) and photos of his parents, Joseph and Rose, in his office-room at the Bellevue Hotel, Boston, Massachusetts.

PART II

Generation *Sputnik*

Election night, November 1, 1952, when John F. Kennedy won a seat to the U.S. Senate from Massachusetts. He would be reelected to the Senate in 1958.

5

Spooked into the Space Race

The notion of a "race" with the Soviet Union permeated Cold War thought, and it applied not only to space but to nuclear arms.

—YANEK MIECZKOWSKI,
EISENHOWER'S SPUTNIK MOMENT (2013)

As John F. Kennedy started his congressional career in early 1947, the Cold War was on, full bore, with the United States and the Soviet Union shoring up their influence in a divided Europe. In following years, advanced intercontinental ballistic missiles (ICBMs) and high-altitude reconnaissance plane innovations would forever change the notion of warfare. Having read so many history books, Kennedy prided himself on being able to make quick summaries of pressing global situations and offer pragmatic recommendations. JFK relied on the power of family ties, trust fund, and personal freedom—he was an island unto himself, questing for the present moment.

For Jack, 1947 was a year of both visibility and irrelevance. An eligible bachelor and nightclub habitué, and possessed with a mercurial attractiveness, he moved into a town house on N Street, in Washington's Georgetown neighborhood. He cut a dashing figure as one of the youngest lawmakers in town, balancing a social life in the fast lane with a minor reputation for occasional eloquence on the House floor. Still, Kennedy was often bored during his first term, full of indifference, realizing quickly that a low-ranking member of the minority party had little of importance to do on Capitol Hill. Perhaps in response, he traveled frequently, not only back to New England but also around the country, speaking to almost any group that would invite him.

In official Washington, dollars and cents ruled the agenda. Having spent a third of a trillion dollars on the war, the federal government had war bonds to repay, both as an obligation and also in the interest of fueling the postwar economy. Another pressing expense was financial assistance to nations devastated during the war. Speaking at Harvard on June 5, 1947, Secretary of State George Marshall presented the outline of a plan that would "help the Europeans help themselves" by pumping more than $13 billion (almost $150 billion in 2019 dollars) into rebuilding war-ravaged Western Europe: stabilizing currencies, budgets, and finances; promoting industrial, agricultural, and cultural production; and facilitating and stimulating international trade relationships. In occupied Japan, a similar program under General Douglas MacArthur was beginning its work of rebuilding the war-ravaged island nation along capitalist lines.

Most Americans believed that the Soviet Union, having lost more than twenty million lives in World War II, must also be focused completely on the task of infrastructure rebuilding—a casual assumption that played to the Soviets' advantage in missile and nuclear bomb development. Because the Kremlin maintained what science-fiction writer Arthur C. Clarke called "an almost impenetrable veil of secrecy," U.S. intelligence services had little idea how rapidly the Soviets were in fact developing their capacity in nuclear weaponry and advanced rocketry. After World War II, Sergei Korolev, principal designer of the future Soviet space program, had been tasked by Stalin to build powerful ICBMs. Initiating a systematic exploitation of German guided missile technology, by 1946 they were already working on ballistic missiles with a range of nearly two thousand miles. Within fifteen years, they would make technological leaps that would lead to the first nuclear-tipped ICBMs and would also propel Russian cosmonauts toward space. Soviet engineer Mikhail Ryazansky also spearheaded a cabal of scientific specialists who pioneered new radar and radio navigation technology. As the Kremlin innovated a large-scale missile program, the United States, von Braun rightfully complained between 1945 and 1951, had "no ballistic missiles worth mentioning."

While the Soviets built, the U.S. military engaged in squabbles fueled by an interservice rivalry. With budgets down from their stratospheric wartime levels, the various armed services jockeyed for every congressional appropriation and threw elbows while they did. The new National Security Act also threatened their autonomy, placing the army, navy, and marines under the authority of a new National Military Establishment (later, the Department of Defense) and separating the U.S. Army Air Forces (AAF) out into its own service, the U.S. Air Force (USAF). Amid these budgetary and administrative struggles, the development of rockets chugged along slowly on multiple tracks across the armed forces as engineers explored their potential as weapons, research tools, launch vehicles for satellites, and vehicles for space exploration.

In the air force, Major General Curtis LeMay, a war hero then serving as deputy chief of air staff for research and development, believed that any future space program should be under the domain of air operations. Born in 1906 in Columbus, Ohio, LeMay had commanded the 305th Operations Group and the Third Air Division in the European Theater during World War II. A dashing fighter pilot able to loop and dive with the best, he ran strategic bombing operations against Japan toward the end of the war. After V-J Day, LeMay was assigned to command the U.S. Air Force in Europe and to deal with the nonstop crisis in Berlin, which was then divided among American, British, French, and Soviet zones of occupation. LeMay commissioned the research-and-development arm of Douglas Aircraft, in Southern California, to answer two fundamental postwar questions: How could satellites benefit the U.S. military? How could space travel advance humanity? He gave the aerospace corporation three weeks to report back.

Douglas Aircraft's RAND unit (which took its name from a contraction of the term *research and development*) had been created immediately following World War II, by a number of military and industry leaders, including General H. H. "Hap" Arnold, who'd served as head of the U.S. Army Air Forces until 1945. Arnold understood that military science was accelerating at a mind-boggling rate, one that would affect all the service branches, but aviation in particular. His insistence that academia, industry, and the military had common goals and would benefit from cooperation was revolutionary during the transition from war to peace, paving the way for decades of expansion in all three sectors.

The 321-page Douglas Aircraft RAND report was delivered on May 2, 1946. Titled *Preliminary Design of an Experimental World-Circling Spaceship*, this document described future military uses for Earth-orbiting artificial satellites, including surveillance and missile guidance. It suggested a raft of civilian possibilities, too, notably in communications and meteorology. The RAND report, the work of fifty researchers, presented two conclusions that offered infinite potential for the air force. First, a satellite vehicle with appropriate instrumentation would become one of the most revolutionary military, scientific, and communication tools in the twentieth century. Second, this type of satellite could inflame the aspirations of mankind. "Whose imagination is not fired by the possibility of voyaging out beyond the limits of our earth, traveling to the Moon, to Venus and Mars?" the report asked. "Such thoughts when put on paper now seem like idle fancy. But, a manmade satellite, circling our globe beyond the limits of the atmosphere is the first step. The other necessary steps would surely follow in rapid succession. Who would be so bold as to say that this might not come within our time?"

The various RAND findings were discussed in detail at a top-secret conference called to consider the idea of uniting the navy and air force in satellite development. The cost of launching a satellite was estimated at around $150 million and to be ready by 1951. Representatives of the two branches, however, couldn't agree, and military officials looking for a compromise soon lost interest. The RAND study was filed away.

Turning away from the RAND satellite recommendations, the navy used its prerogative in late 1946 to order the development of a rocket ultimately named the Viking. Created by Baltimore's Glenn L. Martin Company (later part of Lockheed Martin) in conjunction with the U.S. Naval Research Laboratory, the Viking was built on the basis of the V-2 but was a distinctly American effort, even incorporating some ideas by the late Robert Goddard. "The U.S. Navy wanted no part of the haughty Germans, no matter how talented they were," wrote historian and NASA veteran Doran Baker. The Viking was developed principally to gather

A captured German V-2 rocket just after takeoff at Launching Complex 33 at White Sands Missile Range in New Mexico. The launch was part of an experimental program carried out by the U.S. Army's Upper Atmosphere Research Panel.

upper atmospheric and ionospheric data that would help predict weather and would communicate via satellites. It would prove a huge boon to America's military and commercial aviation industries in coming decades. At the White Sands Proving Ground in New Mexico, test launches of the Viking were able to carry research instruments to altitudes of up to 158 miles.

The U.S. Army was continuing its own missile work, relying heavily on German technology and expertise from Fort Bliss. When the Dora-Mittelbau war crimes trial ensued in 1947 at Dachau, the U.S. Army Ordnance Corps made it clear that the von Braun team (the Peenemünders) had eluded any charges. Unscathed by the Dachau trial, in the fall of 1948 von Braun's team began contemplating the development of Earth-orbiting satellites. The Committee on Guided Missiles requested that von Braun's Fort Bliss–White Sands desert team design a way for the army to pioneer the science of satellite carrier-rocket development. Shortly after Christmas, James Forrestal, now the first "secretary of defense," publicly declared that the Pentagon was looking into the workability of artificial satellites. This was just a preliminary launching by the U.S. government looking into Earth satellite vehicles. Documents from 1949 show that the RAND Corporation had convinced the Pentagon of the satellite's potential for surveillance, reconnaissance, communications, and intimidation, suggesting that the "mere presence in the sky of an artificial satellite would have a strong psychological effect on the potential enemy."

While many researchers in El Paso, Pasadena, Hampton, and at air force bases such as Edwards, Wright-Patterson, and Kirtland were looking to space as the next frontier, others were making history within Earth's atmosphere. On October 14, 1947, a West Virginia test pilot named Chuck Yeager, a master of aerial evasion and dive-bombing, strapped in for a test flight of the Bell X-1 experimental aircraft, nicknamed *Glamorous Glennis*, after his wife. Launched from the bomb bay of a Boeing B-29 Superfortress over the Rogers Dry Lake in California's Mojave Desert, Yeager's X-1 reached an altitude of 43,000 feet and a speed of 700 miles per hour, marking the first time a plane had exceeded the speed of sound in level flight.

Following World War II, the United States went aviation crazy. Even though mass transportation was still the domain of ships, trains, and automobiles, commercial airlines such as Eastern were making inroads with passenger cabins designed for comfort. Cities started building airplane terminals with bonds to be paid out in thirty or forty years. With a plethora of trained pilots, commercial aviation routes were established almost like the old pioneer routes of the nineteenth century's era of westward expansion. During Truman's presidency two reliable routes were opened for transcontinental air travel: the New York–Chicago–San Francisco route in the north and the New York–St. Louis–Los Angeles route to the south. Electronic navigation and anti-icing technology, both developed during the war, soon allowed for all-weather flying, though fog and snow caused groundings. Despite subpar safety standards, every year, more and more Americans were using commercial aviation as their preferred mode of long-distance travel. Still, too often local newspapers ran horrific stories of wings falling off, midair collisions, and crashes on landing. Aviation technology would have to improve before air travel was truly embraced with mass appeal.

On June 24, 1948, long-simmering tensions in divided Berlin came to a boil when Stalin ordered the closure of all land routes to the parts of the city controlled by American, British, and French forces. It was a power play aimed at starving out the Western powers and forcing them to abandon Berlin, located 110 miles within what was the Soviet zone of postwar German occupation, an area soon to become the German Democratic Republic. But the maneuver didn't work. From June 28 to May 11, 1949, the three nations launched the Berlin Airlift, a massive effort that successfully supplied the citizens of Berlin with food, fuel, and medicine. The U.S. Air Force, which continued flying supplies until 1949, scored a major humanitarian triumph with the dropping of foodstuffs and medical provisions, as more than two hundred thousand sorties carried in over 1.5 million tons of supplies to the surrounded city.

The successful U.S. airlift helped Truman defeat his Republican challenger, Thomas Dewey of New York, in November's presidential election, surprising pundits. Kennedy was also reelected, and enough other Democrats took formerly Republican seats to regain control of the House of Representatives. Two years into his political career, JFK had proved surprisingly businesslike in the job, with his congressional offices in Washington and Boston having a reputation for crackerjack staff and unassailable constituent service. He also proved to be a bright, solicitous colleague and was well liked on Capitol Hill. Nobody thought he was a workhorse. But he was his own man, a stand-alone, never part of a faction or clique. Although he usually voted with fellow Democrats, he made no effort to curry favor with such power brokers as Speaker of the House Sam Rayburn or Majority Leader John McCormack—something that was unheard of for a junior representative. It was as if Kennedy's good looks, great wealth, and war-hero status allowed him the privilege of being an island unto himself on Capitol Hill. Speaking frankly, Rayburn called Kennedy "a good boy" but "one of the laziest men I ever talked to."

On the other side of the Capitol Building, in the upper chamber, another New Englander recently arrived on the Hill had chosen a different way of making a name for himself. Even before the U.S. nuclear attack on Hiroshima in 1945, the forward-thinking Connecticut senator Brien McMahon had adopted atomic technology as his area of expertise in Congress. In the years after the war, he deemed the July 1945 test detonation of the first atomic bomb in New Mexico "the most important thing in history since the birth of Jesus Christ." After V-J Day, McMahon believed the United States had a vested national security interest in remaining the world's only atomic superpower. Serving as chair of the Senate Special Committee on Atomic Energy, McMahon became known for the rest of his congressional career as "Mr. Atom Bomb" for his authorship of the Atomic Energy Act of 1946, which created the Atomic Energy Commission to control nuclear weapons development and nuclear power management, stripping this authority from the military. At the same time, McMahon concerned himself with the geopolitical import of atomic weapons, at one point delivering a sober-minded speech in the Senate proposing various diplomatic ways of assuaging Soviet fears that the United States would initiate atomic war.

American officials were confident not only in their atomic monopoly, but also that the United States was sitting on top of the world technologically, politically, and economically. On April 4, 1949, the country had joined Belgium, Canada, Denmark, France, Iceland, Italy, Luxembourg, the Netherlands, Norway, Portugal, and the United Kingdom in creating the North Atlantic Treaty Organization (NATO) to resist Soviet expansionism. Shortly thereafter, the Soviet blockade of Berlin ended with the reopening of access routes from western Germany to the city. Having proved a seemingly limitless ability to resist Stalin's draconian tactics, and having inflicted considerable damage on Soviet-controlled eastern Germany's economy via a retaliatory trade embargo, the newfound Western alliance seemed indomitable. Truman himself, buoyed by his 1948 election victory and the validation of his foreign policy in Berlin, was confident that the USSR lagged behind on every measure of power in the postwar world, including atomic weaponry.

Soviet and ex-Nazi rocket specialists were working on an improved version of the V-2, with a power plant capable of a thrust yield of at least eight hundred thousand pounds—fifteen times greater than the wartime V-2. Because the USSR lacked America's network of strategic air bases around the world, Stalin was betting on intercontinental ballistic missiles to project Soviet power abroad.

In the early postwar years, American intelligence severely underestimated the USSR's atomic capabilities, but that blissful ignorance was soon to be shattered. On the afternoon of September 22, 1949, Senator McMahon received a summons to the White House, where at 3:15 p.m. ushers escorted him into a top-secret meeting with President Truman. Before almost any other American, McMahon heard the disturbing news that, three weeks earlier, at the Semipalatinsk in Kazakhstan, the Soviet Union had detonated its first atomic bomb, nicknamed "Joe 1." A shocked McMahon promised Truman he would not leak the classified briefing to the press. As he drove home that afternoon, he was full of trepidation. Eyeing happy-go-lucky children on playgrounds, watching their purpose-driven parents doing family errands, he felt his heart sink, cognizant that Americans were in their last carefree hours of postwar innocence. The world was about to get more dangerous and complicated.

That evening, Truman paced around the White

House, sipping bourbon. The world order had changed: there were now two nuclear superpowers. But the president's greatest fear wasn't the obvious one: that the Soviets would attack the United States with nuclear weapons. Instead, he intuited that when the public learned that Joseph Stalin had an atomic bomb, the shock could trigger a panic, and recriminations were sure to be unleashed against his administration for having underestimated Soviet capabilities. The next day, when the rest of Congress and the public received the news, lawmakers would have to tread carefully not to frighten the country or fan the flames of war.

Congressman Kennedy deliberated on the Soviet feat in Kazakhstan for two weeks before speaking out via an open letter to President Truman, warning that the Department of Defense had reneged on its obligation to prepare American citizens to survive atomic warfare. Less interested in how the Soviet scientists had achieved nuclear parity, Kennedy seized the issue of civil defense as his calling card. Remembering the woeful lack of U.S. military preparedness in the 1930s, he wrote of "an atomic Pearl Harbor" unless the United States spent "months and even years" planning ways to resist or respond to and survive the catastrophic event of a Soviet atomic attack.

Nobody paid much attention to Kennedy's theatrics. Journalists denigrated the letter as the act of a publicity-hungry young politician who was good at turning a phrase. The Truman administration turned to McMahon to counter Kennedy's alarmism. The Connecticut senator dismissed the young Massachusetts congressman's civil-defense siren call as novice humbuggery. The United States, McMahon retorted, simply couldn't afford to devote the necessary time and expense to the levels of civil defense Kennedy proposed.

Kennedy was playing to the public's still clear memories of the attack on Pearl Harbor of eight years before. Even though a Soviet sneak attack on Western Europe or North America was unlikely, it was plausible, especially considering the tensions over divided Berlin. In coming years JFK endorsed the idea of the federal government's printing pamphlets warning the public about post-explosion radiation hazards, and having schoolchildren learn how to seek shelter if a nuclear flash occurred. But his well-intentioned civil-defense mantra was too simplistic. While building superhighways to escape American cities quickly in the event of a Soviet attack had public works merit, defense strategists such as McMahon knew that actually preventing World War III would be a much more complicated endeavor. It would take the slog of intense U.S.-Soviet diplomacy, and it would also require the United States to build multistaged rockets and hydrogen bombs. And this would mean the United States' developing a sophisticated ICBM program capable of delivering nuclear warheads over long distances. If the USSR could blow America up fifteen times over, America had to build a missile deterrent that would wipe out the USSR fifty times. If the Soviets were mobilizing science and technology in peacetime, so would the Americans. McMahon's thinking framed the Cold War until the USSR's collapse in 1991.

As the national debate unfolded, Kennedy moved beyond his initial civil-defense stance and took a similar but more conventional tack, arguing that the best way to avoid atomic war was for the United States to build up its troop levels; throw billions into modernizing the army, navy, and air force; and stay on a permanent wartime footing.

Unusual for two Democratic leaders, Kennedy and McMahon agreed that President Truman, one of their own, was partly to blame for Stalin's atomic advancements. But their thinking was part of an anticommunist fervor that swept across the country in 1949. When, on October 1, 1949, the Chinese Nationalist government of Chiang Kai-shek fell to the Communists under Mao Zedong, Americans woke up believing that Stalin and Mao were intent on burying U.S. capitalism.

The double whammy of the Soviet nuclear bomb and the Chinese Communist Revolution reinforced Kennedy's tough-minded anticommunist side. The fact that his father despised Stalin made his political stance that much easier. At the same time, Jack realized that it was far wiser, politically, to couch his anti-Sovietism in terms of advancing global peace rather than stoking fears of an impending nuclear war. He knew that communism had to be resisted, whether one sought peace through disarmament or through massive military deterrence. American relations with the Soviets, JFK believed, had to be a mixture of aggression and accord.

In April 1950, while Kennedy was approaching his third congressional race, President Truman received National Security Council Paper Number 68 (NSC-68), a top-secret report that argued for a massive buildup of the U.S. military and the nuclear arsenal to counter the Soviet threat. This document laid the groundwork for American Cold War policy for the next two decades.

On November 7, 1950, Kennedy beat Vincent Celeste to win a third term in Congress. Throughout the Korean War, which had started earlier that year, JFK regularly called for enormous increases in the U.S. military budget. An avatar of constant vigilance against communism, he applauded Truman for authorizing the development and deployment of U.S. thermonuclear missiles, B-52 bombers, supercarriers, tanks, and other heavy weapons. But Kennedy considered Truman's foreign policy feckless, and he regularly criticized the president for weakness against the Soviets, the stalemate in Korea, and a general unwillingness to stop the "onrushing tide of communism from engulfing all of Asia." To the surprise of Democratic liberals, Kennedy distanced himself from the revered George C. Marshall and State Department hands such as John Fairbank and Owen Lattimore for allowing the Nationalist government of China to collapse. Although Kennedy was a traditional liberal on domestic issues such as Social Security, the minimum wage, taxes, and education, he was a pronounced anti-Soviet, pro-military hawk like few other congressmen in his party. Just as nobody—not even Senator Joe McCarthy, a family friend—could accuse JFK of being weak on communism, nobody could accuse him of being a party regular. "I never had the feeling," Kennedy said privately, "that I needed Truman."

In April 1950, von Braun and his group of around 125 German scientists and engineers brought to the United States by Holger Toftoy under Operation Paperclip were transferred from Texas to the Army's Redstone Arsenal, in Huntsville, Alabama, home of the newly renamed Ordnance Guided Missile Center (OGMC). Overall, a thousand personnel were assigned to help von Braun develop what soon became the Redstone rocket. Situated in the Tennessee River Valley, surrounded by rolling hills and caves, Huntsville was a garden paradise compared with arid El Paso. There was no ambiguity about what their mission was building: tactical ballistic missiles of mass destruction, *not* spaceships. Toftoy, then a brigadier general, would soon command von Braun's team in Alabama. (In early 1956, the newly formed Army Ballistic Missile Agency [ABMA], under Major General John Medaris, would take charge of the entire Huntsville operation.)

Speculation was rampant in 1951 that Kennedy would run for Senate the following year. Fresh from a five-week European tour, the young congressman testified before the Senate Foreign Relations and Armed Services Committees on how best to defend Europe against Soviet influence and control. That May, he introduced a bill seeking to restrict U.S. companies from trade with "Red China," and in early fall he embarked on an extended trip to Asia, visiting Hong Kong, India, Vietnam, Japan, Korea, Malaysia, and Thailand. In addition to burnishing his foreign policy credentials, Kennedy's travels confirmed his belief in the necessity of containing communism.

Around Capitol Hill, Kennedy continued to be best known as the House's resident playboy. The *Washington Post* teased that the "current emotional heat wave on Capitol Hill is attributed to bachelor Representative John F. Kennedy of Massachusetts." According to the newspaper, women working in the U.S. Capitol "peek their heads around corridors and have heart palpitations when word spreads that the young lawmaker is approaching." Always smiling boyishly, his teeth a bright white, Kennedy inspired more than workplace crushes; his personal magnetism was an asset in selling the Democratic Party brand. The question hovering over the *PT-109* heartthrob was whether he'd ever marry. Then, in May 1951, he met Jacqueline Bouvier at a dinner party and was smitten—though that didn't stop him from chasing other women. "I knew Jack and he was a playboy," recalled John Lane, a senior aide to Senator McMahon. Lane and Kennedy were similar in age, both Irish Catholics from New England, educated at first-rate colleges, and brimming with the ambition to make good in Washington. "A nice guy," Lane continued, "riding around in the Cadillac convertible with the top down, living it up in Georgetown, because I lived it up there too at the same time."

Though Kennedy lived with a certain esprit and had found a way to use his youthful verve to draw in voters on his own terms, he worked hard to develop a more heavyweight reputation in the halls of Congress. As to his fans among the women at the Capitol, the *Post* added that Kennedy "assiduously dodges them with an inimitable Irish grin and sticks to his legislative duties." Even if his team—or his father—wasn't directly responsible for the addition of that sentence, they could have pointed to the frothy *Post* article as a template for hundreds of Jack Kennedy profiles to come: he was as charming as a matinee idol, but also a staunch anti-Soviet orator. Nevertheless, Kennedy didn't earn the gravitas he sought in the House, where he chafed at the need for legislative patience. He knew you had to be in Congress many years before you had *real* power and influence, or an opportunity to play a historic role on substantive legislation.

Congressman Kennedy remained constantly aware of his father's expectations, which had pointed at the White House since his sons were in swaddling clothes. The pressure to succeed may have been paternal, but the need to act quickly came from Jack himself. Friends who knew him in the House said he treated his life as though death were knocking, making the most out of every day. In this, his health was a major factor: in addition to the digestive and neurological maladies he'd endured since childhood, and which still resulted in occasional hospital stays, he also had near-constant back problems, causing bouts of blinding pain and sometimes necessitating the use of crutches. Additionally, he'd been diagnosed in 1947 with Addison's disease, a dysfunction of the adrenal gland that can cause fatigue, abdominal and muscle pain, and depression, among other symptoms.

In his way, Kennedy seemed to have turned his pessimism about his longevity into optimism for all he wanted to accomplish, as soon as possible. And while it's true his father prodded him to advance in politics, regularly providing money and meddling, Joseph Kennedy's supposedly overbearing influence has been overblown. "Sometimes you read that [Jack] was a reluctant figure being dragooned into politics by his father," recalled Charles Bartlett, a close friend of Jack's who became a syndicated newspaper columnist. "I didn't get that impression at all. I gathered that it was a wholesome, full-blown wish on his part."

Jack Kennedy's political drive was, in part, a reaction to the early deaths of his older brother, Joe Jr., and his sister Kathleen ("Kick"), who'd died in an airplane crash in 1948 at the age of twenty-eight. Close in age, these three Kennedy children had been inseparable in their youth. At thirty-four in 1951, JFK may indeed have felt he was living on borrowed time. Hungering for greatness and steeped in history, he grew eager to move beyond the stultifying House of Representatives.

While Kennedy deliberated his future, the United States was taking steps that would bring it closer to space.

During World War II, three arsenals centered on the town of Huntsville had manufactured and stored ordnance for the U.S. military: the Huntsville Arsenal, the Huntsville Depot, and the Redstone Ordnance Plant. After V-J Day, these facilities were consolidated to form the Redstone Arsenal, which in June 1949 became home to the new Ordnance Rocket Center, the army's headquarters for rocket research and development. Once von Braun had been transferred, he discovered that he loved being in northern Alabama. Huntsville was ideal for von Braun's purposes: the Army already had well-equipped laboratories, large assembly structures, and a dependable firing range. But it wasn't a place where he could fire large booster rockets into the sky. For that his Redstone rockets would be shipped on giant barges on the Tennessee River, which flowed into the Atlantic Ocean. Coinciding with von Braun's move to Alabama was a new rocket launchpad at Cape Canaveral, Florida, where his "babies" would be launched toward the stars. He was frustrated that the U.S. Army was building rockets "at a tempo for peace." Nevertheless, he remained dutiful to the Army and had a born-again Christian conversion, which gave him faith in the future. "We are," von Braun enthused upon resettling, "going to make history here." A further frustration for him was that his novel *Mars Project*, anchored in the uplifting prose of Jules Verne (but larded with empirical science), struggled to find a publisher; it was turned down by nineteen for being "too fantastic." Nevertheless, von Braun remained convinced that landing on the moon was doable in his lifetime. And he believed Mars was reachable in the mid-twenty-first century.

Redstone became the namesake for a new class of von Braun–designed suborbital ballistic missiles that were direct descendants of the V-2 rockets, and Huntsville became an epicenter of Cold War industrial mobilization. In a strange twist, von Braun and his team were tasked with essentially building V-2s with nuclear warheads on them to be shipped to U.S. Army bases in West Germany. Faced with UN forces fighting a North Korean military armed and sponsored by both the USSR and China, the United States had to master military rocketry and begin full-scale production before attempting to make manned space voyages a reality.

Because America had no usable missiles, the Korean War became a showplace for the air force to prove its new-kid-on-the-block importance. More than one thousand U.S. fighter pilots served in the conflict, and the most effective plane proved to be the F-86 Sabre jet, which could fly at forty-five thousand feet and was armed with machine guns for use in aerial engagements. Because the Truman administration couldn't risk escalating the war by bombing the mainland Chinese airstrips where North Korea based its Soviet-made MiG-15 fighter jets, those American pilots were kept busy, and fought magnificently. Thirty-nine U.S. pilots attained ace status, downing five or more enemy planes, and three of these brave fliers would go on to become NASA astronauts who intersected poignantly with Kennedy's career: Neil Armstrong, John Glenn, and Wally Schirra.

Although other politicians typically campaigned for Senate in their forties or fifties and usually boasted a distinguished political record, thirty-five-year-old Jack Kennedy was determined to make his bid for the Senate in 1952, taking on popular Republican senator Henry Cabot Lodge Jr. He could have waited for 1954 and run against the more vulnerable of Massachusetts's senators, Leverett Saltonstall, but he was loath to waste two years. JFK was fired up, and 1952 would be his year. His younger brother Robert would, for the first time, be his campaign manager. Joe Sr. started working the phones and called Connecticut senator Brien McMahon, who was going after the Democratic presidential nomination after Truman announced that he would not seek another term.

"McMahon was home ill when Joe Kennedy called him," recalled John Lane. "I walked in when he was talking to him on the phone. He hangs up and he says, 'Joe is going to enter Jack.' I said, 'Jack Kennedy for the Senate? Really? . . . My God!' But he said, 'I'd rather have a Kennedy in the Senate than a Lodge.'"

Gruff in the way of a friendly bear, McMahon was far unlike Jack Kennedy in temperament, but the self-described Cold War Democrat had staked out much the same ground as Kennedy, and he was someone the younger politician watched closely. In 1952, McMahon was well prepared for the presidency, his only worry being the glass ceiling of his Catholic faith, which had perhaps foiled the candidacy of New York governor Al Smith in 1928. On that score, McMahon thought perhaps he might just be the one to finish the job and win the White House—but it was not to be. Just as he started his campaign, pitching a platform of ensuring world peace through fear of atomic weapons, he was diagnosed with lung cancer and withdrew to wage a more personal battle. He died three months before the election.

Kennedy's Senate opponent, Henry Cabot Lodge Jr., was a moderate Republican from one of New England's most prominent families. Lodge's grandfather had been Theodore Roosevelt's sturdiest ally in the Senate, and one of his brothers, John Davis Lodge, was currently the governor of Connecticut. Henry Lodge himself, then fifty, was a seasoned politician who had resigned from the Senate during his second term to serve in World War II, where he earned a chestful of medals for valor under fire in France and Germany. Hollywood handsome, dapper, and with a fine patrician accent, he was a formidable presence on the American political stage. Next to Lodge, almost any rival would have seemed callow, and the young Jack Kennedy seemed especially green.

But Kennedy, with three successful congressional races under his belt, waged an exemplary fight. The embodiment of poised drive, he traveled tirelessly throughout his home state, appearing before small groups that added up to hundreds of thousands of voters who could go home and say, "I met John Kennedy today." The *Saturday Evening Post* reported that Kennedy was "being spoken of as the hardest campaigner

Massachusetts ever produced." Meanwhile, by backing Dwight Eisenhower, a moderate, over the conservative Robert Taft for the Republican presidential nomination that year, Lodge alienated the right wing of his party; they would stay home on Election Day. In addition, during campaign season, he was often elsewhere in the country, playing the elder statesman. Although both candidates used advertising extensively and spoke whenever possible before large Massachusetts crowds, the personal style of the Kennedy team made a difference, with brother Robert ably serving as Jack's campaign manager and Joe Sr. exercising his phenomenal fund-raising prowess. Channeling the spirit of the times, Kennedy praised the Strategic Air Command decision to deploy Convair B-36 Peacemaker and Boeing B-47 Stratojet long-range nuclear bombers on "Reflex Alert" at overseas bases such as the purpose-built Nouasseur Air Base in French Morocco, placing them within unrefueled striking range of the Kremlin.

The 1952 campaign ground on. In late July, after he'd withdrawn from the Democratic presidential race, Brien McMahon listened to the national convention on radio from his hospital bed in Hartford and heard the Connecticut delegates award him all their votes on the first ballot. Once the other states voted, Governor Adlai Stevenson of Illinois was chosen as the Democratic standard-bearer. From his sickbed, McMahon told the convention leaders by telephone that if Stevenson won, he should immediately instruct the Atomic Energy Commission to mass-produce thousands of hydrogen bombs. A few days later, McMahon was dead. He never got to see his dream of the United States testing its first thermonuclear bomb, which occurred just months later, on November 1, 1952. In some respects, McMahon's death was John Kennedy's gain. As John Lane later recounted, "Kennedy told me later that if McMahon hadn't died he [Kennedy] probably would have never gotten a chance" to run for president in 1960.

On Election Day 1952, Kennedy won a surprise victory in his Senate race even as Republican Dwight Eisenhower won the presidency. The sixty-two-year-old Eisenhower, whose great balding head and fine smile reassured voters of his natural leadership skills, would be the first Republican to occupy the White House in twenty years. A West Point graduate, "Ike" had the distinction of a meteoric rise, climbing over three short years from the rank of colonel in 1941 to that of five-star general, and Supreme Allied Commander, by 1944. After the war, Eisenhower served as army chief of staff from November 1945 until February 1948, then retired and moved into academia as the president of Columbia University. Still looking for his next role as he entered his sixties, he was courted by both political parties, with no less than James Roosevelt, son of Franklin and Eleanor, trying to entice him onto the Democratic presidential ticket. The Republicans won out, and though conservatives hoping to control the party had their doubts about him, Ike brought their party the White House—and they, along with the rest of the nation, waited to see just what kind of president he would turn out to be.

With Harry Truman leaving the presidency with an approval rating of under 30 percent, the Democrats also faced a void of leadership as the Eighty-Third Congress convened in January 1953. Their presidential candidate, Senator Adlai Stevenson, was liked and respected, but his defeat left room for hungrier and louder Democratic voices. Lyndon B. Johnson of Texas, who became the Senate minority leader as Kennedy entered the upper house, was the prime example. Like Kennedy, Johnson had felt stifled in his earlier days in the House; moving on to the Senate in 1948, he thrived there. Even without seniority, Johnson managed to dominate through his mastery of policy and the legislative process. A born politician from Texas Hill Country, he could keep the endless minutiae of issues prioritized simultaneously with the nuances of his colleagues' needs. Nobody in politics worked a telephone or Senate floor (and its back rooms) better than Johnson. And in person, using gesticulating hands and a surprisingly hard grip to his advantage, he came across as a force of nature.

Hubert Humphrey of Minnesota was the unabashed New Dealer among Senate Democrats at the time Kennedy arrived in the upper chamber. A doctrinaire liberal with strong ties to the labor movement, Humphrey had made his reputation at the 1948 Democratic National Convention with a rousing speech on civil rights, for which the Deep South never forgave him. Taking a Senate seat the following year, he championed lost causes and ruffled the feathers of more conservative Democrats. His colleague Robert C. Byrd of West Virginia recalled the time that Humphrey rose during a discussion of Senate committees "to demand the abolition of the Joint Committee on the Reduction of Nonessential Federal Expenditures as a nonessential expenditure." It wasn't as funny then as it sounds now; Humphrey's constant questioning made enemies of many pork-hungry senators that day. By the time of Kennedy's arrival, Humphrey had smoothed his ways and become known as a dogged legislator, someone not as manipulative as Johnson but with a fund of Minnesota decency that could bring him labor and farm support when it was needed.

Jack Kennedy avoided the rookie mistakes that Humphrey had made. After fumbling his opportunities by sidestepping House committee leadership roles from 1946 to 1952, he was more attentive and cooperative in his new Senate role, and felt revitalized working on a variety of bills and projects. Determined not to be confined to championing projects that would help only Massachusetts, he vocally supported the Saint Lawrence Seaway, disregarding his home state's worry that connecting the Great Lakes to the Atlantic would hurt Boston's seaport business. He also enjoyed the Senate's clubbier, more elitist atmosphere, which harked back to his experiences in prep school and the Ivy League. Eloquence was expected, and so were manners. They weren't always delivered, as in the case of the unpredictable Lyndon Johnson, but Kennedy could, without pretension, quote Milton's poetry in his speeches and mention Roman philosophers, certain that he didn't need to explain every reference to his colleagues. While he had always refused to insult audiences by simplifying his rhetoric in public, he seemed inspired to up his game in his speeches in the Senate. For Kennedy, the Senate was about power, and every member had a healthy measure of it. If he had come to the conclusion that the vast majority of representatives were irrelevant, he saw that every senator counted. Toward the end of his life, an old friend asked him what he would like to do after the presidency, and Kennedy replied that he'd like to be a senator or journalist once more.

Because Kennedy's civil-defense advocacy fell flat, he needed a signature issue, and just months after Eisenhower was sworn in, he decided he had one. At issue, remarkably, was the war hero president's disinclination to grow the military. After only six months in the White House, in July 1953, Eisenhower pulled America out of the Korean War, which had cost more than 34,000 American lives. Following what was later dubbed his "New Look" defense strategy, Eisenhower ordered a 25 percent reduction in military funding and demanded accountability from Pentagon planners who had for years slipped their massive pork plans past the accommodating Roosevelt and Truman. The commander who had masterminded D-day, the largest naval offensive operation in the history of the world, didn't believe in maintaining a gargantuan standing army poised for limited Korea-like ground wars. Nor did Eisenhower trust an economy dependent on weapons production for prosperity. Eight years later, in his farewell address upon leaving the presidency, Eisenhower famously warned about the dangerous influence of what he called "the military-industrial complex." Less well remembered is that Eisenhower started his presidency on this same note, with a plea to avoid what he called the "burden of arms. . . . Every gun that is made, every warship launched, every rocket fired signifies, in the final sense, a theft from those who hunger and are not fed, those who are cold and are not clothed."

Eisenhower's rollback of military funding perplexed Jack Kennedy, who complained that the president's New Look, the name given to Ike's national security and defense policy, lacked coherence. This brought Kennedy's voice into the national debate, where his youthful dynamism stood in cosmopolitan contrast with Eisenhower's steadier, Great Plains low-key style. But JFK's barbs lacked bite and weren't a mature public policy issue. For that, Kennedy kept jabbing and searching. On September 7, with the death of Joseph Stalin in March 1953 and the eventual choice of Nikita Khrushchev to become the new Soviet leader Kennedy seized the personal opportunity of this new Cold War dynamic. While Khrushchev won praise in some U.S. foreign policy circles for the de-Stalinization of the USSR, Kennedy railed against him endlessly. At the same time, he burnished his anticommunist credentials with surprisingly robust support for Senator Joseph McCarthy, who claimed that Communists had infiltrated the State Department, the CIA and the U.S. atomic weapons industry.

Although the Eisenhower administration was attempting to shrink the military overall, it supported air power, especially in its most modern incarnations. Because the United States had first-rate intercontinental bombers and had enjoyed indisputable air superiority in the Korean War, the air force had not prioritized development of ICBMs at a level comparable to that of the Soviets. But with the USSR testing a hydrogen bomb on August 12, 1953, and American scientists developing smaller, rocket-mountable nuclear devices, the air force had to catch up. Eisenhower, in his postpresidential memoir *Waging Peace*, criticized Truman for grossly deprioritizing long-range ballistic missile technology during his White House tenure. To put this into perspective: in Truman's last year, defense spending appropriations for ballistic missiles was only $3 million; Eisenhower, by contrast, a few years later had jacked them up to $161 million.

On August 20, eight days after the Soviet hydrogen bomb test, the first of von Braun's Redstone missile was test-fired by the army at Cape Canaveral, Florida. Developed over the previous three years by von Braun's team at the OGMC, the Redstone would go on to be the workhorse of the army's missile program. Von Braun, conditioned to the challenge of funding rocket research ever since his days with Hermann Oberth, believed he could advance his cause by framing U.S. ballistic missile development as necessary for competing with the Soviets. There were other great rocketeers in America (notably, the team assembled by Frank Malina at the Jet Propulsion Laboratory from the 1930s to the 1950s), but they didn't have an engineering wizard-cum–media maven like von Braun to promote their engineering feats.

Von Braun loved to boast publicly that the United States was destined to become a spacefaring nation. He wrote three articles for *Collier's* in 1952 that envisioned launching humans to the moon and a wheel-shaped space station. These articles made a splash. *Collier's* promoted von Braun as an anti-Soviet German engineer and debonair aristocrat who loved the United States. His prose was highly readable. In one article he coined the phrase "across the space frontier" in *Collier's.* Armed with an accumulation of rocket engineering knowledge, he seized the attention of the magazine's four million readers, but his space advocacy came with a stark warning that the United States must "immediately embark on a long-range development program" or lose out to the Soviet Union. In 1953 and 1954, von

Braun wrote articles for *Collier's* about satellites carrying monkeys and Mars exploration.

What von Braun understood was that in a democracy, where taxpayers footed the bill, space exploration would happen only with direct and unvarnished mass appeals for international prestige, beating Russia, and being *first.* His McCarthy-era thesis that the United States must "conquer" space had a Manifest Destiny feel. And his notion that an American-manned space station could be constructed with existing rocket techniques and that, in addition to its uses as a scientific observatory, it could also act as an orbital fortress, dominating Earth as an impregnable launching base for atomic missiles, was alarming. It was up to the United States, von Braun argued, to see that nobody else built one first. "You should know how advertising is everything in America," he told a friend. "The way I'm talking will get people interested [in space exploration]."

Call him an ex-Nazi propagandist, P. T. Barnum–style marketer, or space visionary, but von Braun understood explicitly that space travel had to be couched in the spirit of American exceptionalism. This attitude went against President Eisenhower's belief in holding down expenditures and allowing the private sector to play a large role in shrinking the size of Big Government. Von Braun's saber rattling for the army about conquering the Soviets in space, however, was embraced by those anticommunist Democratic senators with presidential ambitions: Stuart Symington, Lyndon Johnson, and John F. Kennedy. And in 1954 the air force started developing a reconnaissance satellite program, which historians consider the first U.S. government–funded space program.

On September 12, 1953, Jack Kennedy married Jacqueline Bouvier in an elaborate wedding in Newport, Rhode Island. Proud to have descended from French aristocracy, Jackie, as she was called, once stated her ambition as being the "Overall Art Director of the Twentieth Century!" Her sense of fashion and culture leaned highbrow. She had grown up in a wealthy family with residences in Manhattan; East Hampton, Long Island; McLean, Virginia; and Newport. In 1947 she was voted Debutante of the Year at Vassar College. Many Brahmins joked that Jack Kennedy had clearly married up, that Jackie was the best thing ever to happen to the serial dater from Cape Cod. In October, the newlyweds were interviewed from their Boston apartment on Edward R. Murrow's CBS show *Person to Person*, and were turned into Cold War–era celebrities.

While the Kennedys were on their honeymoon, a report that had been commissioned by the Truman administration nineteen months earlier was delivered to Eisenhower. Titled "The Present Status of the Satellite Problem" and prepared by Aristid V. Grosse, a physicist at Temple University who had worked on the Manhattan Project, the report relied heavily on interviews with von Braun and other Huntsville scientists, and laid out the certain propaganda boon that would accrue to the Soviets were they to launch a satellite before the United States. The last paragraph of the report read:

> At the present time our engineering efforts in this field are limited in scope and distributed over various government agencies. It is recommended as a first step in solving the satellite problem that a small but effective committee be set up composed of our top engineers and scientists in the rocket field, with representatives of the Defense and State Departments. This Committee should report to the top levels of our government and should have for its use and evaluation, all data available to our government and industry on this subject. It should report in detail as to what steps should be taken to launch a satellite successfully into outer space and

> to estimate the cost and time required for such a development. It is felt that if such a committee were in existence and a definite decision taken by our government regarding the construction of a satellite, that it would fire the enthusiasm and imagination of our engineers and scientists and effectively increase our success in the whole field of rockets and guided missiles.

This Grosse report didn't immediately jump-start the U.S. satellite program under Eisenhower. But the arguments it raised were reviewed with fresh eyes in 1954. CIA director Allen Dulles also understood that the United States needed to lead the world in satellite technology. A year after the Grosse report, Dulles wrote the Department of Defense that if the Soviets beat the United States in this field, it would be a major Cold War setback. "In addition to the cogent scientific arguments advanced in support of the development of Earth satellites," Dulles said, "there is little doubt but what the nation that first successfully launches the Earth satellite, and thereby introduces the age of space travel, will gain incalculable international prestige and recognition. Our scientific community as well as the nation would gain invaluable respect and confidence should our country be the first to launch the satellite."

In the summer of 1954, Kennedy was suffering chronic pain due to college football and war injuries, along with osteoporosis likely caused by steroid treatment for colitis and failing adrenal glands. Because of his Addison's disease, back surgery was risky, with only a fifty-fifty chance of survival, but Jack chose to undergo the operation at a New York hospital that October. Ironically, when Lyndon Johnson suffered a heart attack on July 2, 1955, and was placed in a Washington-area hospital for seven weeks, Kennedy's stature in the Senate rose considerably on the incorrect presumption that he was healthy and vigorous in comparison to the ailing Senate majority leader.

Speculation swirled that JFK might get the second spot on the Democrats' 1956 presidential ticket. While the chances of defeating Eisenhower were scant (even though he'd suffered cardiac arrest in 1955), several prominent Democrats, including Johnson, chose to run for the top slot. After Adlai Stevenson won the nomination for the second time, Kennedy was among the few who sought to join the ticket, calculating that the VP slot would put him in the national limelight and position him for the top of the ticket in 1960, when the popular Eisenhower would no longer be eligible. JFK's easy good looks and carefree smile made him an attractive guest on television, the exciting new medium for selling a candidate, even if he sometimes came off on the air as cautious and mechanical. After three ballots, Kennedy lost to Senator Estes Kefauver of Tennessee, a loss that deprived Jack of his stepping-stone but proved he could be taken seriously at the presidential level. What handicapped Kennedy was his vote against an Eisenhower bill to bolster farm prices. "This was the first time in his political career that Jack Kennedy had tasted defeat," historian Robert Caro wrote in *The Passage of Power*, "and it was apparent that he didn't like the feeling at all. Yet not only his words but his demeanor, if resigned and disappointed, had been gracious—the demeanor of a young man dignified, even gallant, in defeat."

After the Democratic ticket of Stevenson-Kefauver was defeated that November, Kennedy began preparing his run for the White House in 1960.

In early 1957, Senator Kennedy began publicly criticizing Eisenhower's lagging missile buildup. Although Eisenhower had designated the air force's Atlas ICBM program a top national priority in 1954, JFK believed it was too little, too late. Full of von Braun–like "the Russians are coming" alarm, he contended that Ike had neglected to properly fund both missile and satellite research and development. While Lyndon Johnson had been making many of the same arguments since the Truman administration, his criticisms had been contained largely to government settings, leaving it to Kennedy to take the case to the public. JFK's argument caught fire on August 26, 1957, when the Soviets announced that they'd successfully tested the first nuclear-tipped ICBM capable of reaching the United States within minutes.

Kennedy railed against lackluster administration policies that, he claimed, had led to Soviet domination in ICBM development. Having been raised in a family obsessed with winning at every level, he reduced the complexities of Cold War statesmanship to a simple contest. Who was the first to have something both wanted? Who had more of something both feared? Who was ahead? Who was behind? Kennedy outlined geopolitical strategic initiatives that would allow the United States to surpass the Soviets in military strength (especially in missiles) and, at the same time, he declared, encourage peace.

On June 11, two months before the Soviet ICBM test, at long last the U.S. Air Force had launched a liquid-fueled Atlas rocket. It stayed airborne for only twenty-four seconds before imploding into a curtain of fire—seemingly proving Kennedy's contention that

America was lagging behind the USSR in ICBM capability. But the setback didn't worry Eisenhower much. The glitches and kinks would get worked out. And the United States, he knew, was doing exceedingly well with its solid-fueled missile programs such as the Minuteman, Polaris, and Skybolt.

Advancements in technology were proving helpful to the American effort. The silicon transistor, invented at Bell Laboratories in 1954, revolutionized the world of electronics and silicon became the fundamental component upon which all space computer technology rested. (Before the transistor was perfected, the NACA in Virginia hired women, many African American, to be "human computers" performing difficult calculations by hand.) It was the transistor that helped reduce the size of the computer. In 1946, a single computer with the power of a basic PC today would have been the size of an eighteen-wheel truck and would have needed the same amount of power as a medium-size town. The first silicon transistor in 1954 was about half an inch (gargantuan compared with today, when a cluster of transistors can comfortably fit on the head of a pin), and the first commercially marketed transistor radio, the Regency TR-1, went on the appliance store market for $49.99 in 1954. Visionary aerospace engineers understood long before the general public that the more transistors that engineers could squeeze into a chip, the more speed and power efficiency one could reap.

John Kennedy finally had his national issue in Eisenhower's alleged indifference to technology. Over the next three years he'd turn the "missile gap" itself into a weapon. Experts at the time questioned the relevance of numerical advantage when it came to missiles, given that so many other factors impacted the efficacy of nuclear weapons. They also wondered where Kennedy had gleaned the hard-to-prove data that America was falling behind; by many measures, the "gap" was far smaller than he asserted. But JFK had learned that his message rang loudest when it was boiled down to a score. Realizing that sports analogies resonated with the American electorate, he had given Americans the simplest of yardsticks by which to measure a Cold War world where technological development was racing to the edge of the unknown.

Nevertheless, Kennedy was right that the Soviets had turned rocketry into an unprecedentedly potent military weapon, a Kremlin strategy that stemmed from the immediate aftermath of World War II. In May 1945, the Soviets had been aghast when the United States absconded with the top level of Nazi rocket scientists and their research, minus any coordination with their supposed allies. Even though distrust between the two nations was already brewing on many levels, the American raid of Peenemünde contributed to the Soviets' fear that the United States sought world domination, and it acted like a cattle prod on the Soviet scientific community.

As long as Stalin was alive, the USSR proceeded as though World War II had never ended. Until his death in 1953, Stalin gave priority to the production of atomic weapons, as well as to the rocket science that would combine with it to produce the ICBM. While military technology was treated as a matter of life and death by Stalin, other sectors of the Soviet economy stalled, including consumer goods and other civilian needs. These privations gave rise to a long line of jokes in America, such as the one about a Soviet engineer rushing in to see his boss: "Sir, we have done it, we have built an atomic bomb so compact, it can fit into a suitcase!" The boss looked glum: "But where are we going to get a suitcase?"

Compared with America's scattershot, underfunded initiatives, the Soviets' largely classified efforts in ballistic missile and satellite technology were quite advanced. In April 1955, they'd announced the formation of a Commission for Interplanetary Communications in the Astronomics Council of the USSR Academy of

Sciences, to work on satellite development. Meanwhile, the Soviet missile program was well organized, leveraging captured Nazi scientists, the theoretical legacy of the late Konstantin Tsiolkovsky, and the technical wizardry of their chief rocket designer, Sergei Korolev. Luckily, however, the U.S. Army had perhaps the only rocket engineer more advanced than Korolev: von Braun.

Meanwhile, Eisenhower was most concerned about finding innovative ways to minimize a surprise Soviet missile attack—Pearl Harbor redux, only with weapons of mass destruction. The burning desire to peer into Soviet territory; to identify what Khrushchev's military was up to; to photograph the major military posts, nuclear zones, missile factories, bomber plane runways, and the like, via U.S. strategic reconnaissance, consumed Eisenhower. At that time, in the mid-1950s, only U.S. aircraft overflight and camera-loaded high-altitude balloons were available for the huge job. Neither method was both safe and reliable. So Eisenhower decided to approve the air force's building of high-altitude U-2 spy planes, which could ostensibly fly undetected, even over Moscow, by Soviet antiaircraft technology or advanced radar. And Eisenhower approved the development of space satellites that could, once perfected, spy on the USSR with near-zero danger of being destroyed by Soviet air defenses. These satellites had the big bonus of perhaps not being banned by international law. Space was a new field of human endeavor and lawyers had not firmly established what was legal and what was unlawful once a man-made object left Earth's atmosphere.

For Eisenhower, establishing a doctrine of "freedom of space" was a paramount national security concern. In 1955—the same year von Braun became a naturalized U.S. citizen—Eisenhower plotted the peaceful exploration of space on pure scientific grounds. Three years before, the International Council of Scientific Unions had proposed the "International Geophysical Year" (IGY) to coincide with the high point of the eleven-year cycle of sunspot activity, occurring between July 1957 and December 1958. Uniting scientists from around the world, the IGY would allow for coordinated observation of various geophysical phenomena (including cosmic rays, the aurorae, the ionosphere, solar activity, geomagnetism, and gravity) and would encompass earth sciences such as glaciology, oceanography, meteorology, seismology, and accurate longitude and latitude determinations. Sixty-seven countries would take part, along with four thousand research institutions that would freely share their data. Promoted by the United Nations, the IGY effort won the support of both the United States and the Soviet Union, and both superpowers announced plans to launch satellites in the name of global peace.

Almost obligated to participate, Eisenhower resisted any pressure to show U.S. superiority by rushing a satellite launch, instead proposing to choose among three programs that were already on track. The first was the army's Redstone rocket, which had been refined by von Braun and his Huntsville engineers. The air force had started late but had made impressive progress with its Atlas and Aerobee Hi rockets, while the navy had poured money into its own Project Vanguard, which hoped to launch a satellite using an upgraded Viking rocket. In the summer of 1957, either the army or air force programs, if properly funded and incentivized, could have launched an American satellite into space. The frustrating truth was that the United States had the technology and know-how to be first, but not the will.

Along with workability, optics were a deciding factor in Eisenhower's eventual choice. On July 29, 1955, the president announced that the United States would launch a satellite in IGY. The Pentagon, after careful review, chose the navy's Vanguard rocket because it carried a larger payload than von Braun's Redstones. The air force proposal wasn't taken seriously. The National Academy of Sciences and the Naval Research Laboratory were directed to join forces in planning for a Vanguard satellite launch, emphasizing the peaceful nature of America's space program.

Disgusted that he lost, Eisenhower had established what amounted to a "design contest." "It is a contest to get a satellite into orbit," von Braun fumed, "and we [the army] are way ahead on this." According to *Time*, in the aftermath of the president's 1955 decision, von Braun was "specifically ordered to forget about satellite work."All von Braun could do now was hunger for a president not named Eisenhower; he'd have to wait Ike out to find genuine enthusiasm emanating from the White House for fast-tracking satellites and, eventually, spaceships aimed at the moon. And he quietly worked with the Jet Propulsion Laboratory in Pasadena to be ready for a Redstone (army) launch with a satellite in case the Vanguard (navy) failed.

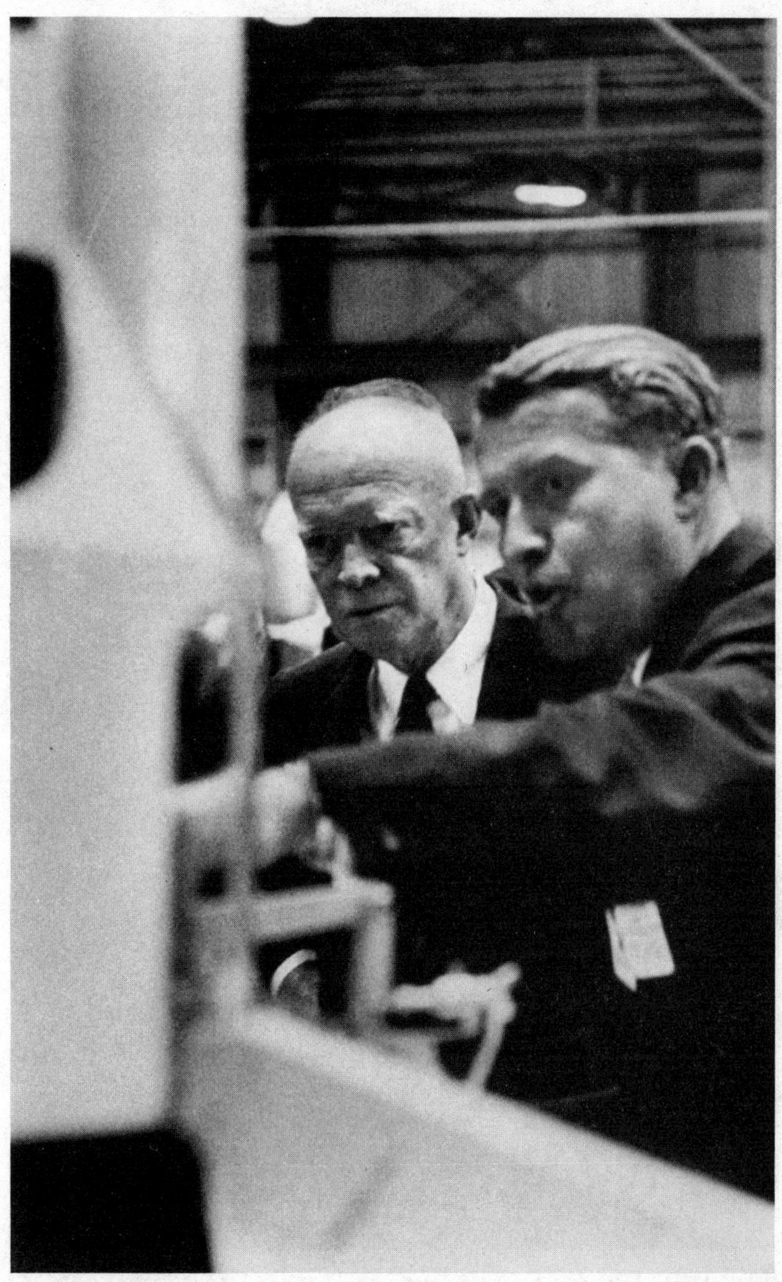

Dr. Wernher von Braun *(right)* with President Dwight D. Eisenhower *(center)* at dedication of Marshall Space Flight Center (Huntsville, Alabama) in 1960. Eisenhower never fully trusted von Braun because of his Nazi past.

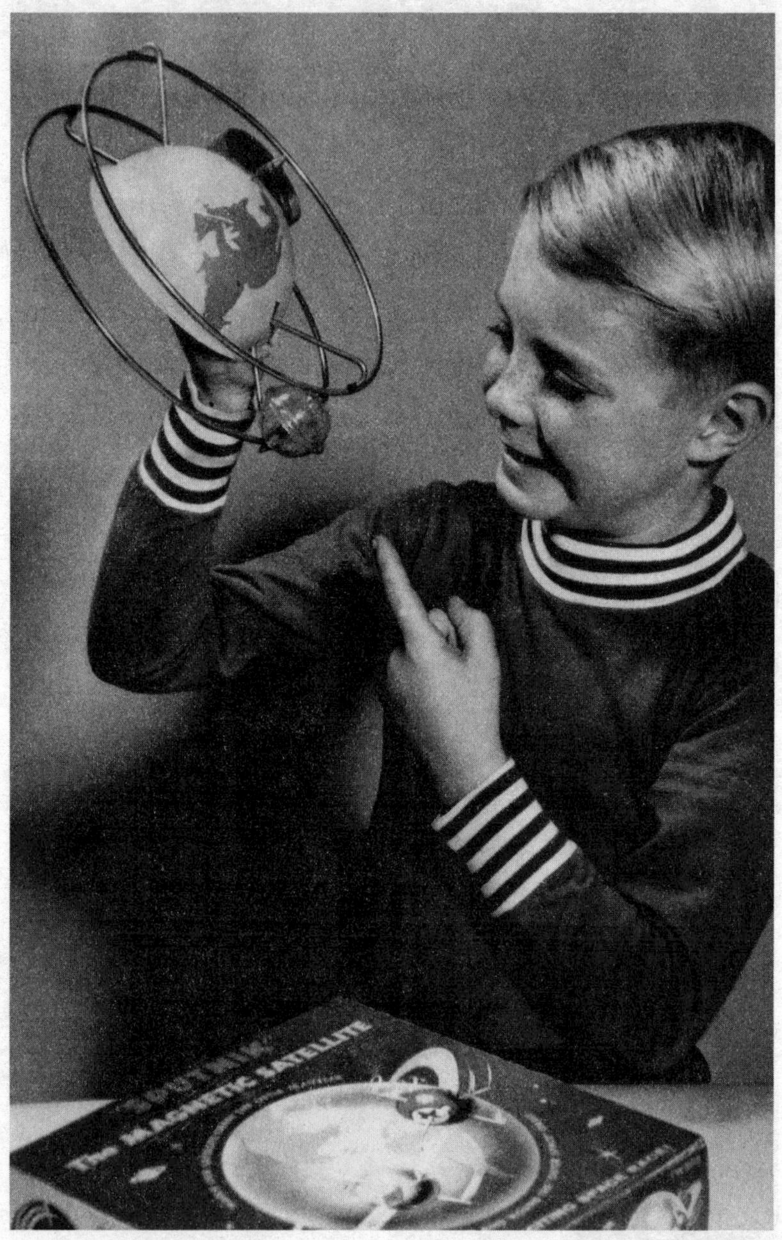

This little boy is playing with a toy *Sputnik* satellite, which was manufactured by Loo/Maggie Magnetic in 1957. The real *Sputnik* was launched into space by the Soviet Union on October 4, 1957, creating a wave of awe and fear.

6
Sputnik Revolution

The Russian word *Sputnik*, used for the satellite, briefly became the most famous word in the world.

—ISAAC ASIMOV,
EXPLORING THE EARTH AND THE COSMOS

Somehow it's fitting that John F. Kennedy and Wernher von Braun first met at Pathé Studios in New York City just after Thanksgiving in 1953. They had been chosen by Henry Luce of *Time* to help nominate Konrad Adenauer, just reelected chancellor of West Germany, as the magazine's Man of the Year, to be announced after Christmas. Both Kennedy and von Braun were exceptional at the art of public discourse and attracting media attention—qualities that helped Luce sell magazines. That afternoon at Pathé Studios, the two looked like clones, wearing Brooks Brothers suits with handkerchiefs in their lapel pockets, white shirts and dark ties, and polished black wingtips on their feet—an Eastern establishment uniform of sorts. Kennedy was the hotshot young Democratic senator from Massachusetts, while von Braun had become the darling of national magazines because of his fantastical determination that Americans were going to the moon and, eventually, Mars. "Senator and Mrs. Kennedy were there and I brought my wife [Maria] along," von Braun recalled. "We had to wait for about an hour until the show was ready for us. During this hour, I had a long conversation with Senator Kennedy, while my wife talked a bit with Mrs. Kennedy."

Right away the New England politician and the army rocketeer got along famously. Jovial and relaxed, enjoying easy banter, proud to already be successful men of the world, they bonded over shared memories of World War II. What surprised von Braun most about his hour chat with Kennedy was that the *PT-109* incident was never mentioned. Instead, Jack explained how his older brother, Joe Jr., died in an aviation mishap that was closely related to the ballistic missile technology von Braun was developing in Peenemünde. "I remember that he said the accident occurred with an obsolescent type of bomb aircraft that had been loaded to the gills with explosives," von Braun recollected in 1964. "The idea was that the plane would be piloted by the Senator's brother up to a certain altitude and then set on autopilot, at which time the pilot, still over friendly territory, would bail out. The autopilot and navigation gear, still somewhat experimental, were rigged in such a way that, upon reaching the destination in level flight, the plane would dive into the target. It seems that the aircraft climbed to the prescribed altitude, but during the transfer to autopilot mode, it suddenly blew up."

What von Braun, a quick study, surmised was that Senator Kennedy was haunted by his brother's death. There was apparently no banter about Operation Paperclip, the limited precision of guided ballistic missiles, or the Soviets' testing their first thermonuclear weapon. For Kennedy, fighting to stop Nazi Germany from acquiring ballistic missile superiority was what Joe had died for. Now the new enemy was the Soviet Union, and JFK, devoid of animosity toward the former Nazi engineer, was expressing his pride that they were both avatars of increased military funding so the United States could achieve long-range missile and satellite supremacy. While President Eisenhower distrusted von Braun for his Hitlerian past, Kennedy adopted the stance that the German rocketeer had simply been swept up in the German nationalism of the 1930s and '40s: von Braun was German, so naturally he had served his government loyally.

After their green-room chat, Kennedy and von Braun went on television to sing the praises of Adenauer, agreeing that West Germany should be admitted into NATO. Von Braun discerned that what got JFK's juices flowing most about missile technology was his belief that his brother hadn't died in vain on that high-risk mission, that ballistic missiles had become *the* coin of the realm of the Cold War square-off. In the mid-fifties, as Kennedy and von Braun lavished praise on Konrad Adenauer, the fast-paced race was on between the United States and the USSR for long-range missile superiority. "The senator pointed at the close relationship of the work his late brother was pursuing with guided missile development and said that missilery had come a long way since those pioneering days," von Braun recounted. "He had been following the work of myself and my associates in missile development with the greatest interest."

World leaders such as Khrushchev, Eisenhower, de Gaulle, Churchill, Macmillan, and Adenauer had been born in the nineteenth century; Kennedy and von Braun were, by contrast, beguiling twentieth-century futurists who understood that the key dividing line in geopolitical terms was no longer BC and AD. It was now pre-V-2 and post-V-2, as well as pre-Hiroshima and post-Hiroshima. Standing armies and naval armadas could soon be annihilated into cinder with long-range nuclear missile volleys. "I would raise a question on something entirely different, and invariably I found him most responsive and concise," von Braun said of Kennedy. "I was greatly impressed by the breadth of his interests and the broad spectrum of his knowledge. In fact, I was so impressed that when we left the studio I told my wife I wouldn't be surprised if Senator Kennedy would one day be President of the United States."

Throughout the fifties, Kennedy kept an interested eye on von Braun. That the Huntsville rocketeer insisted that America could land on the moon in his lifetime lit up JFK's eternal sense of optimism. And von Braun, who had cowritten the book *Conquest of the Moon*, published in 1953, wasn't a lone voice in thinking the American space effort could someday land on the moon. President of British Interplanetary Society Arthur C. Clarke, whose early-1950s books such as

The Sands of Mars were popular, publicly insisted that a lunar voyage was doable. Clarke was born in 1917, the same year as Kennedy, and it was part of their abiding faith that in the near future, Buck Rogers and Flash Gordon would metamorphose into real space travelers. At least that was the generational conceit. "When I published my first space novel [*Prelude to Space*] in the early 1950s, I very optimistically imagined a lunar landing in 1978," Clarke recalled. "I didn't really believe it would be done so soon, but I wanted to boost my morale by pretending."

It is difficult to determine what Kennedy knew or really thought about ballistic missiles and space exploration from 1953 to 1957 because he was so reflexively critical of Eisenhower's defense and space policy for his own politicial purposes. In truth, Eisenhower's space program was not as ineffective as Kennedy had made it seem. In the summer of 1957, six months into Dwight D. Eisenhower's second term, the president had asked the National Security Council (NSC) to review the space program of the United States to ensure that the level of investment and progress being made was sufficient. He intended to field the first ICBMs and reconnaissance satellites by the time he left office.

These capabilities in the new high ground of space would confirm that the United States could compete effectively with the Soviet Union in the Cold War. Eisenhower discovered that between 1953 and 1957 the nation had spent $1.18 billion on military space activities, mostly on ballistic missile and reconnaissance satellite development. "The cost of continuing these programs from FY 1957 through FY 1963," the NSC reported, "would amount to approximately $36.1 billion, for a grand total of $47 billion."

Von Braun, however, considered 1947 to 1957 the lost decade in U.S. space technology research. Like Kennedy he was concerned that the Kremlin had gained long-range missile supremacy over the United States. This fear seemingly became a reality on the night of October 4, 1957, when radio operators working at the Naval Research Laboratory in Washington, DC, picked up a steady A-flat beep emanating from an onboard transmitter, which they identified as a signal from a Soviet satellite in Earth's orbit. Eisenhower's incremental approach was given a comeuppance. The device, named *Sputnik* ("traveling companion of the world"), had been launched from a secluded steppe in southern Kazakhstan aboard a modified Semyorka ("Seventh") ICBM, and it carried equipment to analyze the density and temperature of the upper atmosphere. Amateur radio operators also picked up the signal, and a few of them (residents of Ohio, Indiana, and California) ran outside and were able to see with the naked eye a strange object streaking across the skies. The next morning, the Naval Research Laboratory announced that the Soviet satellite had successfully orbited Earth. The Kremlin confirmed the stunning report, adding details such as the weight of this "artificial moon" (184 pounds) and its shape (a ball about twenty-two inches in diameter, with four antennae protruding from it). Far heavier and faster than American intelligence agencies had predicted, *Sputnik* was nevertheless a very simple satellite. The U.S. Army, in fact, had the very design capability to match or outperform the Soviet feat.

On October 5, the *New York Times* devoted nearly half its front page to the satellite, under a banner headline of three boldface lines: "Soviet Fires Earth Satellite into Space / It Is Circling the Globe at 18,000 M.P.H. / Sphere Tracked in Four Crossings Over U.S." Other newspapers were equally breathless. America's pride had been deflated by a satellite composed of a battery, a radio transmitter, and a fanlike cooling device, orbiting the Earth elliptically every ninety minutes at an altitude of between 140 and 560 miles. All the U.S. government could do was ask the seventy thousand members of the American Radio Relay League, a society of ham radio buffs, to help them track the Soviet beeps.

For a few days, the world was captivated by this polished sphere of aluminum, magnesium, and titanium that had pushed beyond the bounds of Earth, extending the hand of man and forever changing perceptions of the possible and impossible. As the philosopher Hannah Arendt wrote in the prologue to *The Human Condition*, published the following year, *Sputnik* was an event "second in importance to no other, not even the splitting of the atom." Tracking the satellite became a cause for global excitement, and science museums and planetariums reported a surge in telephone calls asking how to spot the man-made miracle. Without the tension of the Cold War, the Soviet launch might have been hailed throughout the United States as a scientific accomplishment on par with Marie Curie's discovering of radium or Guglielmo Marconi's inventing the radio. But with America and the USSR locked in geopolitical combat, feelings were mixed. To some, *Sputnik* felt less like an uplift of humanity and more like a body

The first Soviet artificial satellite was launched on October 4, 1957, frightening many Americans into believing that they were losing the Cold War. On January 2, 1959, the Soviets launched *Luna 1 (above),* the first spacecraft to pass close to the moon, to Senator John F. Kennedy's great consternation.

blow, prompting fear that if the Soviets could launch a satellite into space via a superlong-distance R-7 missile, then perhaps they could also annihilate parts of the United States with nuclear-armed ICBM rockets.

While *Sputnik* was newsworthy and startled some Americans, the Soviet feat also generated enthusiasm for the prospects for the United States to launch a counter-satellite. Three days after *Sputnik* went up, social anthropologists Margaret Mead and Rhoda Metraux began to collect data gauging various American responses. They asked colleagues and friends around the United States to conduct surveys asking three open-ended questions among diverse age, gender, race, economic, and social groups:

> What do you think about the satellite?
> How do you explain Russia's getting their satellite up first?
> What do you think we can do to make up for it?

Between October 7 and October 18, Mead and Metraux collected 2,991 adults' responses. Analyzed as a unit, these responses suggest the need for a revision to the shopworn idea of a post-*Sputnik* shock reverberating across America. An exceptionally small number said that the Soviet launch of *Sputnik* was an unexpected event; an even smaller number registered no knowledge of the launch. Of those who had scant knowledge, the rejoinder of one twenty-two-year-old woman from Austin, Texas, was characteristic: "It was a surprise to hear that the satellite was launched successfully. . . . I was skeptical that such a project would ever materialize. Now that it has, it shows that science is still progressing." Another respondent, a forty-year-old Kentucky man, put it this way: "It's been a scientific possibility for some time. . . . Russia had said she would launch it, so it did not come as a surprise."

There were, however, humanitarian skeptics about *Sputnik*'s value to human civilization. Biologist Rachel Carson, whose 1951 nonfiction masterpiece, *The Sea Around Us*, had won the National Book Award, thought *Sputnik* was going to intensify Cold War tensions. With a cold and critical eye, she worried that humans would be poor stewards of space. "In the pre-Sputnik days, it was easy to dismiss so much as science-fiction fantasies," she lamented to a friend. "Now the most farfetched schemes seem entirely possible of achievement. And man seems actually likely to take into his hands—ill-prepared as he is psychologically—many of the functions of 'God.'"

Overall, it seems, the more frantic response to *Sputnik* that ensued was a politically constructed event aimed at specific political ends. Democrats piled onto the national security issue of losing the Cold War, stoking a paranoia that quickly edged toward panic. Just days after *Sputnik* appeared in the night sky, Senator Henry "Scoop" Jackson of Washington decried "a week of shame and danger," as if World War III had just erupted. A hawkish liberal and anticommunist with the piercing eyes of a navy admiral, Jackson never hesitated to insist that Americans needed a far stronger military to beat the Soviets—a position that benefited aviation giant Boeing, which was then headquartered in his home state of Washington. Before long, other Democratic leaders joined Jackson, including Jack Kennedy, in hammering home the message that *Sputnik* was a devastating blow to America's postwar global prestige as well as to homeland security.

Michigan governor G. Mennen Williams, a potential Democratic candidate for president in 1960, joined the criticism of Eisenhower with a satiric poem that mocked Ike's penchant for golf:

Oh Little Sputnik, *flying high*
With made-in-Moscow beep,
You tell the world it's a Commie sky
And Uncle Sam's asleep.

You say on fairways and on rough
The Kremlin knows it all,
We hope our golfer knows enough
To get us on the ball.

Senate Majority Leader Lyndon Johnson became the most high-profile Democratic critic. The night *Sputnik* was launched, Johnson heard the news on CBS Radio while at his ranch along the Pedernales River in the Texas Hill Country. At once he knew that "a new era of history [had] dawned on the world." In his memoirs, he reflected about what a frightening moment it was. "In the Open West you learn to live closely with the sky," he wrote. "It is part of your life. But now, somehow, in some new way, the sky seemed almost alien. I also remember the profound shock of realizing that it might be possible for another nation to achieve technological superiority over this great country of ours."

George Reedy, a high-powered Democratic strategist, wrote to LBJ on October 17, 1957, about how they could effectively use *Sputnik* to the party's advantage: "The issue is one which, if properly handled, would blast the Republicans out of the water, unify the Democratic Party, and elect you as President. You should plan to plunge heavily into this one," a gung-ho Reedy suggested. "As long as you stick to the facts and do not get partisan, you will not be out on any limb." Reedy urged Johnson and other Democratic lawmakers to establish the legitimacy, breadth, and dynamism of *Sputnik* as a serious homeland security threat. Outlining ways in which that might be accomplished, Reedy noted the Democrats needed to generate a considerable public outcry:

> Nevertheless, as the facts sink home, the American people are bound to become increasingly uneasy. It is unpleasant to feel that there is something floating around in the air which the Russians can put up and we can't. The American people do not like to be "second best." Furthermore, the various dopey stories about the *Sputnik* mapping the ground with infra-red rays and about the possibility of one flying overhead with a television camera are bound to have an effect. People will soon imagine some Russian sitting in *Sputnik* with a pair of binoculars and reading their mail over their shoulders. Folks will start getting together in the evening over a case of beer and some field glasses watching for *Sputnik* and ignoring their television. And when two or three of the satellites get into the ionosphere, what is now curiosity may turn into something close to panic.

As a political pro, Johnson knew at once that Reedy was right, that *Sputnik* presented a huge opportunity for the Democrats to score points. In one Texas stem-winder, he delivered a blistering attack on Eisenhower that tapped into the snowballing misapprehensions in the country. "The Roman Empire controlled the world because it could build roads," he said. "Later—when men moved to the sea—the British Empire was dominant because it had ships. In the Air Age, we were powerful because we had airplanes. Now the Communists have established a foothold in outer space. It is not very reassuring to be told that next year we will put a better satellite into the air. Perhaps it will even have chrome trim and automatic windshield wipers. . . . Soon, the Russians will be dropping bombs on us from space like kids dropping rocks onto cars from freeway overpasses."

Johnson's alarmist reaction to *Sputnik* was heard loud and clear on Capitol Hill, and is frequently cited by Cold War scholars as the tipping point in the push to create a unified American space program. While the public had not been immediately afraid of the ramifications of *Sputnik*, by the time Johnson was done with them, they were. Eight years after the Soviets detonated an atomic weapon and China turned to communism, and four years after the bloody Korean War ended in what was essentially a stalemate, the United States was now lagging behind the Kremlin in space technology. For Johnson, it was too much losing to bear—in military power, in science and technology, and in global prestige. Each of the Soviet satellite's 1,440 orbits (lasting twenty-two days) suddenly felt like a tether tying America down to lonely planet Earth while the USSR was soaring around space, gallantly reaching for the stars.

Amid the chorus calling *Sputnik* a national humiliation, President Eisenhower took the launch in stride. Acting just a little surprised, he noted that space had been around a long time, and it would continue to be. To the president, it didn't matter who happened to be first to circle Earth with an artificial moon. Refusing to buy into the whole superpower paradigm and insisting that self-evident democratic virtues over communism needed no spin, he rejected the idea that "prestige" was the defining motivation in U.S. foreign policy. But in truth, Eisenhower the politician realized that the *Sputnik* feat was unsettling. His own Joint Chiefs of Staff were demanding that space supremacy not be ceded without a full-bore fight. As Neil deGrasse Tyson would put it decades later in *Space Chronicles*, the launch of *Sputnik* "spooked" the United States into the space race.

Many national security analysts worried about Eisenhower's perceived lassitude about *Sputnik*, but such criticism only dented the president's personal popularity. Even if he were a little anemic and a trifle dull, he still owned the "I Like Ike" brand. And in truth, the U.S. satellite program in late 1957 was essentially on a par with or even more advanced than the Soviet one, but perception mattered mightily in the U.S. Cold War dynamic. Just as Harry Truman and the Democrats had gotten beaten up for not winning the Korean War quickly, Republicans were now being brutalized over *Sputnik*. With Eisenhower ineligible for another term, it gave potential candidates in the 1960 presidential election the chance to be perceived as tough anticommunists and to shed the intellectual, New Deal, pacifistic yoke of Adlai Stevenson's 1956 campaign.

Stoking disaffection, Lyndon Johnson took deliberate action in the Senate. In concert with the powerful Richard Russell, a feisty Georgian who served on the Armed Services Committee, he opened an official investigation to gauge the reasons behind America's allegedly lagging satellite program. By late November, Johnson was chairing a series of Senate hearings (officially an "Inquiry into Satellite and Missile Programs" but known in the press as the Johnson Hearings), seizing on the space gap as he eyed the 1960 nomination.

Jack Kennedy, also gearing up for 1960, made his first calibrated remarks about what he called "the age of *Sputnik*" on October 18, 1957, in a speech at the University of Florida, in Gainesville. Full of sarcasm, he mocked Eisenhower's shrug of apathy about losing the space race, claiming that the president's lackadaisical attitude was that the Soviets may have launched the first satellite, but the United States had the Edsel, Ford Motor Company's "car of the future" for 1958, whose subsequent commercial flop made Kennedy's jab even more biting in retrospect.

Surpassing the Soviets in satellite technology, JFK said, would require the federal government to act for the long term by funding a civilian space agency. Furthermore, schoolchildren needed to be better educated in such fields as engineering, physics, and mathematics. "It is rather difficult to reverse these trends," Kennedy complained of the apparent lag in American technology, "when the teaching of the physical sciences and mathematics in our own secondary schools has declined; when about half of those with talent in these fields who graduate from high school are either unable [to go] or uninterested in going to college; and when, of the half who enter college, scarcely 40 percent graduate." If U.S. schools didn't toughen science standards, he worried, the space deficit with the USSR was only going to grow. Later, JFK escalated his criticism by warning that the United States was "losing the satellite and missile race because of complacent miscalculations, penny-pinching, budget cutbacks, incredibly confused mismanagement, and wasteful rivalries and jealousies."

For the several hundred dedicated, high-level scientists and engineers toiling in America's disparate and fractured space rocketry programs, the first *Sputnik* launch was deeply upsetting. When a British reporter informed von Braun about *Sputnik*, the U.S. Army engineer visibly winced, stung by the blow to his ego and to his adopted country's prestige. A colleague saw von Braun at a social event that same October night; like Kennedy, von Braun despaired about the Soviet achievement and "started to talk as if he had suddenly been vaccinated with a Victrola needle." Refusing to be good-tempered, he was devoid of amusement; in his driving urgency to unburden his feelings, his "words tumbled over one another." If only "President Eisenhower would understand," he lamented, "that the future of technology was in space!" It infuriated him even more that the outgoing Secretary of Defense Charlie Wilson, his nemesis, derided *Sputnik* as "a useless hunk of iron." Sherman Adams, Eisenhower's chief of staff, told the press that the United States wasn't interested "in an outer space basketball game."

Less than a month after the *Sputnik 1* success, on November 3, the Soviets launched *Sputnik 2*, once again throwing the Americans into a panic. This time, the satellite carried the first living creature ever to orbit Earth: Laika, a dark beige mixed-breed dog, whose experience was monitored for vital data and whose image was beamed back to Earth via an onboard television camera. With a dog aboard, the media began calling *Sputnik 2* "Muttnik" or "Poochnik." Space buffs all across the planet looked upward hoping to see what novelist Jack Kerouac, from the lawn of his little cottage in Orlando, Florida, described as "a brown star racing northward."

Laika immediately became the mascot of the entire Soviet nation, her image adorning commemorative buttons, plates, posters, and other souvenirs. The Soviets possessed the technology to send the brave mutt into space, starting with the power to lift a satellite weighing more than 1,100 pounds, but they lacked one notable aeronautical necessity: a proper means for reentry. Immobilized inside the rocket capsule, Laika died early in the mission from dehydration, overheating, and stress.

The sacrifice of Laika indicated that the Soviets

were skipping steps in their research and experiments, ignoring reentry concerns in order to make a conspicuous step forward toward manned flight. The Soviet space dog had proved that a living creature could survive for a spell in a weightless environment, but by failing to bring Laika home alive, the USSR opened itself to accusations of brutality and callousness. *Sputnik 2* was mocked by some as a dog coffin floating in space.

Still, for Soviet leader Nikita Khrushchev, *Sputnik 2*'s 163-day voyage was yet another coup against the Americans, proving Soviet mastery of space technology and validating his strategy of prioritizing missiles over conventional weapons, itself a reversal of Stalin's plans to build a massive Soviet navy. In January 1958, *Time* declared Khrushchev its 1957 "Man of the Year," the cover photograph showing him holding a model of the *Sputnik* satellite in his hands. Strategically boastful in his success, Khrushchev bluffed that he had an arsenal of ICBMs at his disposal (though few were in fact operable) and that the USSR had "outstripped the leading capitalist country—the United States—in the field of scientific and technical progress."

While Khrushchev crowed and Laika became a posthumous shoo-in for the Cold War Hall of Fame, Eisenhower was coming off as disconnected from the moment, not rattled at or worried over Laika, and seemingly uninterested in the Space Age technology that was rapidly changing today into the future. He had other concerns. Not only had the president been hospitalized for seven weeks after a 1955 heart attack, but he had also undergone intestinal surgery related to Crohn's disease in 1956, so his health suggested to some that he might not be fully engaged or firmly in command. Facing a brave new tomorrow that Americans could see streaking through their nighttime sky, Eisenhower, with his low-key reaction to *Sputnik*, began to seem out of step.

Predictably, *Sputnik 2* only amplified the storm of controversy surrounding America's lagging position in space exploration. Nuclear weapons designer Edward Teller claimed that the United States had lost "a battle more important and greater than Pearl Harbor." Racing against Lyndon Johnson to be the more ardent Cold Warrior, Jack Kennedy doubled down, lashing out even more forcefully against Eisenhower than he had in Gainesville. Dave Powers, his personal aide, recalled his boss resolving, after vying for the vice presidency in 1956, to become "a total politician," and the record bears this out: Kennedy did noticeably shift into a higher anti-Soviet gear in late 1957 and 1958. He'd always been mindful of his own constituents and every potential voter beyond Massachusetts, but he hadn't paid as much attention to the needs of the Democratic Party leadership or his colleagues. Now he transformed into a crisply professional senator who didn't miss a photo op or, it seemed, make a wrong political move. His youth explained the public nature of his evolution. Politicians who start in their forties have already been to finishing school in one way or another, but JFK was still growing up in the public eye. A late bloomer to start with, he had been a kid yet at thirty. Turning forty in 1957, however, he was very different, a mature and sophisticated Senate star.

Kennedy's take-charge reaction to the *Sputnik* crisis was indicative of the total politician he had become. Initially, he gave no hard-and-fast answers on the space supremacy issue. Instead, he waited in the shadows of politicians such as Lyndon Johnson, Stuart Symington, and Henry Jackson, each of whom wanted to make it his national issue. Slowly and deliberately, Kennedy waited for the right time to take his own, even more aggressive stab at Eisenhower's seeming distractedness, insouciance, and weak national security leadership. Scheduled to address a group in Topeka, Kansas, on November 8, five days after the launch of *Sputnik 2*, JFK co-opted the title of an Eisenhower speech scheduled for the very next night, "Science and Security." It was the Massachusetts senator's first address solely on the topic of space and what he called "the solemn consequences of *Sputnik*."

With unusual vigor, Kennedy slammed Eisenhower in the five-star general's home state in practically every sentence, complaining of dithering, complacency, and weak-mindedness on space technology and science education. Almost mechanically, he named the areas in which he thought Eisenhower had failed, including insufficient funding for satellites and a lack of focus that resulted in squabbling and a lack of cooperation among military services and private contractors, who were often conducting the same research and development; to JFK, this duplicative effort slowed America's overall progress toward space. Buying entirely into the "space race" concept that the president rejected, Kennedy complained, "It is now apparent that we could have been first with the satellite, but failed to see any reason for doing so, failed to see the scientific, military and propaganda advantage it would give to the Soviets if they were first."

The leadership styles of Eisenhower and Kennedy couldn't have been more different. Ike, the strategizing general, was used to working out the details in private and releasing information on a need-to-know basis, without any fanfare. Always operating calmly and as silently as growing grass, the president eschewed soaring rhetoric aimed at gaining favorable press—or spreading doom. JFK, who had depended mainly on charisma in his political career, was frustrated that Eisenhower seemed uninterested in sharing the adventure of space with the American people. A close student of Winston Churchill, Kennedy believed that leadership was about galvanizing a slumbering public (via speeches, articles, and radio addresses) to achieve great things. "The President must tell us exactly where we stand today," the senator from Massachusetts insisted, referring to the satellite program, "and where we go from here."

The next day, the Associated Press account of Kennedy's Topeka sermonette appeared in newspapers around the country. "The Massachusetts Democrat said this country had never stood in so critical a position in world affairs," reported the AP. Fully engaging in the post-*Sputnik* crisis from his Kansas podium, JFK had thrown down the gauntlet, pressing for an urgent response. While Eisenhower had recently mocked the Democratic inclination to "go frantic" over the *Sputniks*, Kennedy was in essence pleading with Ike to "go frantic" for the good of the country. As Johnson pressed forward with his Senate investigation, Kennedy reiterated his Florida focus on the gap in science education, stating that "the race for advantage in the Cold War is . . . a race of education and research."

Kennedy soon got a boost from *Life* magazine, which printed a long feature comparing American and Soviet performance at the high school level. The article focused on a Moscow school where the students bore down on demanding subjects such as physics and algebra six days a week, contrasting it with a Chicago school where goofing off was prevalent. By the end of the piece, readers were left surprised that American graduates could even operate a bicycle or sewing machine, much less launch a satellite or two into outer space. Staying on the technology deficit issue, JFK was an early proponent of the National Defense Education Act, which increased funding for what are now called STEM disciplines in schools and created the first federal student loan program.

Soon after the launch of *Sputnik 2*, Soviet premier Nikolai Bulganin was asked by the *New York Times* when his country would launch a third satellite. "It's the Americans' turn now," he replied, devilishly yanking Eisenhower's chain. "Let them send up their satellite." Bulganin knew that the different U.S. armed services' rocket programs were all struggling, including the one being run by von Braun.

Von Braun's rocket team in Huntsville built the Jupiter-C as a payload-bearing army rocket suitable for putting a satellite in orbit. It was used instead as a nuclear-bomb-equipped ballistic missile, but it was not a particularly useful weapon. This missile took many features from the V-2, added an engine from the Navaho test missile, and incorporated some of the electronic components from other rocket test programs. Its first launch took place at Cape Canaveral, Florida, on August 20, 1953, and its capability as an IRBM was tested on May 16, 1958, when combat-ready troops first test-fired the rocket itself. The Jupiter-C was then deployed to U.S. units in Italy and Turkey and served until 1963.

The Eisenhower administration's deployment of von Braun's Jupiter missile destabilized the situation with the Soviet Union. It is not an uplifting story. One serious problem of Jupiter missiles was their range, necessitating that they be deployed near the border of the Soviet Union. Secretary of State John Foster Dulles argued that the placement of the IRBMs so close to the Soviet border invited attacks and therefore was a provocative act that destabilized the Cold War balance. Moreover, the technology was such that it took hours to ready the missiles for launch—they had to be deployed at aboveground launch complexes—and could be destroyed by a Soviet-trained sniper with a high-powered rifle. "It would have been better to dump them in the ocean," Eisenhower complained, when the Jupiters were deployed to Europe, "instead of trying to dump them on our allies."

If Eisenhower was dubious of the Jupiter's worth as a missile, von Braun and his team were frustrated by the refusal to compete with the Soviets in satellite launches. He was truly befuddled by how Washington had downgraded the technology of tomorrow. Back on October 4, 1957, President Eisenhower had been at his farm in Gettysburg, Pennsylvania, when he learned about *Sputnik* at 6:30 p.m., following a frustrating day of talks with Arkansas governor Orval Faubus, who was refusing to desegregate public schools in Little Rock. Ike regarded his decision to federalize the Arkansas National Guard in order to enforce the order far more important than *Sputnik*, and he tasked White House press secretary James Hagerty with communicating his feeling that the Soviet feat "did not come as a surprise" and that America wasn't "in a race" with the Kremlin.

In a speech, Eisenhower credited the Soviet success to "all the German scientists" who had been captured at the end of World War II and taken east to Moscow and then beyond. This comment embarrassed von Braun. Still, the German rocket scientists now working for the Soviets had been given greater resources and opportunities behind the Iron Curtain than von Braun and his colleagues had in the United States. To von Braun, it was the height of irony.

In the middle of the *Sputnik* uproar, von Braun protested to Neil McElroy, who had replaced Charlie Wilson as secretary of defense, that the army's satellite program wasn't being supported with the urgency the historical moment demanded. "Vanguard will never make it," von Braun said, railing against the navy effort. "We [the Army] have the hardware on the shelf. For God's sake turn us loose and let us do something. We can put up the satellite in sixty days." Major General John Medaris, von Braun's supportive commander at the ABMA, quietly insisted on ninety days, and two weeks later, McElroy approved the Jupiter-C/Redstone rocket-testing plan. Those on the Huntsville team were the direct beneficiaries of *Sputnik*. Moving quickly, von Braun reserved a late-January launch date at the Cape Canaveral range and began frantic preparations.

By the time of the *Sputnik* launches, von Braun had grown in public acclaim. Not only had *Collier's* magazine given him a chance to publish his vision of space exploration, but Walt Disney had provided the German-born engineer with a television platform for Sunday-night talks about the development of the Redstone and Jupiter rockets in Huntsville, and about the cosmos in general. Although his Nazi past still occasioned some resentment, it wasn't a debilitating public liability. Indeed, von Braun, and his role in American life, had become an embodiment of West Germany reformed by the American way.

Projecting strength with a forceful voice and conniving eyes, von Braun exuded, as writer Norman Mailer put it, "a confusing aura of strength and vulnerability, of calm and agitation, cruelty and concern, phlegm and sensitivity." From 1947 to 1957, von Braun's rocketry prowess and space technology genius were celebrated as national educational assets on TV and in glossy magazines. After *Sputnik*, Kennedy and others viewed von Braun as the indispensable man in America's fight against Soviet space dominance. When von Braun had first arrived in Huntsville, the army arsenal had a few handsome stone- and tile-faced office buildings and test laboratories, but not much more. The automobile showrooms and movie theaters in town were where the local excitement resided. But after *Sputnik*, cathedral-like buildings were erected, and at night the lights around the base gave the complex the feel of an aerospace fairyland.

In the year after *Sputnik*, von Braun adorned the covers of both *Time* and *Life*. Swooning reporters treated him as half rocket scientist, half charming and cultured poet/raconteur, imbued with a deep affinity for all things American. His originality, his commitment to rocketry, and his steady faith in going to the moon were considered priceless national assets. Fan mail poured into Huntsville after these publications, and women wrote von Braun mash notes. His aggressive belief that the United States could outperform the USSR in space made him a heroic figure in Cold War America. "I get about ten letters a day," he told *Life*. "About half come from youngsters who want advice on how to become rocketeers. We tell them hit math and physics heavily. One lady wrote that God doesn't want man to leave the Earth and was willing to bet me $10 we wouldn't make it. I answered that as far as I knew, the Bible said nothing about space flight, but it was clearly against gambling."

Not all the letters *Time* received were positive about von Braun's service to U.S. missile development. There were taunts about his being a Nazi and having no moral conscience. "I cannot share the enthusiasm for Von Braun," Samuel E. Lessere of Clearwater, Florida, wrote to the magazine. "All that can be said for him is that he's willing to do for us what he was willing to do for Hitler." Nowhere was anti–von Braun sentiment greater than in Soviet-dominated East Germany, where press reports excoriated him as a Nazi war criminal.

Invariably, whenever von Braun was forced to explain his work for the Third Reich, his stock defense was that just as the Wright brothers had signed a contract with the U.S. War Department during World War I, he had tied his kite to the German Army in World War II. But this defense wasn't credible. Von Braun had mastered the simpleton act, claiming he had no idea how brutal Hitler was to Jews. "I wasn't truly aware that atrocities were being committed in Germany against anyone," he offered. "I knew that many prominent Jewish, Catholic, and Protestant leaders had been jailed for their opposition to the government. I also suspected from the fact that I had lost sight of my own Jewish friends, that many Jews had either fled the country or were held in concentration camps. But being jailed and being butchered are two different things." Ultimately, hoping to cleanse his legacy in the last decade of his life, von Braun confessed that he was "ashamed of having been associated with a regime that was capable of such brutality."

Even though the fall of 1957 had been a busy time in America—the first meaningful civil rights bill since Reconstruction had been enacted in Congress; federal troops were sent into Little Rock, Arkansas, to protect nine black students entering its newly integrated Central High School; and the United States conducted its first underground nuclear test, code-named "Rainier," in the Nevada desert—a consensus was emerging by Thanksgiving that the *Sputnik*s had made a counterstatement U.S. satellite launch vitally urgent. After conferring with officials from the Naval Research Laboratory (NRL), the White House announced that the navy's Vanguard program would launch "small satellite spheres" by late in the year. The message, however, was somewhat garbled. The Vanguard team had actually been planning to launch *test* vehicles by the end of 1957, in anticipation of sending an actual satellite into orbit the following spring. With the added pressure to complete the package in December, a certain sense of anxiety set in at the NRL. Joe Siri, an MIT graduate in charge of Vanguard's Theory and Analysis Office, developed a plan to use one of the army's Jupiter-C rockets in combination with Vanguard satellite engineering. Technically sound but bureaucratically impossible, given the level of interservice rivalry among the navy, army, and air force rocket programs, the idea and its reception were emblematic of what one NRL official termed "those rather trying days" after *Sputnik*.

Senator Kennedy regarded the "mismanagement" of the country's space research offices as a failure of Eisenhower's leadership. In his Topeka speech, he rightfully condemned "the costly, harmful rivalries and jealousies between the three services and between various companies." He despaired at seeing them "duplicating each other's efforts, competing for funds, for personnel, for scientific facilities and brainpower and surreptitiously undercutting the other two to Congress and the press." What Kennedy didn't tell the audience was that both *Sputnik* and the American Vanguard (navy) satellite programs were the result of the international scientific effort known as the International Geophysical Year (IGY), which was planned, known to the public, and embraced by him.

Immediately after the news of *Sputnik 2*, the Department of Defense lost no time in acting on an earlier Eisenhower directive to fully authorize the army's Redstone as an alternative to the navy's Vanguard. During the first week of December 1957, the NRL hoped it would succeed with the scheduled launch of its Test Vehicle (TV-3) Vanguard rocket, with a small satellite on board. The navy got first crack at counterbalancing the Soviets' space achievement, while the army waited in the wings.

The same week was a watershed for Senator Kennedy, who adorned the cover of *Time* as the "Man Out Front." The profile inside the magazine invoked the familiar aspects of his legend—his heroics in the *PT-109* incident, the athleticism of touch football games at the Kennedy compound on Cape Cod, his personal appeal to women, his firm anti-Soviet resolve, and his elegant wife Jackie's appeal to everyone. *Time* acknowledged his "unannounced but unabashed run for the Democratic Party's nomination for President in 1960." According to the lore at *Time*, effusive pro-Kennedy letters poured in at a record-setting pace, from readers wholeheartedly embracing "the Democratic whiz kid of 1957."

On Friday, December 6, Vanguard engineers, scientists, and officials convened at Cape Canaveral, midway between Jacksonville and Palm Beach along Florida's east coast, for the first Vanguard launch. Used as a navy and air force aeronautic test site since 1949, Cape Canaveral, a twenty-mile-long marshland of sand, scrub, and seagrass, had in July 1950 seen the launch of a rocket called *Bumper 8*, a modified V-2. The launch site, known as the Long Range Proving Ground Base, was far enough away from heavily populated areas to protect citizens against engineering disaster, yet easily accessible from the larger barrier island, closer to the mainland, where test personnel were headquartered. The reason an East Coast location was desirable for U.S. rocket launches was simple physics: traveling eastward provided a projective boost from the Earth's spin. In addition, if there were a mishap, the Atlantic Ocean could absorb a crash without killing civilians on the ground. Conversely, a West Coast location would either send rockets over populated areas or have to grapple with launching against Earth's rotation. The southern Florida location put the launch site close to the equator. The tangential velocity of a point on the surface of the Earth is at its maximum at the equator, so to launch from there would start the flight just a little closer to orbit. Both the army and navy agreed that the key determining factor in choosing a location for a rocket launch site was its relation to the easiest way of reaching orbit. And all that Florida sunshine meant fewer bad weather delays.

Cape Canaveral was hunting and fishing country; the adjoining Banana River, actually a lagoon, was a wonderful place to catch largemouth bass and bream. Eventually, this marshland would become the headquarters for America's first space port, a patch of thicket and brush into which the federal government would invest billions of dollars and around which aerospace industries would sprout like mushrooms. With the hot Atlantic breeze holding sway and no blazing metropolitan electric lights to disturb government-paid stargazers, this subtropical wilderness would be where the United States would try to reach the moon. But first the Americans needed to launch a satellite to match the Soviets'.

The NRL had already completed suborbital rocket tests at Cape Canaveral by the time the three-stage Vanguard TV-3 rocket was ready for its December 6 test. Seventy-two feet in height, the rocket would be propelled by two liquid-fueled sections (or stages), while the third stage, carrying a payload of one small satellite, would be propelled by solid fuel. As Robert Goddard had predicted back in the 1920s, much of the challenge in creating the rocket lay in managing the heat that was generated. The capacity to produce thrust in each stage was staggering, but the stages had to be kept separate.

Two months after *Sputnik*, a herd of news-hungry reporters was on hand at Cape Canaveral to cover the landmark Vanguard TV-3 event. Television cameras carried images nationwide on CBS and NBC, beginning a tradition of live coverage for rocket launches that would last into the 1970s. The countdown gave drama to the moment, with millions watching as the Vanguard TV-3 took off, rose four feet into the air, and then sank back to Earth, the tons of fuel on board exploding into an expensive fireball. All that was left of America's dream of an answer to *Sputnik* was the battered satellite, which had been thrown to the side when the rocket hit the ground, rolling around on the launchpad in a hissing cloud of black soot. It didn't take long for the launch to be derided as "Kaputnik," "Stay-putnik," and "Flopnik."

To scientists working under normal circumstances, a test is a success whether it brings the desired result or not; if something is learned, the test is considered worthwhile. Much was learned from the TV-3, but with all the pressure to launch an American satellite, it was seen by the rest of the nation as a low point in the annals of U.S. ingenuity. Various explanations were offered for the failure. One possible culprit was the change of a key fuel-delivery system component from steel to aluminum, without the recommended temperature-resistant coating. Another was a possible fuel leak caused by workmen having climbed the fuel line as though it were a rope to tend to details out of their reach.

Senator John Kennedy didn't comment on Flopnik.

The day after the failed launch, he had his own problem to handle in the wake of the *Time* cover story, when Washington journalist Drew Pearson appeared on ABC News and asserted unequivocally that Kennedy had not written *Profiles in Courage*, the book for which he'd won a Pulitzer Prize earlier in 1957. The primary author, Pearson said, was Kennedy's speechwriter, Ted Sorensen. "You know," Pearson said on national television, in disparaging Kennedy's actions, "some of his colleagues [in the Senate] say, 'Jack, I wish you had a little less profile and more courage.'"

It was a rare strike against the Massachusetts golden boy and his carefully honed image of playing aboveboard. What the public didn't see was how the Kennedy family fought the accusation relentlessly, with lawsuits, affidavits, bare-knuckled threats, and intense pressure that resulted in a retraction. With help from the Kennedy team, the charge was mitigated enough that Jack could leave discussions of it to his surrogates as he continued to travel throughout the country on speaking engagements.

While Kennedy's primary focus was on politics, Senate Majority Leader Lyndon Johnson began 1958 with a seminal February 5 speech on the subject of missiles, rockets, and space, attempting to get America's scattershot space program on track. In *Profiles in*

Courage, Kennedy had honored the legislative genius of past senators such as George Norris of Nebraska and Robert Taft of Ohio, but Johnson was a living example. The Texas senator's bold address, written by aide Horace Busby, coldly described the apparent Soviet superiority in missiles and satellites; however, his sentiment was anything but defeatist. Instead, Johnson challenged his colleagues with his belief that "we must work as though no other Congress would ever have an opportunity to meet this challenge, for, in fact, none will have an opportunity comparable."

To Johnson, it was sickening that the Kremlin was using *Sputnik* in its global propaganda, rubbing its technological superiority into Eisenhower's amiable face. At the same time, LBJ faulted the president for his baby-steps approach to organizing the U.S. space effort. By contrast, Johnson advocated America's taking giant steps, and in the process became the Senate's leading voice on space issues. Some have questioned whether he studied the subject himself or simply delegated it to aides. Others thought Johnson was trying to ride the coattails of popular sentiment. But as space came into keen geopolitical focus, LBJ was intent on leading the parade, intuiting the political, scientific, and even humanitarian opportunities America could reap.

One day after Johnson's speech, he set up a new Special Committee on Space and Astronautics in the U.S. Senate, dedicated to aeronautical and space science. He would serve as chairman—in part because no other senator knew a damn thing about aeronautics and astronautics, while he at least was conversant in the lingo and acronyms. Regardless, LBJ felt it was high time for greater federal organization to be applied to the rocket engineers in Huntsville, Hampton, and Pasadena who *did* understand the subject. Johnson's committee immediately began to help plan for the new agency to pull expertise from the various military services into one civilian-run outfit. That spring and summer, Johnson worked tirelessly to create a sleek rocket of an agency from the spare parts of America's disparate space programs. Reedy, a longtime Johnson aide, later called it a rare instance "where the initiative for a very major law and a very major change was the initiative of Congress."

Kennedy campaigned continually in early 1958, making three speeches per week on average, in all corners of the country. He didn't focus on space, except inasmuch as the *Sputnik* satellites reflected "the flaunting of the Soviets of their ability to rain death on any hostile neighbor." He coined the term "*Sputnik* diplomacy" for the dark side of space exploration, as he saw it. And he intimated that *Sputnik* meant that the Soviet Union had the capacity to deliver an atomic weapon or an ICBM capable of blowing up Boston, New York City, or Washington, DC. For Kennedy, the United States was now in a take-no-prisoners fight to survive as the premier superpower.

While Johnson tried to streamline America's space policy in a bipartisan fashion, von Braun's Jupiter-C/Redstone team was focused on launching an *Explorer 1* satellite on a missile. At the same time, the navy's Vanguard program designed a space rocket aptly named TV-3BU ("backup), but the vehicle was having glitches. If this second Vanguard failed, the army might have a chance with Jupiter. Von Braun assured his boss at ABMA that the Jupiter-C/Redstone would work—a promise taken literally by ranking officers, who were themselves under pressure from Washington to deliver results.

On yet another space venture, a navy–air force partnership was developing a space-worthy vehicle that had nothing in common with the multistage rockets designed to lift satellites into orbit. The X-15 was a totally experimental research aircraft program meant to provide data in high-speed aircraft and spacecraft design. It was a winged, rocket-powered airplane that counted among its masterminds Walter Dornberger, von Braun's former boss at Peenemünde, who had come to the United States courtesy of Operation Paperclip and developed guided missiles for the air force before becoming Bell Aircraft Corporation's in-house genius.

Promising manned flight from the very start, the X-15 was built to detach from a B-52 mother ship at an altitude of about seven miles, a process that would substitute for the first stage of a Viking-type rocket. Once released, the X-15 would use its own rocket propulsion to attain speeds in excess of 3,600 miles per hour and altitudes of up to 62 miles (100 kilometers), which is regarded as the borderline of space. Pilots, who would be weightless at that elevation, could release the entire cockpit as an ejection capsule in an emergency situation.

Captain Iven C. Kincheloe, an air force test pilot, was scheduled to fly the X-15 across the space barrier sometime in 1959, which would make him America's first astronaut. A graduate of Purdue University, Kincheloe had joined the air force with an obsession for flight. After completing 131 missions during the Korean War, he campaigned for a career as a test pilot. "Boy, did he lap up publicity and attention," the technical director of the X-15 program, Paul Bikle, recalled. Scheduled for the second trip was Joseph A. Walker, another test pilot with air force experience. A no-nonsense Pennsylvanian, Walker was always eager to be the top pilot whenever he flew. According to the X-15 master plan, the third flight would be piloted by a navy pilot, John McKay of Virginia, while the fourth would be piloted by an Ohioan who was more reserved and seemingly less ambitious than the rest. An apocryphal quote circulated, attributed to Armstrong, that perfectly captured his low-key demeanor. "God gave man a fixed number of heartbeats," Neil Armstrong liked to say, before adding that he wasn't going to gobble up his quota by overexerting himself.

Armstrong had become one of the most respected jet pilots of the Korean War generation, part of an "ace club" that also included John Glenn and Wally Schirra. Born on August 5, 1930, near Wapakoneta, Ohio, he developed a single-minded passion for aviation while still a child, fantasizing that if he held his breath long enough, he could float up and hover above the ground like a figure in a Marc Chagall painting. His mother described making many excursions to the Cleveland Municipal Airport so her son could watch the planes take off and land. "He was so fascinated," Viola Armstrong recalled. "He was never ready to leave."

Once Neil took his first airplane ride, in a Ford

Trimotor at age six, he was hooked for good. He enthusiastically built models to the point that he even constructed a seven-foot-long wind tunnel in his parents' basement to test his creations. The subjects of aeronautics and astronomy could never be exhausted.

After Armstrong received his degree from Purdue in 1955, the former naval aviator returned to aeronautics as a test pilot, settling at Edwards Air Force Base, in Lancaster, California. Admired for his composure and detached powers of observation, he was designated as one of the twelve pilots on hand during the X-15's development. Bill Dana, another pilot at Edwards, said that Armstrong "had a mind that absorbed things like a sponge and a memory that remembered them like a photograph." Milt Thompson, who worked on the rocket-powered X-15 project in a variety of capacities, concurred, saying that "Neil was probably the most intelligent of all the X-15 pilots, in a technical sense. He was extremely well qualified to fly the X-15." Armstrong's reserve, however, meant that it was hard to say much more than that he was a smart man and a fine pilot. "I worked and flew with Neil for over six years," Thompson concluded. "I knew Neil, but I did not know him."

While at Purdue completing an education that had been interrupted by the Korean War, Armstrong had been a visible figure on campus, joining a fraternity and the band. An excellent pianist, he was just extroverted enough to write and perform in a musical comedy for his fraternity. As he became more serious about his career in aeronautics, however, he grew even more noticeably private. "All in all," he said later about his thinking in high school, "for someone who was immersed in, fascinated by, and dedicated to flight, I was disappointed by the wrinkle in history that brought me along one generation late. I had missed all the great times and adventures in flight."

By 1955, however, Armstrong knew well what great opportunities were newly available for a pilot and aeronautical engineer.

***Time* magazine's** judges had it right when they picked Khrushchev as Man of the Year for 1957. The *Sputnik* launches had changed the world. Overnight, mankind knew that it had a future in space, and soon media stories would be credibly discussing the feasibility of going to the moon.

For the Soviets, the launches were an unalloyed public relations win, a sigh of relief, and probably their greatest achievement of the Cold War. But they also proved a quieter but still significant win for the United States and the global aerospace community. With no protests from the other UN member states erupting in response, the Soviet satellites set a precedent in international law, one that opened space to unfettered exploration. For Eisenhower, that was *Sputnik*'s major upside: while the Kremlin had beaten the Americans to space, their satellites' use of America's high-altitude airspace tacitly gave the United States reciprocal rights to the Soviets' own high-altitude airspace. This played into Ike's "Open Skies" insistence that the best use for space was peaceful scientific research and satellite communications—and, much less publicly, CIA overhead reconnaissance: the president had already approved the top secret Corona program, the development of spy satellites that were capable of letting U.S. intelligence agencies know exactly where Soviet missiles were positioned and tanks were massed.

In the coming years, spy satellites trained on the USSR would improve America's strategic position in the Cold War enormously. Between 1956 and 1960, high-altitude U-2 spy planes conducted numerous missions over Soviet territory, photographing everything from military bases and atomic power facilities to the ICBM launch compound in Kazakhstan, with the captured images sent to the CIA's National Photographic Interpretation Center to be scrutinized by experts. The U-2 program displayed the kind of nuanced thinking that had made Eisenhower a success as D-day's Supreme Allied Commander, but the program's secrecy meant it didn't generate the kind of public, tit-for-tat one-upmanship that made for positive newspaper headlines. In pure power-politics terms, Kennedy was a lucky beneficiary of Eisenhower's decoys and discretion.

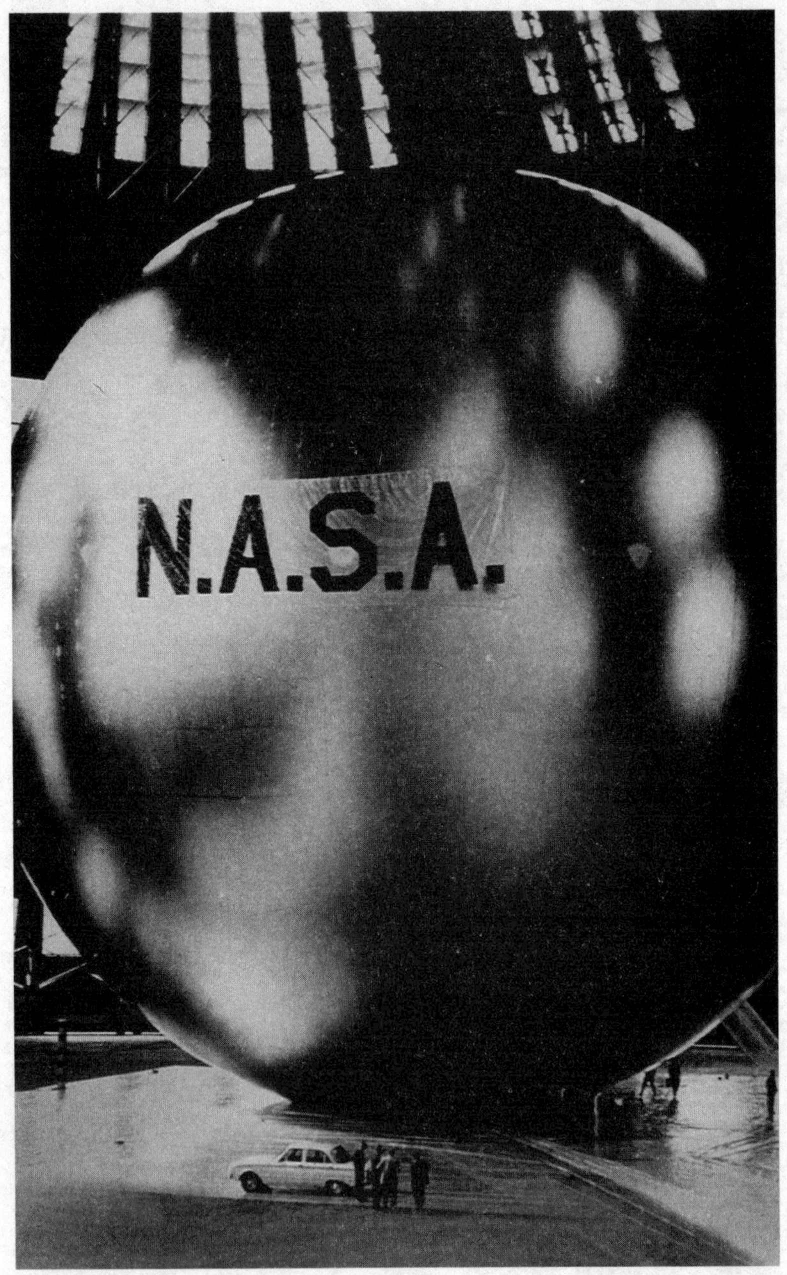

President Dwight D. Eisenhower deserves credit for establishing NASA in 1958. The new agency inherited the large organization that already existed with the NACA, and a workforce of more than eight thousand people.

7

Missile Gaps and the Creation of NASA

Events since last October show all too clearly the public appeal and the Cold War importance of satellites. The importance to the Nation of a successful space program cannot be overestimated.

—W. H. PICKERING (DIRECTOR OF CALTECH'S JET PROPULSION LABORATORY) TO W. V. HOUSTON (PRESIDENT OF THE RICE INSTITUTE), APRIL 19, 1958

At the beginning of 1958, the U.S. Navy's Vanguard program was still mired in the short-term effort to put a satellite into space. By then, Senator Kennedy had taken to calling the International Geophysical Year the "Geo Fizzle Year," and his disillusionment with Eisenhower's systematic approach to the space race was now a regular talking point. At a Harvard Club dinner in March, he even subtly mocked Vanguard's first disastrous launch effort. "When the elevator at the Washington Monument caught fire in December and smoke poured out of the building," he deadpanned, "one drunk staggered by and declared, 'They will never get it off the ground.'"

That winter, both U.S. military rocket programs, the army's Redstone project and the navy's Vanguard, were so hungry for a satellite success that there was a race to launch first at Cape Canaveral. One launchpad was quickly repaired after sustaining damage in the December 6 Vanguard TV-3 explosion. The Vanguard team had a launch of their TV-3BU backup scheduled for February 3. The indefatigable von Braun, raring for the Army Ballistic Missile Agency (ABMA) in Huntsville to be first, had fulfilled his promise to Secretary of Defense Neil McElroy to deliver a modified Redstone rocket (the Jupiter-C) within ninety days. The pressure was on.

On January 31, 1958, von Braun was in the Communications Room at the Pentagon while his primary boss, Major General John Medaris, oversaw the launch from Cape Canaveral. He had gotten permission to launch the Jupiter-C as a counterstatement to *Sputnik 2.* According to Medaris's journal, he "cautioned Dr. von Braun that there must be no public claims or discussion by employees of this agency which would falsely give the impression that we are in the satellite business." This was the only day the army team had to test the white, cone-shaped Jupiter-C; after that, they'd have to cede the Cape Canaveral range to the navy's Vanguard, which might then beat them into space. Von Braun knew that the Jupiter-C launch vehicle, a descendant of the Redstone, was ready, but the weather still had to cooperate.

Fickle weather was the bane of rocketeers the world over. Sudden wind or a lazy drizzle could scrub a launch. Fortunately for the army's Redstone team, after two rainy days, the skies cleared that January evening. Cape Canaveral's chief meteorologist, to von Braun's relief, gave the launch a late-evening thumbs-up. At 10:48 p.m., the Jupiter-C rose exactly as planned, disappearing into the upper atmosphere and then beyond. The stages fell away on cue, leaving America's first orbiting satellite, *Explorer 1*, circling the Earth. Von Braun was euphoric when he heard confirmation of *Explorer*'s radio signal, after it first circled Earth. At 1:00 a.m. on February 1, two hours after the launch, he emerged in a state of jubilation to announce his ABMA team's success to the sleepy press corps, who had been given no advance warning of the launch. It was the triumph of the Huntsville team. Pleased with the breaking news, Secretary of the Army Wilber Brucker and Army Chief of Staff Maxwell Taylor congratulated von Braun and touted the launch as a great American achievement. Speaking on behalf of all the Operation Paperclip rocketeers, von Braun said, "It makes us feel that we paid back part of the debt of gratitude we owe this country."

The next morning, newspapers and broadcasts on radio and TV ballyhooed the success of *Explorer 1*, which had brought relief and considerable delight to millions of Americans. Most major U.S. newspapers suggested that the *Explorer* was the first step toward American space supremacy. The *San Francisco Chronicle* headline read, "First U.S. Moon Circling Globe" while the *Orlando Sentinel* boasted, "Moon Over the Cape." President Eisenhower was ecstatic upon hearing the news at his cottage at Georgia's Augusta National Golf Club. Repeating the word *wonderful* three times when told of the success, he admitted, "I sure feel a lot better now," though he also gave orders to "not make too big a hullabaloo over this."

Three days after the *Explorer* launch, the navy's Vanguard TV-3BU fared only a little better than its predecessor, breaking up in midair long before reaching space. But in the wake of *Explorer*'s success, the U.S. government wasn't overly concerned. Major General Medaris boasted that the trifecta of the U.S. Army, academic science wizards, and the aerospace industry was indeed the winner of the interservice rivalry with the navy. "This is the beginning," von Braun said, "in a long-range program to conquer outer space."

Measuring just 6.4 inches in diameter, *Explorer* wasn't very sophisticated or large—Nikita Khrushchev famously derided it as the "grapefruit satellite"—but it proved a trouper, orbiting Earth until 1970, while both *Sputnik*s disintegrated upon reentry just months after launch. Not to be outdone in the competitive one-upmanship mode, the Soviets decided to launch a complex geophysical laboratory for their next satellite, and the TASS News Agency in Moscow claimed that the Kremlin's advanced space initiative was the most ambitious yet, designing rockets that would go near the moon as soon as 1959. The U.S.-Soviet space race was accelerating.

Encouraged by *Explorer* but still wary of pouring funds into space, Congress began looking into upgrading the peaceful aspects of American space exploration. Likewise, President Eisenhower, in an effort to harness government activities in the technology field, ordered

Defense Secretary McElroy to streamline space activities across the army, air force, and navy. As a result, the Advanced Research Projects Agency (ARPA) was created in February 1958 with a budget of $520 million. According to McElroy, ARPA's mandate was to coordinate cutting-edge research in artificial satellites, planet probes, and other space-related endeavors. It would act as a technology hub, overseeing existing research laboratories within each branch of the military while also sponsoring scientific work in private business and academia—all with the goal of beating the Soviets in overall space innovation.

The first director of ARPA, a civilian named Roy Johnson, ran the agency on a lean budget. A former General Electric executive, he had a dubious grasp of science but surrounded himself with aerospace experts and astrophysicists who appreciated that the new agency could help them achieve space feats. Hunting for a big-league success, Johnson lobbied around U.S. government power corridors on behalf of his bagful of space imperatives.

One proposal, dubbed Project Orion, suggested using a series of atomic explosions to propel a spacecraft at high velocity, with great fuel efficiency. At first "it looked screwball," Johnson told a congressional committee in 1959. "It doesn't look so screwball today. . . .

You use little bombs, and you use a lot of them. The trick is the creation of a spring mechanism on the platform." The platform would provide a barrier between the nuclear bombs and the spaceship, somehow protecting astronauts onboard from the bombs' concussions. While thermal nuclear propulsion in rocketry was the kind of research undertaken through ARPA contracts, the agency ultimately decided against bomb-powered spaceflight.

Having excoriated the Eisenhower administration the previous autumn over mismanagement of national security, Kennedy was not actively involved with ARPA's plans to reorganize American space exploration. When, on March 17, Project Vanguard finally sent a navy satellite into orbit, Kennedy retired his space jokes. Running for reelection to the Senate, he instead switched gears, homing in on a recent National Intelligence Estimate that forecast the Kremlin would have a "first operational capability with up to ten prototype ICBMs" somewhere between mid-1958 and mid-1959. Such a nuclear arsenal in the hands of a leader like Khrushchev, who regularly threatened to "bury capitalism," was represented to Kennedy as the most dangerous threat the United States had ever had to face—the possibility of total annihilation.

JFK also began taking control of some of the less

decorous aspects of his personal narrative. Speaking that March at the Gridiron Club, during a glittering evening of status-conscious Washingtonian satire for which he prepared as though it were a Harvard final exam, Kennedy pulled from his jacket pocket what he said was a telegram, claiming it had just arrived from his "generous daddy." Poking fun at himself, he then pretended to read his father's faux message aloud: "Dear Jack, don't buy a single vote more than is necessary. I'll be damned if I'm going to pay for a landslide." From that point on, he drew roars from the audience, often at his own expense. It was a hilarious night overall, and a watershed for Kennedy. After living ten years in Georgetown, he was at last part of the Washington establishment on his own terms: an independent Democrat, beholden to nobody.

In fact, Kennedy's Gridiron jest belied the actual plan for his Senate campaign. Knowing full well that Joe Sr. would pour nearly unlimited funds into the coming presidential campaign, Jack was adamant that his Senate reelection campaign be bare bones. Not wanting the image that "Daddy buys my votes" to have a glimmer of truth in 1958, he made sure his campaign team was disciplined and effective in its own right. Hard as it might have been on his health, he returned to Massachusetts regularly and circulated around the nation tirelessly, delivering speech after speech.

On February 6, 1958, as noted, the Senate voted to establish a Special Committee on Space and Astronautics to map a more aggressive U.S. space policy. Lyndon Johnson had steamrolled the resolution through the Senate and succeeded with a 78–1 vote. The lone dissenter was Allen Ellender of Louisiana, a Democrat who rejected all new committees as a matter of principle. The new Senate committee was chaired by Johnson. Soon thereafter the House created its own Select Committee on Astronautics and Space Exploration. Seizing on the post-*Sputnik* moment, both these legislative bodies were geared to advocate civilian space relevancy over those in the military. In the deliberations, Kennedy was merely a yes vote, while Johnson seized on space with assiduity. "LBJ was eager to get out front in space because it was the new national toy," White House speechwriter Bryce Harlow believed. "He was trying to get to become President of the United States . . . so LBJ wanted to get in front of the space race so that everybody would say, 'oh, that's our leader."

In the Senate, both Kennedy and Johnson vigorously championed a bill that was gathering steam: the

National Defense Education Act (NDEA). Promising funding to encourage studies in the sciences and related subjects, the proposal was controversial in that it initiated direct federal aid for public schools, as well as universities and individual students. Some saw this as the beginning of the end for local control of public schools. The book *Why Johnny Can't Read and What You Can Do About It*, published by Harper and Row, had hit the best-seller list, and Kennedy rode the wave. Nathan Pusey, the president of Harvard, joined JFK in insisting that a larger percentage of the GNP be directed toward education. Kennedy believed that science education was the critical component of American success in the space race. Today's high-schoolers would be tomorrow's physicists and astronomers. He supported the bill and also proved his ability to balance fairness with his ardent anticommunism by arguing against a provision requiring loan applicants to sign a loyalty oath. Once the NDEA was enacted, it indeed laid the groundwork for advancements in space by putting hundreds of thousands of students on track to what might be called science/space technology literacy.

At ARPA, Roy Johnson received a proposal in March 1958 that caught his fancy. Sporting the catchy name "Man-in-Space-Soonest" (MISS), the idea originated at the air force's Air Research and Development

Command, one of the many groups that had sprouted up throughout the military over which ARPA was supposed to ride herd. Within days, Johnson expressed his support for the $133 million program, telling the press, "The Air Force has a long-term development responsibility for manned spaceflight capability with the primary objective of accomplishing satellite flight as soon as technology permits." The plan called for an eleven-step protocol, with each step constituting the launch of a more refined and sophisticated space vehicle, leading to the ultimate goal of a "Manned Space Flight to the Moon and Return." The total projected cost of landing an American on the moon by 1965 was $1.5 billion (off by $24 billion and four years, as it turned out).

Even before the initial ARPA disbursement came through, the air force began work on MISS. Rudimentary plans were debated, scientists were contracted, and astronauts were recruited. Accordingly, the USAF asked Dr. Edward Teller, the so-called father of the hydrogen bomb, and several other members of the scientific technology elite to study the issue of human spaceflight and make recommendations for the future. Teller's group concluded that the air force could place a human in orbit within two years, and urged that the department pursue this effort. Teller understood, however, that there was essentially no military reason for undertaking this mission and chose not to tie his recommendation to any specific rationale, falling back on a basic belief that the first nation to do so would accrue national prestige in a general manner. In early 1958, Lieutenant General Donald Putt, the USAF deputy chief of staff for development, informed NACA director Hugh Dryden of the air force's intention of aggressively pursuing "a research vehicle program having as its objective the earliest possible manned orbital flight which will contribute substantially and essentially to follow-on scientific and military space systems." Putt asked Dryden to collaborate in this effort, but with the NACA as a decidedly junior partner. Dryden agreed, but insisted on a nonmilitary lead in the effort. In the fall of 1958, the newly established NASA gained authority of MISS. It laid the groundwork for Project Mercury.

While space historians have often ignored or dismissed MISS as an underfunded effort, it accomplished three important things over the summer of 1958. First, it identified an easily quantified objective: a manned flight into space. Second, the eleven-step guideline provided a framework for realistic discussions about Americans in space. Third and most important for manned space in the coming decades, MISS proactively named the men who would venture into space.

That summer, nine names made the air force MISS list, all of them high-altitude test pilots who also fit the profile of being smaller, lighter men. Four fearless X-15 test pilots were on the roster: Iven Kincheloe, Joe Walker, Jack McKay, and Neil Armstrong. Also listed was Scott Crossfield, an aviation engineer from California who from 1946 to 1950 had worked in the University of Washington's Kirsten Wind Tunnel, and who was slated to fly the X-15 even while helping to engineer it. Bob White, a New York City native, was equally at home in the cockpit and at the drawing board, having worked as a systems engineer at the Rome Air Development Center in upstate New York before seeking a billet at Edwards Air Force Base as a test pilot. Bob Rushworth was a stalwart of air force combat and experimental aviation. Alvin White (no relation to Bob White) was a World War II hero with a degree from the University of California, Berkeley. Bill Bridgeman, at forty-two the oldest of the MISS candidates, had risen from obscurity, flying relatively slow airplanes in the war and immediately afterward, and then becoming a flashy, record-breaking test pilot by 1951.

Among the X-15 pilots themselves, Joe Walker

of Pennsylvania was already a legend of sorts. During World War II he earned the Distinguished Flying Cross by flying weather reconnaissance flights for the army air corps. Once the war ended, Walker left the air corps and joined the NACA's Aircraft Engine Research Laboratory, in Cleveland. There he met his hero Jimmy Doolittle. Something of a daredevil, Walker would fly test planes to unprecedented altitudes. While Chuck Yeager became famous in 1947 for being the first person to break the sound barrier, it was Walker who logged more hours and went higher and faster on the X-15 than any other American pilot. On his inaugural X-15 flight, he was surprised by the sheer brute thrust of its rocket engines. "Oh my God!" he screamed into the radio as his body vibrated madly. The flight controller comically responded, "Yes? You called?" By 1951, Walker was relocated to the High-Speed Flight Research Station, in Edwards, California. With experience in practically every new type of chase plane or fighter jet, Walker was an obvious choice for the air force's MISS group in 1958.

Denying that he wanted to break the speed records, Neil Armstrong was skeptical about the MISS program. The decorated Korean War pilot was not especially interested in sleek rockets or spaceships, insisting that he preferred planes with wings, which the X-15 had. Armstrong volunteered, and with the other candidate astronauts was put through several days of medical testing at Edwards Air Force Base. The patriotic Armstrong, in his non-flashy Ohio way, spoke for the rest of the corps when he said that if the country needed a Man-in-Space-Soonest, he was "in the line-up."

With this lineup, MISS put faces on those American test pilots who, it was thought at the time, would someday conquer space. But this MISS program wasn't publicized. President Eisenhower refused to seek headlines in mid-1958. That however, would change, with personal publicity soon becoming a hallmark of American space exploration. The air force promoted only a few articles about the MISS roster. But once *Time*, *Life*, *Collier's*, and *Newsweek* turned von Braun into the Space Maestro of the 1950s, astronauts—as the aviators turned space travelers were now called—were also poised on the brink of fame.

If Eisenhower hadn't been convinced previously of the need for an American presence in outer space, the six months since *Sputnik* had brought him a long way around from his naturally conservative fiscal beliefs—his modus operandi, he often claimed, was saving taxpayers "every possible dime." In sending the legislation for a new space agency to Congress in April 1958, the president admitted that public relations had become the master. "The highest priority should go of course to space research with a military application," he explained, "but for national morale, and to some extent national prestige, this should likewise be pushed through a separate agency." Infuriating army, navy, and air force brass, the new space agency was to be run by civilians. Trying to calm the military's disgruntlement, the president promised generals and admirals that every branch of the armed services would still play a major role in all space-related White House decisions.

On April 2, 1958, President Eisenhower spoke before a joint session of Congress to champion the creation of a civilian National Aeronautics and Space Administration (NASA), which would incorporate the existing National Advisory Committee for Aeronautics (NACA), headquartered in Washington, DC, with laboratories in Hampton, Cleveland, and the San Francisco Bay area. Lyndon Johnson of Texas wasted no time in embracing Eisenhower's call. Working with Senator Styles Bridges (R-New Hampshire), Johnson cosponsored the Senate version of the NASA bill and resisted arm-twisting efforts to militarize the adolescent American space program as the two senators worked on the legislation in bipartisan tandem with the Eisenhower administration.

The bill's controversial proposal for civilian control

of NASA led to months of negotiations on Capitol Hill, as the small contingent of space advocates in Congress was overwhelmed by the horde of legislators who demanded a say in military matters. Ultimately, the exact nature of NASA's work was delineated as "exercising control over aeronautical and space activities sponsored by the United States," with the military services retaining control over "activities peculiar to or primarily associated with the development of weapons systems, military operations, or the defense of the United States."

Washington lawmakers sought expert opinions from the brightest minds working the aerospace beat for ideas on how NASA should function. Many thoughtful space leaders were put off by Lyndon Johnson's claim that the "conquest of space" was a Cold War military necessity. "General Donald Putt recently called for an Air Force base on the moon," director of Caltech's Jet Propulsion Laboratory W. H. Pickering scoffed at the U.S. Air Force's anti-NASA reluctance. "These and similar statements, I believe, are nonsense. The direct military value of a space program is almost zero. Satellites may have some value for reconnaissance or communications, though from a Russian point of view this would appear to be quite minor. Space vehicles capable of journeys to the moon or beyond appear to me to be of no military significance."

General Jimmy Doolittle of the U.S. Air Force worried that the establishment of NASA meant the dissolution of the NACA. Since World War I, the Langley Research Center had been a pioneer in aviation research. Having served as chairman of the NACA right up until 1956, Doolittle knew virtually every employee there by name. "While we [the NACA] knew that missiles would have a very important place, while we knew that space must be explored, we were hesitant to turn over to the missile people and their supporters all of the funds that we had been receiving for the development of the airplane and associated equipment," Doolittle wrote. "In retrospect, I think we all agree that we were wrong."

When Eisenhower signed the National Aeronautics and Space Act of 1958 on July 29, some generals and admirals dissented. Didn't Eisenhower understand the horrific consequences of *Sputnik*? Did he really believe all that International Geophysical Year nonsense? Space was rapidly being militarized by the Soviets; therefore, America should follow suit. Speaking for many career soldiers, Lieutenant General Gilbert Trudeau, the head of U.S. Army Research and Development, expressed utter dismay. "I just can't believe," he said, "that anyone would take the capability of the most capable element in the Nation to explore space and do away with it." Trudeau was just one of a chorus of pissed-off military officers. Dozens of Pentagon denizens insisted that outside civilian control over their aerospace research spelled doom. Of course, the Department of Defense and its predecessor, the War Department, had exerted such civilian power since the founding of the nation in 1789, and the Pentagon was stocked with civilian administrators. But that was different in one respect: while military personnel were trained to use intimidation and force to maintain the security of the United States and its interests abroad, NASA lacked that motivation, not being subordinated to military objectives. As Lyndon Johnson put it, space exploration for science's sake was the civilian agency's purported mandate (at least for public consumption).

Senator John F. Kennedy was still regularly using space exploration as a wedge issue with which to attack President Eisenhower. In a speech in western Pennsylvania on April 18, 1958, he delivered the bitter pill that "Americans were no longer the paramount power in arms, aid, trade or appeal to the underdeveloped world. We are acting largely only in reflex to Soviet initiative." With the verve of a political powerhouse who had finally hit his stride, JFK suggested instead that "the United States for once appeal to the world with constructive solutions and our own vision of the future." To Kennedy, trying to beat the USSR in the near vacuum of outer space, satellite by satellite, was untenable. Instead, the senator from Massachusetts asserted, the United States needed to rally toward the cause of beating the Soviets in intermediate-range ballistic missile (IRBM) and ICBM technology and accelerating the production schedules of the Atlas, Thor, Jupiter, and Titan rockets.

At the same time that Kennedy was delivering his anti-Eisenhower campaign speeches, an influential report appeared to support his positions. *Deterrence and Survival in the Nuclear Age* (commonly referred to as the Gaither Report, after H. Rowan Gaither, the chairman of a blue-ribbon commission whose aim was to offer Eisenhower advice on how to deal with the Soviet Union following *Sputnik*) called for a tougher stance with Moscow, more money for military research and missile development, and increased U.S. conventional forces. It also recommended that the United States develop an invulnerable second-strike force, warning of a threat that could become critical by 1959 or early 1960 because of the "unexpected Soviet development of the ICBM." Kennedy embraced the Gaither Report as his personal white paper on the campaign trail and in press releases, while arguing his concern that the Western democracies possessed little reliable information about the USSR's military and technological strengths and weaknesses. Even the most elementary facts were unavailable—the locations of railroads and bridges, the locations of factories and the nature and volume of their production, the size and readiness of the Soviet army, navy, and air force. The Soviet Union, JFK complained, was tightly wrapped in ominous secrecy.

On August 14, 1958, Kennedy put a new phrase into the American lexicon, earning himself a prominent citation in the *Oxford English Dictionary.* Speaking on the floor of the Senate, he said, "Our nation could have afforded, and can afford now, the steps necessary to close the missile gap." That "missile gap" became Kennedy's calling card, wielded as a campaign cudgel even though its contention—that the Soviets had hundreds of ICBMs whereas the United States hadn't deployed a single one—proved to be a fiction. In the wake of JFK's assertions, the air force and the CIA sparred over how advanced the Soviets actually were. Air force analysts backed Kennedy's notion that the USSR had stockpiles of ICBMs. The CIA disagreed, insisting that there were fewer than a dozen. (Declassified documents would later show the CIA was right: at the time in question, the Kremlin had only four ICBMs.)

Kennedy's missile gap was a direct descendant of the "bomber gap," the mid-1950s fear that the Soviets had a strategic bomber force bigger than America's. But U-2 reconnaissance flights over the USSR soon proved this to be a fable. Just as Kennedy was propagating the "missile gap" fallacy, U-2 photographs proved that the Soviets were behind in ICBM development. JFK wasn't given this CIA intelligence until the summer of 1960. Like most U.S. senators in 1958, he had been briefed on the Corona intelligence satellite program, which Eisenhower approved that year. None of this mattered to him. His "missile gap" spiel was a winner. At heart, Kennedy was set in the technocratic idea that the federal government needed to play a huge role in spurring social change at home and abroad through the bankrolling of technological innovation and military modernization.

On October 1, 1958, President Eisenhower signed an executive order to put the National Aeronautics and Space Act into effect by transferring "Certain Functions from the Department of Defense to the National Aeronautics and Space Administration." But sorting out civilian versus military control remained a tricky proposition. After all, rockets and satellites used very similar technology, infrastructure, and personnel regardless of whether they were deployed for military, scientific research, or communications purposes. Plus, the military was deeply territorial of the field. Von Braun and ABMA's commander, Major General John Medaris, insisted that control of outer space was the sine qua non of national greatness and that the army had to lead the way. And when Medaris was asked in 1958 if outer space was going to become an active Cold War battleground, he snapped, "As sure as anything in the world."

Medaris kept fighting, asserting to the Defense Department that giant boosters were needed for army rockets (though he couldn't cite a single military reason) and calling Eisenhower's plan to have NASA take over ABMA's Redstone and Jupiter-C rocket development "rather disastrous." This disapproval was shared by most top army brass, including such World War II air power generals as Carl Spaatz and Nathan F. Twining, who were dismayed that Eisenhower, of all people, seemed intent on punishing the army. Some, such as Medaris, eventually retired in protest, but all their carping was for naught. On October 21, 1959, after months of bickering, Eisenhower announced he would transfer ABMA, its massive new Saturn project, and von Braun's Huntsville rocket team from the army to NASA in the coming year. Finally, space exploration and space technology were no longer only the purview of the U.S. military. Like its predecessor, the NACA,

NASA would pursue astronomy and astronautics in the interest of scientific knowledge and engineering excellence, rather than building ICBMs for war. Inventing NASA didn't mean that peaceful exploration of space and military usage could be divided with ease in the organization: the overlap between civilian and military spheres remained. The legislation that had created NASA specified that it make open to defense agencies "discoveries that have military value or significance," and defense agencies were in turn to provide "information as to discoveries which have value of significance" to NASA.

When NASA opened its doors in the fall of 1958, it had eight thousand employees, an annual budget of $100 million, and a main office in a brownstone in Northwest Washington (once the home of First Lady Dolley Madison), near the White House. Of course, the new agency was built on the foundation of the NACA, which had been set up back in 1915. The NACA's Langley Memorial Aeronautical Laboratory, in Virginia; Ames Aeronautical Laboratory, near San Francisco; Jet Propulsion Laboratory in Pasadena; and Lewis Flight Propulsion Laboratory, in Cleveland, were all incorporated into NASA, along with the two NACA test-flight facilities: the High-Speed Flight Research Stations at

Edwards Air Force Base, in California (for high-speed-flight research), and Wallops Island, Virginia (for research rocket launches).

Furthermore, as noted, on July 1, 1960, the old ABMA facility in Huntsville was officially incorporated into NASA as the George C. Marshall Space Flight Center, dedicated to providing launch vehicles for space exploration. Eisenhower toured von Braun's Space Science Laboratory and inspected a Saturn I rocket model, which would soon become the largest rocket ever built at the time. The two men were distrustful of each other, but for the sake of NASA they were cordial. Speaking before a huge crowd of ten thousand Alabama workers, Eisenhower praised von Braun's work. "No doubt this mighty rocket system makes its presence known loudly—possibly too loudly—in Huntsville," he said. "But it is a significant forward step in our conquest of space and for growth in human comprehension."

Outside military circles, the creation of NASA wasn't headline news. Root questions about its specific mandates were relegated to the inside pages of those newspapers that even noticed them. Within NASA, administrators had worked hard to guarantee a smooth transition. "Employees had been reassured for several weeks by the NACA headquarters and by Langley management that they were to come to work as always and do the same things they had been doing," historian James Hansen explained. "Their jobs already had much to do with the nation's quickly accelerating efforts to catch up with the Soviet Union and launch America into space. As NASA personnel, they were to keep up the good work." On October 11, 1958, little more than a week after it became operational, the infant NASA launched *Pioneer 1*, a three-stage Thor-Able rocket carrying a scientific instrument package intended to measure cosmic radiation between Earth and the moon and to collect information about the lunar surface. Although it was intended to prove that America was in the space race for real, *Pioneer 1* failed to achieve lunar orbit, and plunged back into Earth's atmosphere after forty-three hours, having transmitted back a small quantity of useful scientific information.

To fill the job of NASA's first chief administrator, Eisenhower approached Air Force General Jimmy Doolittle, who had helped guide the United States through the *Sputnik*, Vanguard, and *Explorer* events. However, Doolittle refused, opting instead for a more lucrative position as chairman of the board of TRW Space Technology Laboratories. The NACA director Hugh Dryden, a respected aerodynamicist, was also in the running, until his appearance before the House Select Committee on Astronautics and Space Exploration proved a bust; he offered that sending a man into space "has about the same technical value as the circus stunt of shooting a young lady from a cannon."

Following the false starts of Doolittle and Dryden, Eisenhower nominated T. Keith Glennan, president of the Case Institute of Technology (today's Case Western Reserve University), in Cleveland. From 1950 to 1952, Glennan was head of the Atomic Energy Commission (a new federal agency) and had excelled. His appeal to Eisenhower was his fiscally conservative viewpoint. Glennan feared that the United States was rushing into a fast-water "socialist stream," and he rejected the proposition that "the federal government is a 365-day Santa Claus and that the national treasure is an inexhaustible storehouse of largesse." He was easily confirmed. Furthermore, Dryden disappointed hardened Cold Warriors on the committee when he said that "the prospective space programs are not such as to leapfrog the Soviets immediately or very soon." Following *Sputnik*, they wanted a NASA administrator who wanted to beat the Soviets and seize U.S. leadership in space exploration.

Hugh Dryden became NASA's first deputy administrator. Just three months after NASA was founded, Glennan and Dryden proclaimed Project Mercury America's first manned space mission and named

Robert Gilruth, a talented test engineer, to lead a new Space Task Group at the renamed Langley Research Center dedicated to jump-starting the mission. (The group's base of operations would remain at Langley until NASA relocated manned space research to Houston in 1962.) NASA promised that a Mercury astronaut would be rocketed into space within three years.

Historian Walter A. McDougall wrote that once NASA was established, the big question was a Hamlet-like pondering over whether "to race, or not to race" the Soviets in space. Eisenhower was against both the militarization of space and trying to one-up the USSR feat by feat. The dilemma the president faced—one that Senator Kennedy exploited—was that each time the Kremlin put space points on the board, American critics charged that the Cold War was being lost. "One purpose of Eisenhower's strategic posture was to restrain those elements in government and society willing to jettison limited government and financial restraint in order to prove American superiority," McDougall wrote. "Racing with the Soviets for space spectaculars ran against his grain."

Regardless of Eisenhower's cautionary approach to the space race between the United States and the USSR, the competition was on. By the fall of 1958, America had launched four orbiting satellites to the Soviet Union's three. The Soviet satellites were heavier, which was a credit to their advanced rocketry. While *Explorer 1* was lighter, it was able to perform many of the same functions, which was a credit to American ingenuity. The U.S. satellites also uniformly orbited at a higher altitude than their Soviet counterparts. In anticipation of manned flight, the Soviet space program had performed more animal experiments in their laboratories, while the Americans felt confident that they had amassed more supporting data on the challenges of humans in space. On October 5, 1958, the *New York Times* correctly opined that "the balance sheet of a year of effort since *Sputnik I* would seem to indicate that the United States was not as far behind at the time of the launching of the first satellite as was then imagined."

The Soviets had enjoyed an early advantage because their space-related activities were streamlined, with the central government in Moscow overseeing all developments in IRBMs, ICBMs, satellites, and space exploration. But the Kremlin didn't yet realize what an advantage unfettered capitalism would be to the Americans' new space agency. Established with the cooperaton of the army, air force, and navy, NASA became the lucky beneficiary of an astounding, interconnected network of industrial contractors and aerospace firms that had invested in the development of long-range missiles even before the civilian agency's creation, and of an even wider net of innovators and suppliers that would join the effort as the space race wore on. For example, there was simply no entity in the Soviet Union as dynamic as North American Aviation's Rocketdyne Division, the leading designer and manufacturer of liquid-fueled engines for most of the army and air force missiles during the Truman and Eisenhower years. When tight Communist control competed with free-market capitalism on the playing field of large-scale innovation, the Americans outshone the Soviets, and space wasn't the only beneficiary.

At Langley, the Space Task Group was beginning another NASA tradition: leveraging space-related research for the benefit of commercial air transport. NASA partnered with aviation companies such as Boeing of Seattle, Convair's Astronautics division of San Diego, General Electric of Philadelphia, the Martin Company of Baltimore, and McDonnell Douglas of St. Louis. Having shed their NACA smocks for NASA ones, Langley engineers were still modernizing in the realm of flight research and wind tunnel testing, solving a variety of problems related to the transonic flow regime (mach.8-1.2) through the implementation of swept-wing design. During World War II American pilots had controllability problems with some planes' suddenly diving down and accelerating to transonic speeds; these problems had to be rectified at the NACA. They also invented the idea of grooved runways, which offer better grip for aircraft tires in heavy rain. Any way one peered into the looking glass, the truth was that most aerospace-related research and development had military applications. John F. Kennedy understood that if the nation's goal was beating the Soviets in space, U.S. military and civilian aims had to be integrated.

At the end of 1958, NASA launched *Pioneer 3* (the first U.S. satellite to ascend to an altitude of 63,580 miles), while the air force achieved the first long-distance flight of an ICBM (anAtlas 12B, which flew more than 6,300 miles). Three weeks later, another Atlas ferried a communications relay satellite into orbit as part of Project SCORE (Signal Communication by Orbiting Relay Equipment), which broadcast President Eisenhower's Christmas message to the world—the first voice sent from space. The development that gave America its greatest edge over the Soviets in 1958 came courtesy of the private sector. Working separately, electrical engineers Jack Kilby of Texas Instruments and Robert Noyce of Fairchild Semiconductor invented the monolithic integrated circuit, also known as the microchip—with that innovation, transistors, resistors, capacitors, and connecting wiring, all previously separate components, could be placed onto a single small "chip" of semiconductor material. This tiny, integrated circuit would soon lead to the development of portable, efficient, and affordable high-speed communication systems, revolutionizing both space exploration and terrestrial technologies.

In preparation for manned Project Mercury spaceflight, NASA modified the U.S. Navy's jet aircraft suits (the inside lined with neoprene-coated nylon, the exterior aluminized nylon) for surviving galactic conditions. Recognizing that a hard-shell suit was unworkable, NASA designers made soft silver suits that could furnish oxygen, regulate temperature, enhance flexible movement, generate communications, and shield against solar radiation. At NASA, the hunt for astronauts was on.

Many names that should have been shoo-ins were absent from the lottery. Iven Kincheloe, once on track to become America's first astronaut with the air force's now-defunct MISS program, had been killed the previous summer on a test flight after ejecting too late from a crashing plane for his parachute to open. Two other men in the MISS group were above forty, and deemed too old. Neil Armstrong, for his part, chose not to apply, remaining loyal to his work on the X-15. Chuck Yeager, the first man to break the sound barrier, also declined to apply. His lack of a college degree would have made him ineligible in any case. NASA's insistence on academic credentials reflected the dual role envisioned for NASA's astronauts: not just "a man in a can," as some aviators had said disparagingly, but contributing to the ongoing engineering of their flights in the manner of X-15 engineer-pilot Scott Crossfield.

While Armstrong passed on the Mercury program, another Korean War veteran from Ohio, John Glenn, looked on it as a "tonic." Considered an overgrown Boy Scout by other test pilots, full of gentlemanly manners and a quarterback's drive, Glenn believed from the outset that NASA would move the United States into space in an organized way—one that would also advance his career. Glenn's guiding light was merit: a challenge fought for and achieved by sheer willpower and self-conquest over natural limitations.

At the time of Pearl Harbor, Glenn was a twenty-year-old student at Muskingum College in New Concord, Ohio. Hungry for combat action, he left his engineering studies to join the armed forces. After stints with the U.S. Army Air Corps and naval aviation, he was given a commission in the marines and acquitted himself well, flying Corsair fighters in the Pacific Theater. After the war, he remained in the Marine Corps, accepting relatively dull assignments with equanimity. Even after the United States entered the Korean War, Glenn initially remained stuck stateside in administrative posts, a "non-entity," as a friend termed him. Finally ordered into the war in early 1953, he flew Panther jets with the marines' "Tomcat" squadron before being seconded to the air force, where he began flying transonic fighter jets for the Twenty-Fifth Fighter-Interceptor Squadron and was credited with three kills in dogfights.

Glenn returned home from Korea a highly decorated hero, and was able to fly the most advanced jet airplanes of the era. Hoping to get a serious aviation education, he sought admission to the test pilot school at Naval Air Station Patuxent River, in Maryland. As Glenn was aware, he was ill prepared for this elite institution in one respect: he still lacked a college degree. Even without completing his bachelor's at Muskingum, he used enthusiastic recommendations from his superiors in Korea to win a coveted place at Patuxent River. And then, after pulling every string to get in, he nearly quit. He just didn't speak the same language as the college-trained pilots at the school. The word used for Glenn at such times was *dogged.* Patching together courses at local colleges and getting tutoring when he could, he managed to stay abreast of his training and graduate from Patuxent with distinction.

Ironically, Glenn's persistence in learning about aeronautical engineering, combined with his certification as one of the navy's newest test pilots, earned him a desk job in Washington examining airplane designs. At pains to understand how all aircraft worked, he heard talk of a proposed supersonic cross-country flight to stress-test the Pratt and Whitney J-57 jet engine, and he volunteered for the assignment. On the morning of July 16, 1957, he took off from Los Alamitos Naval Air Station in California, piloting his Vought F-8U Crusader at a speed of over 725 miles per hour and touching down at New York's Floyd Bennett Field 3 hours, 23 minutes, and 8.4 seconds later, setting a new transcontinental record. This supersonic event turned Glenn into a minor celebrity, earning him an appearance on the television show *Name That Tune*, partnered with the child star Eddie Hodges. When NASA started thinking in earnest about its first astronaut class at the end of 1958, Glenn was certain that he was "in a pretty good position" to join it. Among navy aviators, he was known as the best man to land safely if something went wrong with a craft's controls or if the wind didn't cooperate on final approach.

Glenn's record-breaking flight had positioned him well for Mercury, but he had other deficits besides his education: he was a little heavy and was near NASA's age limit. On the plus side, he was also one of the few marines with an interest in the program—important because corps brass had informed NASA that they expected at least one marine to be chosen for the Mercury team. There was also a certain fund of decency in Glenn's overall character that military leaders admired. There was never a calculated love of battle, grandstanding, medals, or glory. Instead, Glenn exhibited, often with a self-deprecating smile, a Midwestern devotion to duty, honor, and country. While waiting to hear from NASA, he lost forty-one pounds, intent on being in shape should he get the call. He also tried, without success, to turn his hodgepodge of college credits into a degree from Muskingum.

While Glenn was dieting madly, Kennedy was preparing for Election Day 1958 in a curious way. From Labor Day to the first Tuesday in November, he spent only seventeen days in Massachusetts, mostly to rest and relax at Cape Cod. Two-thirds of his time were spent in other states or in Washington, DC. He had become a national figure, and that became part of his senatorial appeal in his home state. He oversaw a well-oiled staff that offered up surrogates, rather than the candidate himself, to do his campaigning. JFK continually referred to himself not as the senator from Massachusetts, but as the "senator of New England." As it turned out, Kennedy could do no wrong in Massachusetts, coasting to a second Senate term with 74 percent of the vote. Winning the race so handily seemed just another step on the way to the 1960 presidential election. Doubters of his political chops were forced to admit to having underestimated his talent.

What might be called Kennedy's presidential years, in fact, started right after the 1958 election. Free and easy with the press, he was unquestionably aiming for the White House and positioned as the front-runner. His political persona rose so steeply that it was no longer easy to see the line between the man and the image he had created. Long gone were the days when he arrived to give a speech disheveled and harried, quickly tucking in his shirt on the way to the dais. Now he was the crisp Ivy Leaguer, radiating self-esteem and eminently comfortable with himself. Nearly all politicians choose which part of their personalities to project and which to leave at home, but Kennedy had edited himself with unusual precision, becoming one of the most unique and recognizable public figures of the late 1950s. Perhaps he was "sold like a box of Wheaties," as Adlai Stevenson had said about attempts to market

Kennedy as a national leader. If so, there were few constituent complaints about him from the Commonwealth of Massachusetts.

But in truth, voters didn't know the real JFK. Even though he exuded confidence and political combativeness, he camouflaged a lot, including his precarious balance of vigor and infirmity, which necessitated prodigious use of prescription medicine. But the well-rounded image Kennedy projected as the nation's potential next president was convincing because it was a real part of the truth about himself. His questioning intelligence and keen alertness were integral to his being. As the 1958 election receded, he faced two years in which to reach Americans and sell them on the hawkish humanitarian he had become. Nobody feared he would ever capitulate to the Soviets. JFK may have been a product, but in yet another contradiction, he also seemed the least artificial senator in Washington. "I have never seen anybody in my life develop like Jack Kennedy did as a personality and as a speaker, and as an attractive person, over the last seven, eight years of his life," Democratic senator George Smathers from Florida recalled. "It was a miracle transformation."

Smathers was in a position to know. First elected to the U.S. Senate in 1950, he was widely considered Kennedy's best friend in Washington. Tall and handsome, and the former captain of the University of Florida basketball team, he often partied with Kennedy in both Georgetown and Palm Beach. Smathers was pleased that Cape Canaveral was in his state—he would talk a blue streak with Kennedy about aerospace industries of tomorrow—but when it came to civil rights, unlike Kennedy, he was a determined segregationist, one of the Dixiecrats who had signed the 1956 Southern Manifesto, which denounced the U.S. Supreme Court's *Brown v. Board of Education* ruling as a "clear abuse of judicial power."

As for NASA, Kennedy's view was exactly in sync with Smathers's: no more Soviet "firsts" in space. But Kennedy didn't try to bring space business to Massachusetts the way Smathers did in Florida. In fact, he practically rushed into 1959 largely oblivious to the beehive of activity around Project Mercury, though one bit of space-related news undoubtedly caught his attention because of its implications for the upcoming Democratic primary battle. In November 1959, at President Eisenhower's request, Lyndon Johnson gave an important speech on space at the United Nations. Kennedy believed that by accepting Ike's request, Johnson had aligned himself with the president and with policies that had birthed Kennedy's favorite subject: the alleged missile gap. At his own public appearances, JFK argued for a smarter approach to the Soviet rivalry. "It is not necessary that we match the Russians missile for missile, invention for invention," he said in a Detroit speech. "If the Russians succeed in sending a man to Hell, there is no need for each of our defense agencies to clamor the next morning for a new appropriation to match them. But neither can this challenge be met by men of little minds and little vision—by those who fix weapons policies as a part of our budgetary policies. The Democratic Party rejects the principle of a cheap, second best defense—and it intends to see that we have the money and brainpower necessary to do the job."

Influenced by a 1958 CBS News two-hour television special titled *Shooting for the Moon* (hosted by Walter Cronkite and starring Wernher von Braun), Kennedy leaned toward a beefed-up NASA but hedged his bets in public, not wanting to alienate the army, where some still had bitter emotions over Eisenhower's ABMA transfer. Nevertheless, he was understandably proud that the United States notched some productive successes in early 1959, including the launch of communications and weather satellites and, on March 3, the launch of *Pioneer 4*, which made the first successful flyby of the moon by a U.S. spacecraft. These were important steps toward a goal on which nearly every American public official in the post-*Sputnik* era agreed: getting an American astronaut into space soonest. What mattered to Kennedy was that NASA wasn't window dressing for a lack of commitment in space exploration; he wanted to ensure that the new agency was well funded and results oriented.

Every time NASA administrator Keith Glennan circulated around the Senate looking for increasing NASA appropriations, Kennedy essentially said, "Well, of course, uncap the faucet." JFK understood that in the realm of global prestige, NASA astronauts were going to be seen as knights of American exceptionalism—when a Mercury astronaut eventually broke the shackles of Earth to soar into space, average citizens in India, Venezuela, or Portugal weren't going to debate whether NASA (civilian) or the U.S. Air Force (military) deserved the credit. The buzz would be that America had pioneered into the galaxy, proving definitively that democratic capitalism was superior to state-run communism. And Kennedy agreed with the House Select Committee on Astronautics and Space Exploration conclusion that "outer space is fast becoming the heart and soul of advanced military science. It constitutes at once the threat and the defense of man's existence on earth."

More so than President Eisenhower or even Lyndon Johnson, Kennedy was a prestige maven when it came to space-related issues: it was about *winning* boasting rights. Refusing to be first in space, JFK would say, telegraphed the wrong signals to Third World countries debating the political virtues of democracy over communism. Given this beat-the-Soviets attitude, he offered blanket endorsements of all things NASA related and marketed a doomsday scenario due to the missile gap with the Soviets. According to Kennedy, the USSR could destroy "85 percent of our industry, 43 of our 50 largest cities, and most of the nation's population." As historian Yanek Mieczkowski explained, for Kennedy the term *missile gap* encompassed "Sputnik, the Gaither Report, military decline, vanishing prestige, and deep-seated worry that the U.S. under Eisenhower had reached second-place status." When convenient, Kennedy used *missile gap* as a general term for the chronic Eisenhower malaise for falling behind the Soviets. But he knew he also needed an optimistic catchphrase in which to bundle his better-days-are-a-comin' rhetoric. As NASA stories circulated in the public press, he circled around "the New Frontier" as his uplifting New Deal/Fair Deal–type moniker.

Space and ballistic missile technology became the rage in the 1950s and 1960s. Here radar echoes were absorbed in an anechoic chamber so that engineers could bounce echoless beams off the nose cone of an ICBM model.

8

Mercury Seven to the Rescue

Greetings, my friend. We are all interested in the future, for that is where you and I are going to spend the rest of our lives.

—CELEBRITY PSYCHIC CRISWELL, AT THE BEGINNING OF THE ASTRO-FICTION FILM *PLAN 9 FROM OUTER SPACE* (1959)

The thought of manned spaceflight, especially of landing on the moon, seized the American imagination in the days of consternation following *Sputnik*. Once NASA was established, astronaut mania swept the land. Although the timing of the Mercury rollout emanated from Cold War concerns, the public enthrallment with space grew as much out of its frontier heritage and football fanaticism. In popular publications, test pilots were called "space cowboys" or "space jocks." Space exploration was marketed by *Collier's* and *National Geographic* as a heart-racing adventure performed by brave test pilots willing to risk their lives to be pioneers in space. The days of the Wrights' lowly twelve-horsepower engine had been replaced by the loud thrust of space-bound rockets. Questing held a "mystical lure of the unknown," historian Ray Allen Billington wrote about postwar America, because it answered the "call of the primitive, the dominance of the explorer impulse."

NASA wasn't inventing the notion of the space frontier in the late 1950s for it was already part of the national DNA. Somehow going to the moon seemed to be part of America's destiny. Science-fiction novels were called "space-opera Westerns," and Disneyland in California had Frontierland (Wild West) next to Tomorrowland (space). When von Braun contributed a series of articles to *Collier's*, his first effort was titled "Crossing the Last Frontier." In the wind-up to announcing the Mercury Seven astronauts NASA used words such as *frontier*, *adventure*, *pioneer*, *challenge*, and *explorers* to stoke the public enthusiasm. Designer David Clark and pilot Scott Crossfield convinced NASA to make the astronaut space suits silver to give them a futuristic look.

While dozens of top military aviators made the NASA "consideration" list to become Mercury astronauts in early 1959, only seven would be selected. Those eventually chosen to "conquer space" needed three characteristics shared by Daniel Boone, Davy Crockett, Theodore Roosevelt, Jimmy Doolittle, and Charles Lindbergh: drive, self-reliance, and guts. (The only European NASA regularly evoked in its public relations blitz was Christopher Columbus, an honorary American for "discovering the New World" in 1492.)

Because the winning seven would be going into space *alone*, in one-man capsules, the phenomenon reminded some citizens of Buck Rogers and the Lone Ranger. For others Mercury astronauts were great team players, like baseball stars Al Kaline and Ernie Banks. NASA basically tested, then overtested, its 160 serious applicants to discover if they had a genuine "pioneering spirit," or "the right stuff," as Tom Wolfe so memorably titled his New Journalism classic about Project Mercury.

On April 9, 1959, the so-called Mercury Seven, the test-proven astronauts chosen after a two-month selection process, were introduced to America at NASA headquarters in Washington, DC. Although Eisenhower had initially been inclined to keep the identities of the seven men low-key, if not secret, NASA's announcement that afternoon became a PR coup, complete with simple but effective stagecraft: as Administrator Keith Glennan addressed a packed press briefing in Washington, DC, a curtain was pulled open, revealing the seven astronauts, all clad in civilian clothes befitting NASA's status as a civilian agency. "It is my pleasure to introduce to you," Glennan said, "Malcolm S. Carpenter, Leroy G. Cooper, John H. Glenn, Jr., Virgil I. Grissom, Walter M. Schirra, Jr., Alan B. Shepard, Jr., and Donald K. Slayton . . . the nation's Mercury Astronauts!" When asked who wanted to be the first space traveler, each man raised his hand, eliciting loud laughs even from the hard-bitten reporters.

Appealing to a youthful audience hungry for adventure, the handsome astronauts—three from the air force (Cooper, Grissom, and Slayton), three navy (Carpenter, Schirra, and Shepard), and one marine (Glenn), all possessing stoically all-American faces atop lean and rangy frames—received a standing ovation and an outpouring of adulation, becoming overnight heroes.

Their camaraderie was palpable. Ranging in age from Glenn (at thirty-seven) to Gordon Cooper (thirty-two) and all standing shorter than five feet eleven, they were almost interchangeable. All were white and male—a given in this chauvinist, pre–1960s civil rights/women's movement era. All had a patriotic avidity for space adventure, held college diplomas (Glenn using the combination of Muskingum and Patuxent as a fudge); were seasoned jet test pilots, with a proven record of aviation proficiency; could barrel-roll or figure-eight loop; knew aircraft mechanics inside out; possessed an unwavering devotion to beating the Soviets; and had the mental and physical requirements to handle zero gravity. Even though pilots were killed in crashes caused by mechanical or structural malfunctions, none of the chosen astronauts obsesssed about mortality. They were masters of the sky, prepared to be masters of space. "We didn't know what was going to happen," Glenn later recalled. "We were making up the music as we went along."

The press gushed enthusiasms for these new instant space cadet heroes in no uncertain terms. Leading the charge was the *New York Times*'s James "Scotty" Reston, who was enthralled by all things Project Mercury. "Those gloomy students of the American character who think we've lost the hop on our fast ball should have been around here this week when seven young American men dropped into Washington on their way to outer space," he marveled. "Somehow they had managed to survive the imagined terrors of our affluent society, our waist-high culture, our hidden persuaders, power elite and organization men, and here they were, aged 32 to 37 and all married, in the first stages of training for the first manned rocket flights into space. . . . [W]hat made them so exciting was not that they said anything new, but that they said all the old things with such fierce conviction."

Among the other shared attributes of the "Magnificent Seven," as the press soon dubbed them, was their ability to survive Dr. Randy Lovelace's endurance tests in New Mexico, which included swallowing a two-foot rubber hose, parachuting at night, cycling in place past the point of lassitude, and having jets of ice water gushed into their eardrums at ten-second intervals. Apollo 11 astronaut Michael Collins best described the Lovelace experience as being "poked, prodded, pummeled, and pierced" in a hellish torrent where "no orifice is inviolate, no privacy respected." Psychologists also administered thirteen varied "personality and motivation" tests to the chosen astronauts. And at the Lewis Research Center, the

Mercury astronauts completed disorientation flights on a three-axis space simulator.

What made the Mercury Seven story so powerful was the bedrock faith the men had in one another. Even though these aviators were all super-achievers with competitive drives and immoderate egos, they bonded like brothers. To avoid duplication, each of the men took on special responsibilities. Cooper and Slayton mastered the art of booster-monitoring the Redstone rockets (army missiles built by Chrysler) and the Atlas (air force missiles by Convair). Shepard put his navy background to work interacting with the branch's spacecraft recovery forces, based in Norfolk. The most trusted astronaut on cockpit layout issues was Glenn. Always tinkering with gadgetry, Grissom was the flight control maestro. Schirra was responsible for the life-support systems, including oxygen intake procedures while in space. And Carpenter, who liked radio communications, was the chief navigator of the Seven. "I'd go so far as to say that the most significant achievement of the space program was a concept of teamwork," Schirra believed. "A guy like Chuck Yeager is thus really out of place in my profession. I hesitate to snipe at Yeager, but he asks for it. He boasts about not being a team player."

The public ate it up. Barely a week went by without

a major story praising the Mercury Seven and predicting grand American achievements in space. *Life* magazine bought the exclusive rights to their personal stories. There was great national pride in the openness of NASA compared with the secrecy of the Soviet space program, and a sense of shared adventure that trickled down to the army of technocrats, physicists, engineers, and rocket scientists underpinning the Mercury program. In the culture at large, space was the place. Jules Verne's *From the Earth to the Moon* became popular again. Architect and designer Eero Saarinen, best known for the Gateway Arch in St. Louis, created curvy womb chairs that made sitters feel space-capsule snug. The aesthetic interior of new art museums, the so-called ice-white cube look, grew out of NASA culture. "Scientist alone is true poet he gives us the moon," Beat Generation bard Allen Ginsberg prophesied in *Kaddish and Other Poems.* "He promises the stars he'll make us a new universe if it comes to that."

With the media frenzy came rekindled interest in the life and legacy of Dr. Robert Goddard, whose patented innovations had been used in designing engines for the Atlas, Thor, Jupiter, Redstone, and Vanguard rockets—the oomph that boosted NASA to space. On May 1, 1959, the NASA facility in Greenbelt, Maryland, was named the Goddard Space Flight Center in long-overdue appreciation of this undersung genius. As Goddard's biographer Milton Lehman put it, the rocketeer had "opened the door to the Space Age."

The von Braun team in Huntsville was ecstatic about Project Mercury, which would fulfill their long-held dream of achieving manned spaceflight. While NASA officials thought of the Mercury Seven astronauts as dexterous test pilots in silvery flight suits, von Braun saw them as field scientists exploring the contours of outer space, courtesy of his Mercury-Redstone launch vehicles. "Man is still the best computer that we can put aboard a spacecraft," he said, "and the only one that can be mass produced with unskilled labor." But he also knew the Seven needed to be supported by technology that didn't yet exist. For instance, he envisioned a control protocol whereby astronauts circling Earth could communicate with the ground via a global network of NASA tracking systems; this was soon achieved. Computers in 1959 existed only as huge mainframes that filled rooms. What von Braun envisioned, and NASA contracted from IBM, were smaller computers for the Goddard Space Flight Center that would provide "mission critical" data analysis for NASA in a hurry. These new IBM transistorized computers helped NASA determine "trajectory dynamics" during the launch and early orbit phases.

On the day the Mercury astronauts were introduced, John Kennedy was giving a speech in Milwaukee, but no distance from Washington could erase the impact Mercury would have on both the country and his own personal brand. Space was America's Cold War Manifest Destiny, and the Mercury astronauts were its rough-and-ready trailblazers, following in the footsteps of Kennedy's own World War II generation and almost two centuries of American adventurers before. Later that spring, journalist Ben Bradlee, JFK's Georgetown neighbor, wrote in his private diary that "Kennedy identifies enthusiastically with the astronauts, the glamour surrounding them and the courage and skill it takes to do their jobs." While Kennedy the seafarer identified with the mythos of the Mercury Seven, Kennedy the politician understood that to be identified with them, to be part of that magnificent fraternity, would be a bonus in his pursuit of the presidency.

On the other side of the political aisle, Vice President Richard Nixon was raising his own presidential stock by getting up in communism's face. In late

July 1959, while viewing a display of modern kitchen conveniences at the American National Exhibition in Moscow, Nixon goaded Soviet premier Khrushchev into a schoolyard quarrel on the virtues of democracy versus communism, an exchange that became known as the Kitchen Debate. Two months later, while touring the United States for thirteen days, Khrushchev was denied access to Disneyland, where Tomorrowland space rides were popular attractions with thrill-seekers.

As their leaders squabbled and positioned, scientists advanced. On January 2, 1959, the first of the Soviets' *Luna* (or *Lunik*) spacecraft was launched from Baikonur Cosmodrome in Kazakhstan, on a trajectory intended to impact the moon. A malfunction caused *Luna 1* to miss by some 3,600 miles, but the Soviets recast the mission as a success for becoming the first man-made object to escape Earth's gravity and enter into a heliocentric orbit. Eight months later, on September 14, the three-thousand-pound *Luna 2* probe completed its predecessor's mission by crash-landing between the Archimedes and Autolycus craters on the lunar surface, becoming the first man-made device to connect with another planetary body. The event prompted elation from people around the world; for instance, the *New York Times* treated the news as the biggest story of the day. The data in those signals showed, among other things, that the moon has no significant magnetic field.

A month before *Luna 2*, the United States had scored its own coup when the newly launched *Explorer 6* satellite sent back the first-ever photograph of Earth from orbit, showing a sunlit area of the central Pacific Ocean from an altitude of 17,000 miles. Two months later, the Soviet *Luna 3* made further history by sending back the first photographic images of the far side of the moon. Meanwhile, operating away from public glare in Huntsville, von Braun had developed a rocket whose first stage could deliver 1.5 million pounds of thrust—an amazing start for a possible American moonshot. He called the rocket Saturn.

In Washington, the Eisenhower administration continued to push for global "freedom of space," conducting U-2 reconnaissance flights over the USSR, and gearing up NASA for manned-space exploration. The president's science advisors—prominent experts such as Caltech president Lee DuBridge, MIT president James Killian (who preferred robots to men in space), and engineer and inventor Vannevar Bush (who told a congressional committee that rockets couldn't span oceans)—warned him not to be goaded by the Kremlin into rushing the manned spaceflight program. The administration's methodical approach soon became fodder for Kennedy on the campaign trail, where he dismissed Eisenhower, Nixon, Bush, Glennan, Killian, and DuBridge as flat-out behind the times. America, JFK was soon saying, had a "space gap" with the USSR, which in the next year would choose twenty "cosmonauts" for its own manned spaceflight program, almost triple Project Mercury's total. Under the continuing leadership of rocket designer Sergei Korolev, Soviet engineers had also devised a completely automated spacecraft in which a cosmonaut would ride as a passenger instead of as an active pilot.

Kennedy essentially agreed with a snarky *Newsweek* article that mocked Eisenhower's space policy recipe: "start late, downgrade Russian feats, fragment authority, pinch pennies, think small, and shirk decisions." Advanced technology, he believed, was a primary indicator of the economic health of a nation. While Ike methodically slow-walked into the future, Kennedy wanted a decisive American victory in space. As a senator, he pushed for NASA's getting caught up with Russian space technology. Not yet saying directly that America would put an astronaut on the moon by 1970, he nonetheless guaranteed crowds that if *he* were U.S. president, the nation would not just be "first but, first and, first when, first if, but first *period*."

Wives of the seven Mercury astronauts became NASA celebrities in their own right. *Top row:* Jo Schirra and Louise Shepard. *Middle row:* Annie Glenn and Marjorie Slayton. *Bottom row:* Trudy Cooper, Rene Carpenter, and Betty Grissom.

Senator John F. Kennedy during his 1960 campaign for the U.S. presidency.

9
Kennedy for President

The American, by nature, is optimistic. He is experimental, an inventor and a builder who builds best when called upon to build greatly.

—JOHN FITZGERALD KENNEDY ANNOUNCING HIS CANDIDACY FOR PRESIDENT OF THE UNITED STATES (JANUARY 2, 1960)

When John F. Kennedy formally announced his candidacy for president of the United States on January 2, 1960, he boasted that he would "rebuild the stature of American science and education." He showcased himself as the oracle of public education. There was genuine excitement in the Democratic Party that the precocious Kennedy was a relatively fresh face. Party leaders relished the distinction of having a presidential aspirant under fifty. This was also welcome news at NASA, which was struggling to move out of what it called its "formative stage, accelerating its space research and development." In his more technology-oriented speeches in early 1960, Kennedy came across as a potential inspirational president willing to prioritize ICBMs and NASA funding more vigorously than Eisenhower had.

A strange sense of momentum and destiny swirled about Kennedy. Hugh Sidey of *Time* called him "a serious man on a serious mission." Extremely in demand and the darling of the press, JFK was receiving five hundred speaking invitations per month, according to his secretary, Evelyn Lincoln. His visionary rhetoric about innovation and technology threw into sharp relief the more staid approach of Vice President Nixon, who was gunning for the Republican nomination. Joseph Kennedy Sr. bought a Convair 240 twin-engine airliner for Jack's personal use, marking the first time a White House aspirant had a plane at his constant disposal. Named the *Caroline*, after Jack and Jackie's two-year-old daughter, JFK's Convair became an extension of his personality, helping him cover some one hundred thousand campaign miles. (Unbeknownst to

Kennedy, rocketeer extraordinaire Hermann Oberth was then working as a technical consultant for Convair, on the Atlas rocket program. Von Braun befriended his old hero, who was on loan to the United States from West Germany, even coaxing him to join the Huntsville rocket team at the Marshall Space Flight Center, for a stint working at the Space Sciences Laboratory there.)

The *Caroline* gave Kennedy an edge over his political rivals, letting him choose his itinerary based on political benefit rather than train and airline schedules. He was helped by a top-notch campaign staff that mixed battle-proven Democratic operatives with Ivy League academics, all held together by the steely ambition of two generations of Kennedys and their spouses. Also, the public hadn't missed the fact that Kennedy's focused, well-organized campaign had essentially begun with his 1958 Senate reelection, an extreme rarity in a time when campaigns typically began within the same calendar year as Election Day.

Although, in the past, sitting U.S. senators had had a hard time jumping from Capitol Hill to the White House—only one, Warren Harding, in 1920, had successfully made the journey—Kennedy's three major rivals for the Democratic nomination were all among his Senate colleagues: airpower maven Stuart Symington of Missouri, liberal Hubert Humphrey of Minnesota, and political operator Lyndon Johnson of Texas. Also hovering around the race was the 1952 and 1956 nominee, Adlai Stevenson, who, though not actively seeking the nomination, let it be known that he'd accept a convention draft. Kennedy himself leaned Cold War center-left—more Scoop Jackson than Adlai Stevenson—but his true ideological home was simply the winners' club; he had never lost an election. While Symington, on moral principle, refused to embrace racism and flatly refused to speak to segregated audiences in the South, Kennedy would and did. Like FDR before him, JFK didn't want to lose rural white Southern voters over civil rights, so he pandered around the issue rather than risk alienating too many white Protestants in the region. Later in 1960, though, he telephoned Martin Luther King Jr. in jail, sending a strong symbolic signal to African Americans that he was on their side of the freedom struggle.

Yet, throughout hundreds of resolute speeches and rah-rah articles, what made Kennedy stand out among all the Democratic candidates of 1960 was his streak of vivid Catholic idealism, which shone brightly enough to mask ideological divisions. Right-wing voters could focus on Kennedy's anti-Soviet "missile gap" tough talk, while more liberal voters heard his high-minded defense of New Deal programs such as Social Security.

Another of JFK's gifts was his ability to attract smart people to serve as personal assistants and aides. First and foremost was Theodore Sorensen from Nebraska, whose family had a distinguished history in progressive politics. Handsome, quick-minded, and a brilliant speechwriter, Sorensen knew how to make campaigning fun for Kennedy. He was also, incongruously, a bookish intellectual mired in the carnival of grassroots Democratic politics, an erudite fix-it man with an ironic sense of existential philosophy, and one of the few people who knew about Kennedy's health problems, including his two major spine operations in the mid-fifties and his use of steroids to combat his Addison's disease.

Kennedy was thoroughly in his element on the campaign trail. His well-crafted eloquence, private plane, picture-perfect family, fund-raising chops, circumlocutory manner, singlemindedness, and willingness to speak anywhere reflected a strategy that separated him from the other Democratic senators who were running. Compared with the other professional pols, JFK was a breath of fresh air—unflappable, graceful, and comfortable around people. Even though he was often in the throes of health-related pain, the demands of his twelve-hour workday never debilitated him. "Jack was always out kissing babies, while I was passing bills," Lyndon Johnson crossly complained, "including his bills." Later in life, Johnson, in an oral history, censured Kennedy for being a "pathetic" congressman and senator who had mastered the suspect art of purposelessness. It puzzled the driven Texas operator how Kennedy could be so detached and yet so beloved.

Johnson had a point: Kennedy's lack of a legislative record was his biggest liability during the 1960 campaign. But LBJ was wrong that JFK (whom he regularly referred to as "the boy") was a lightweight. Kennedy was a resolute man on a mission, and while Johnson, Humphrey, and Symington might have been better senators, Kennedy was the better presidential candidate. Like Elizabeth Hardwick said of philosopher William James, Kennedy was a "sort of Irishman among the Brahmins," which allowed him to be apropos on every occasion. And although Adlai Stevenson still had an adoring Democratic following, there was a palpable sense that his time had passed; by contrast, JFK was the flush, exuberant candidate of the bold future.

As 1960 began, NASA was still trying to fully absorb the army's Saturn rocket program in Huntsville. Although Eisenhower had ordered the transfer of von Braun and his rocketeers from the army to NASA, resentment abounded. Wanting to avoid brawls with the service branches, NASA administrator Keith Glennan cultivated bonds with the air force. By favoring the air force publicly during a time when NASA's Mercury program was dominating the headlines, Glennan made the army jealous enough to begin to fall in line.

Recognizing that his fate was in the hands of NASA, von Braun visited Glennan at his Washington home and aggressively prodded his new boss to adopt his ambitious plan of going to the moon in a three-stage rocket. He was envisioning a Saturn rocket that would dwarf other rockets in complexity. Glennan was vague and reserved, giving von Braun general assurances. "Wernher finally ended the discussion," Glennan recalled, "by saying, 'Look, all we want is a very rich and very benevolent uncle.' What a personality!" Glennan promised von Braun that the rocket programs in Huntsville would be fully funded by NASA, but von Braun still went away wondering just what kind of uncle Glennan would turn out to be. And Glennan saw von Braun as a blowtorch personality hungry to have the army beat the air force in interrelated fields of ballistic missiles, satellites, and space vehicles."I had not realized," a perturbed Glennan wrote, "how much a pet of the Army's von Braun and his operation had become."

Glennan may have been benevolent, but he didn't

want to be profligate, and the idea of planning a moon mission before humans had yet to send an astronaut into space smacked of the kind of Cold War one-upmanship he distrusted. Like Eisenhower, Glennan preferred to advance more slowly and methodically, funding programs that offered "intrinsic merit" rather than showering money on fast-track programs that promised the United States a chance to be "first." It was this kind of conservative approach that persuaded Ike, on January 14, 1960, to grant top priority to "high-thrust space vehicles."

In early 1960 the term *New Frontier* was ubiquitous in space-related television and print stories. It floated around, and Kennedy grabbed it as his own. NASA publicists were successfully shopping the idea of selling Project Mercury as a Western frontier opera. "I am profoundly worried," von Braun told the *Washington Post* in a March 1960 article, "as to what has happened to the American frontier spirit." On the NBC-TV program *World Wide '60*s, a May 1960 segment "Report from Outer Space," Glennan seized upon the Wild West analogy, saying that "space is the greatest new frontier to be breached by man in over four hundred years. Backing away from the opportunity would be a denial of our heritage." That kind of Glennan "new frontier" space rhetoric appealed mightily to Kennedy in the throes of an election season. Not that Glennan had coined the term *New Frontier* in politics. Back in 1934, then-secretary of agriculture Henry Wallace published an economic manifesto titled *New Frontiers*; he argued that the "old frontier" of raw individualism had to be grown into a "new frontier" of "cooperation" anchored around large projects such as the Boulder Dam and Tennessee Valley Authority. By choosing the singular "New Frontier" as his own overarching campaign slogan, Kennedy was cleverly giving a nod to Wallace, who had run for president in 1948 as the Progressive Party nominee. At the same time, the term had futuristic techonology and innovation connotations, something that appealed to those ready to beat the Soviets in space exploration.

Exhibit A of how Kennedy processed space imperatives came in a serendipitous way. In February 1960, William Everdell, a Princeton freshman who described himself as a Republican, wrote Kennedy to ask for his solutions to the mixed messaging in America's space program. In the midst of his perpetual campaign, Kennedy normally didn't have time to answer letters personally, but while traveling in Wisconsin, he composed a detailed response to Everdell, who later became a professional historian.

In a humorous opening, Kennedy thanked Everdell

for a letter that stood out because of his "undeviated Republicanism, Princetonian self-assurance and uncomplicated handwriting." Then the letter addressed the serious issues the college student had raised:

> *Whatever the scale and pace of the American space effort, it should [be] and is a scientific program. In this interval when we lack adequate propulsion units, we should not attempt to cover this weakness with stunts. And when this weakness is overcome, our ventures should remain seriously scientific in their purposes.*
>
> *Since the exploration of space is, scientifically, a relatively new venture, it is rational to expect payoffs we cannot calculate, as in the early stage of any major scientific breakthrough. This has two consequences. First, the basic scientific component of our program should be financed and encouraged to the hilt. It is out of the work in basic research that possibilities of leap-frogging the Russians are likely to emerge. And without leap-frogging I fear we shall be getting their exhaust in our face for quite a long time. Second, projects for exploration should at this early stage be viewed with a bias toward hope rather than skepticism. We can count on good payoff from a high proportion of our probes, at this*

stage. Thus, on a scientific basis alone, the program should be generously financed.

With respect to the competitive and psychological aspects of the space program, it is evident that we have suffered damage to American prestige and will continue to suffer damage for some time. But, our recent loss of international prestige results from an accumulation of real or believed deficiencies in the American performance on the world scene: military, diplomatic, and economic. It is not simply a consequence of our lag in the exploration of space vis-à-vis the Soviet Union. The space lag has in fact, had a disproportionate impact because it is one of a group of lags and gaps.

At the time Kennedy wrote this letter, Eisenhower was considering having NASA announce that a successful manned Mercury mission would close out America's foray into manned space exploration. But as this letter to William Everdell indicates, Kennedy considered Mercury not an end, but a beginning. Incrementalism would keep Americans perpetually behind the Soviets in space, while bold space exploration and the seismic scientific upswings it could engender could be a new polestar of American technological superiority and national prosperity. While others disparaged the space race, JFK embraced it on grounds of national pride. While others concentrated on the military applications of rocketry, he saw the scientific benefit of aiming for the stars. The key was to find the goal that would not merely match but leapfrog the Soviet program.

Unbeknownst to Kennedy, events were about to prove that the United States was not quite as technologically behind as he claimed, and that it might soon be in a position to make a historic leap.

On May 1, 1960, the more clandestine aspects of Cold War strategy—espionage and technology—were laid open to public view when an American U-2 spy plane was shot down in Soviet airspace. The U-2, built by Lockheed in 1955, was a top-secret, ultra-high-altitude single-jet aircraft able to gather intelligence day or night from an elevation of seventy thousand feet, where pilots needed to wear partially pressurized suits able to deliver an oxygen supply, much like astronauts' space suits. U.S. pilot Francis Gary Powers was taking high-altitude reconnaissance photos when the Soviets detected the plane and fired three surface-to-air missiles, one of which hit its mark and caused Powers to crash in the Sverdlovsk Oblast, in the Ural Mountains.

Assuming Powers had not survived the crash, the Eisenhower administration initially tried to cover up the U-2's mission, putting out a press release claiming that the pilot, on a weather-gathering mission in Turkish airspace, had become incapacitated due to oxygen deprivation and had crashed in Soviet territory. Khrushchev soon revealed, however, that Powers was very much alive and had been captured by the Soviets, who also recovered part of his U-2. With egg on his face, Eisenhower was forced to admit guilt while Powers was convicted of espionage at a well-publicized trial in Moscow.

The incident caused already tense U.S.–Soviet relations to deteriorate further, but it also provided clear evidence that the United States was not only keeping up with but perhaps even exceeding Soviet technology. At NASA facilities in Huntsville, Hampton, Pasadena, and Cape Canaveral, engineers cheered Lockheed's triumph in producing an aeronautical marvel like the U-2, and confidence grew that American ingenuity would soon prevail over the Soviets in the space race. The technology, in fact, was already in development. Although Project Mercury was still a year away from its goal of putting a solo American in space using von Braun's Mercury-Redstone launch vehicle, draft-room planning was already under way for a follow-up program that would attempt to launch a three-man team with his new Saturn rocket.

In 1959, von Braun had named his new rocket program Saturn because the army had been forced to transfer what he called "its cherished Jupiter missile" to the air force. "Saturn," he simply said, "was the next outer planet in the solar system." But in naming the proposed manned missions the Saturn rockets would support, Dr. Abe Silverstein, NASA's third in command as chief of Space Flight Programs, had more transcendent themes in mind. Having named Project Mercury after the fleet-footed messenger of the Roman gods, he turned again to a book on Greco-Roman mythology for inspiration, and the winning name jumped out at him: Apollo, the god of music, medicine, prophecy, light, and progress. "I thought the image of the god Apollo riding his chariot across the sun," Silverstein recalled, "gave the best representation of the grand scale of the proposed program." Silverstein proposed the name to his superiors, and on July 28, 1960, NASA deputy administrator Hugh Dryden told a planning conference audience of government, aerospace, and academic representatives that "the next spacecraft beyond Mercury will be called Project Apollo." Over the following days, conference attendees would learn more about Apollo's potential missions, which included ferrying astronauts to a space station, orbiting the moon, and eventually making a manned lunar landing.

In July 1960, Jack Kennedy secured his party's nomination for president at the Democratic National Convention in Los Angeles. Even detractors admitted that the Massachusetts senator had run a flawless, media-savvy campaign, but the bad blood that had developed between him and Johnson on the campaign trail had spilled over into the convention. Suspicious that JFK was spreading rumors that LBJ's serious heart attack in 1955 made him unfit for the White House, the retaliatory Texan tried to beat him to the punch, telling a *Chicago Daily News* reporter that Jack was a "little scrawny fellow with rickets" who suffered from Addison's disease and took regular injections of cortisone—a nasty string of invective, but not entirely inaccurate. Bobby Kennedy, an expert at putting out brush fires, lambasted the Johnson campaign for spreading smears against the *PT-109* hero. The counterattack worked, and LBJ was forced to apologize. With cool efficiency, Kennedy then chose Johnson as his running mate, in part to balance the ticket—the northeastern Irish Catholic offset by the Protestant Texas Hill Country rancher. In personality, education, religion, style, region, and expertise, the two were near-complete opposites, causing soon-to-be Republican nominee Richard Nixon to characterize the relationship as "an uneasy and joyless marriage of convenience." When it came to the importance of the United States' being first in space exploration, however, they were on the exact same page.

In his acceptance speech on July 15, Kennedy declared, with the shining ring of both antiquity and the future, that the "New Frontier" had arrived. Somehow his aspirational words threw open a lever. Unlike the New Deal, his administration wouldn't focus on the federal government's helping people, but rather on people helping the United States achieve new greatness. Kennedy's animating spirit and his call for a federally funded technological revolution boded well for NASA. "The New Frontier of which I speak is not a set of promises," Kennedy said. "It is a set of challenges. It sums up not what I intend to *offer* the American people, but what I intend to ask of them. It appeals to their pride, not to their pocketbook—it holds out the promise of more sacrifice instead of more security."

Whatever the New Frontier would be exactly, it seemed it would be anchored, in part, in a new age of space exploration—a concept that appealed to the younger generation immensely. The Mercury Seven astronauts being ballyhooed in *Life* and *Time* magazines were all relatively young Americans. Southern California, where the Democrats held their convention, had become the epicenter of a burgeoning youth movement that liked rock and roll, surfing, convertibles, and space. Such giant aerospace corporations such as Northrop Corporation (now Northrop Grumman), Douglas Aircraft Company (later McDonnell Douglas), TRW Inc. (now Northrop Grumman), Lockheed Corporation (currently Lockheed Martin), North American Aviation (now Rockwell), and Hughes Aircraft Company (later Raytheon and Boeing) were based in the metropolitan Los Angeles area. Pasadena was also the home of Caltech (the Jet Propulsion Laboratory), while Santa Monica was where the RAND think tank was headquartered.

And Kennedy himself was the Hollywood-like leader of a new generation that was starting to dominate American society. Surrounding Kennedy was a team of action-oriented intellectuals, mostly in their thirties, like speechwriter Ted Sorensen, Robert F. Kennedy, Harris Wofford, Pierre Salinger, and Ken O'Donnell. Like the Mercury Seven astronauts, the Kennedy team exuded a sense of self-possession, glamour, celebrity, and patriotism, and embodied the political rise of the World War II junior officers. "Even before Kennedy took office," historian William Leuchtenburg noted, "they were all riding high, those days, on their own bravado, their own idealistic temerity of purpose, young people had begun to appropriate him for their own, and he, in turn, had shown an affinity for them."

Science and space were crowd-pleasers in Los Angeles, and they continued to play a role in the campaign after the Republicans, at their convention in Chicago, nominated Richard Nixon, as expected. Nixon in turn selected Henry Cabot Lodge Jr., Kennedy's former opponent from Massachusetts, as his running mate. With both tickets supporting the already popular NASA, space exploration itself was not a particular bone of contention, but the optics of the space race were, with Kennedy declaring that the United States had lost its lead due to Eisenhower's and Nixon's fecklessness.

While the "space gap" would be a Kennedy talking point throughout the campaign, it took a backseat to the alleged "missile gap" that he continued to hawk. In late July, CIA director Allen Dulles had had enough. Presenting Kennedy (July 23) and Johnson (July 28) with classified intelligence gathered from U-2 reconnaissance and on-the-ground sources, he assured them that the missile gap was a Cold War myth, and that the United States was in fact considerably ahead of its enemy in ICBMs. Kennedy, however, chose to disregard CIA and Pentagon evidence and continued mining the missile gap for political advantage and to advocate for accelerated deployment of Polaris and Minuteman missiles. Nixon and Lodge, who had received the same briefing from Dulles following their nomination, fumed. "I could expose that phony in ten minutes by displaying our high-altitude photographs and explaining the quality of information we are getting," Nixon raged. "I can't do that without destroying our source, and Kennedy, the bastard, knows I can't."

According to Ted Sorensen, soon after the convention, Kennedy read Allan C. Fisher Jr.'s article "Explaining Tomorrow with the Space Agency" in *National Geographic.* While much of the article evaluated NASA's accomplishments over its two years of existence, there were prose riffs about how space was the "strongest of oceans" and Earth a "relatively small island" in a "solar sea," which Kennedy liked. "This is year three of the Space Age, and a vast new environment, the solar system," Fisher wrote, "lies open to mankind's assault." Although Kennedy and Sorensen didn't get to write a speech premised on the *National Geographic* article, they would use it down the line as a conceptual springboard for New Frontier space speeches. In particular, JFK was attracted to a story von Braun told Fisher about Benjamin Franklin watching the rise of history's first gas-filled balloon in Paris. A skeptic asked Franklin what possible use the balloon was to mankind. Franklin answered, "What good is a baby?" When thinking about NASA's justification for going to the moon, Kennedy would sometimes retell this Franklin anecdote.

On the campaign trail, the forty-three-year-old Kennedy used space the same way he used the missile gap: as a metaphor for American technology falling behind that of the Soviets. "The people of the world respect achievement," Kennedy told a crowd in Portland, Oregon, on September 7. "For most of the twentieth century they admired American science and American education, which was second to none. But they are not at all certain about which way the future lies. The first vehicle in outer space was called *Sputnik*, not *Vanguard*. The first country to place its national emblem on the moon was the Soviet Union, not the United States. The first canine passengers in space who safely returned were named Strelka and Belka, not Rover or Fido, or even Checkers," the last a reference to Nixon's infamous cocker spaniel.

Forced to defend the Eisenhower administration's technological accomplishments without benefit of its best intel, Nixon instead touted NASA's successful 1959 launch of *Echo 1*, a one-hundred-foot, self-inflatable aluminized balloon satellite. A predecessor of active-repeater communications satellites like 1962's *Telstar*, *Echo 1* functioned as a kind of orbital mirror, bouncing radio-television beams off the Earth's surface to facilitate long-range communications. While a successful idea, *Echo 1* paled by the standards of mid-

1960, when the Soviets launched *Korabl–Sputnik 2* and its canine passengers, the aforementioned Belka and Strelka. Unlike their martyred predecessor, Laika, these Russian dogs survived, returning to Earth after a day in space, during which they were heard barking as they apparently spotted *Echo 1* through a window. One year after the mission, Strelka gave birth to a litter of puppies, one of whom, named Pushinka ("Fluffy"), was gifted to Jacqueline Kennedy as a flattering gesture of goodwill by Khrushchev.

While Kennedy was able to distort the facts to support his missile gap and space gap stances, he had less initial success in addressing the fact of his religion. Bigots from coast to coast argued that a Catholic in the White House wouldn't uphold America's constitutional separation of church and state, instead placing the dictates of the Vatican above the national interest. The Minnesota Baptist Convention declared that both Catholicism and communism were "serious threats" to America. Dr. Norman Vincent Peale, author of *The Power of Positive Thinking* and a crony of Nixon's, told *Time* that having a Roman Catholic as president was unacceptable. To rebut these smears, Kennedy spoke on September 12 before the Greater Houston Ministerial Association, insisting that his Catholicism shouldn't be a major issue in the election. Pointing out that nobody had cared about his religious affiliation when he served in the navy during World War II, Kennedy pivoted to say that too many American "old people" couldn't "pay their doctor bills" and families were being forced to "give up their farms." Brushing aside the religion issue, Kennedy then turned visionary, saying that he saw an America "with too many slums, with too few schools, and too late to the moon and outer space." The speech was televised live and proved to be a campaign turning point.

On stump speeches around the country, Kennedy spoke of leadership being about unruffled boldness and tenacious daring for greatness. "I am tired of reading every morning what Mr. Khrushchev is doing," Kennedy said in Syracuse, New York. "I want to read what the President of the United States is doing." Singling out Nixon by name, JFK lamented that Republicans were tepid on funding technology gambits. "We have been repeatedly reassured by Mr. Nixon—in glowing, sugar-coated terms—that we have nothing to worry about in arms, science and space," Kennedy observed. But Nixon, he said, had "a tendency to react instead of act."

During a series of four televised debates—the first presidential debates *ever* in American history—both candidates promised a new era and a sharp turn from

Eisenhower's methodical style, but Kennedy seemed to embody change. While Nixon presented himself as an old-style politician of finesse, JFK appealed to voters who wanted a man of action, in touch with the moment and possessed of the charisma and vision to move democracy forward by thwarting communism. "I look up and see the Soviet flag on the Moon," Kennedy goaded Nixon at the October 21 debate. "Polls on our prestige and influence around the world have shown such a sharp drop that up till now the State Department has been unwilling to release them." Pushing this theme of GOP complacency further, Kennedy mocked the vice president as a weak-kneed Cold Warrior. "You yourself said to Khrushchev, 'You may be ahead of us in rocket thrust, but we're ahead of you in color television,'" Kennedy chided Nixon, hoping to get a rise. "I will take my television in black and white. I want to be ahead of them in rocket thrust."

Optics mattered in the end. A majority of Americans watching on TV believed the calm and collected Kennedy won the debate; Nixon, by contrast, had kept glancing at the camera, as if it were an invader. Those listening on the radio gave the nod to Nixon, where his makeup-free, sweaty lip didn't offend.

The palpable enthusiasm for Kennedy continued through October, yet no one could predict what would happen on Election Day. Kennedy unleashed Lyndon Johnson to sharply criticize the Republicans' downgrading of space through incompetence, complacency, foot-dragging, indifference, and budgetary restrictions. Speaking as both the Democratic vice presidential candidate and chairman of the Special Committee on Aeronautical and Space Sciences (aka Space Committee), Johnson charged that it was only the Democrats' ceaseless prodding that had brought NASA to the brink of success with Project Mercury and other endeavors. "The Republican presidential candidate has not at any time assisted in this prodding," LBJ said of Nixon. "It was only when he himself was prodded by the forthcoming election that he recognized publicly the importance of active work in this field."

On Election Day, polls showed the two candidates in a dead heat. Votes were counted long into the night. Eventually, Kennedy won the nail-biter with 303 electoral votes to Nixon's 219, and by only 0.1 percent of the popular vote. The next day, NASA administrator Glennan lamented in his diary, "It seems to be all over," knowing full well that JFK would want one of his New Frontiersmen to lead NASA. That November, Glennan continued his work at NASA, scheduling research and rocket launch dates, and awaited the new president's first move on the civilian space program.

The four televised Kennedy-Nixon debates of October 1960 were watched by millions of Americans. On October 21, the upstart Kennedy chided Vice President Nixon that the United States was losing the space race to Khrushchev. "I look up and see the Soviet flag on the moon. Polls on our prestige and influence around the world have shown such a sharp drop that up till now the State Department has been unwilling to release them."

PART III
Moonbound

Ham, the astrochimp, in 1961, celebrating his space feat.

10
Skyward with James Webb

Webb's legacy is measured by Apollo, which stands today as a virtually unparalleled example of U.S. technological brilliance, as the moment when the human species took its first journey to another world.

—W. HENRY LAMBRIGHT, *POWERING APOLLO: JAMES E. WEBB OF NASA* (1995)

The poet Marianne Moore famously said that a poem should not *mean* but *be*. That was John F. Kennedy's approach to his inaugural address, one of the most memorable in American history. En route from Palm Beach to Washington that January 1961, Kennedy worked with Ted Sorensen to draft, on a yellow legal pad, a new opening for his upcoming speech. Once at his Georgetown house, he grappled with polishing the poetic words that would illustrate the passing of the torch from a generation of leaders born in the nineteenth century to one that embraced the new Space Age.

Inauguration Day dawned with a frigid wind ripping through the nation's capital, the ground covered in snow from a storm the previous night. But as if on cue, the sun shone boldly down as a coatless Kennedy took the oath of office as the thirty-fifth president. The youngest man ever elected to the White House, he looked vibrant and vigorous as he stood to speak, confident in the power of oratory to move the masses toward public service. Kennedy's high-toned thirteen-hundred-word oration lived up to expectations. "Let the word go forth from this time and place, to friend and foe alike," he proclaimed, "that the torch has been passed to a new generation of Americans—born in this century, tempered by war, disciplined by a hard and bitter peace, proud of our ancient heritage—and unwilling to witness or permit the slow undoing of those human rights to which this nation has always been committed, and to which we are committed today at home and around the world."

The most memorable line in the address was arguably "Ask not what your country can do for you, ask what you can do for your country." This urging of a new sense of national service and sacrifice was nothing short of magical. Speaker of the House Sam Rayburn boasted that the speech was "better than anything Franklin Roosevelt said at his best—it was better than Lincoln." Having proudly voted for Kennedy, von Braun, who attended the inauguration, was transfixed by the powerful oratory. Watching on TV sets in Cape Canaveral, Huntsville, Hampton, and Pasadena, NASA employees, like everybody else, had their morale uplifted by oratory worthy of Abraham Lincoln. "I was so proud of Jack," Jacqueline Kennedy gushed. "There was so much I wanted to say! But I could scarcely embrace him in front of all those people so I remember I just put my hand on his cheek and said 'Jack, you were so wonderful.'"

The campaign behind him, Jack Kennedy faced responsibility for righting the ship he'd charged Dwight Eisenhower with destabilizing. If there really was a missile gap and a space gap, they were now his problems to fix.

That February, however, the *New York Times*

reported that Kennedy's own defense study had concluded that there was no missile gap with the Soviets. All JFK's taunts at Ike had been overdrawn. The facts were that when Eisenhower left the White House, the United States had one hundred sixty operational ICBMs to a paltry four R-7s in the Soviet arsenal. Kennedy now gladly accepted this reality. Later in 1961, Corona satellite intelligence indicated that Khrushchev, expectations aside, had only six ICBMs. Kennedy and Johnson's campaign swipes that Eisenhower and Nixon were the architects of a national security missile strategy of "drift, delay, and dilution" had clearly been off base. Furthermore, America's three coastal launch sites (Cape Canaveral, Florida; Vandenberg Air Force Base, California; and Wallops Island, Virginia) were already operating around the clock to further spaceflight advancement. Each of these sites had technical facilities, a control center, and the most modern of launchpads.

What Kennedy learned was that Eisenhower had actually done an able job of building up U.S. defenses. When Ike was inaugurated for his first term in 1953, the air force still used piston-driven bombers, and navy strategy focused on basing ships around the Pacific. By the time he handed the reins over to JFK, the United States had developed reconnaissance and communication satellite capabilities. And there were nuclear submarines—including the USS *George Washington,* deployed the very month Kennedy was elected and able to carry sixteen nuclear Polaris missiles. Five generations of rockets—starting with the early Vanguard, and then onward with ICBMs like Atlas and Titan—were born in the Eisenhower years. The army had teams designing heavy-launch Saturn rockets, while the air force, not to be outdone, had made headwind with its Space Launching System (SLS), experimenting with a myriad of launch configurations using solid-fuel boosters and hydrogen/oxygen upper stages. The Strategic Air Command had more than fifteen hundred jet bombers capable of dropping hydrogen bombs on America's enemies. And in just a few weeks, the United States would fly a three-stage Minuteman from Cape Canaveral.

The more Kennedy learned from conversations with top brass at the Pentagon, including Robert McNamara, his new secretary of defense, the more obvious it was that the United States hadn't ceded space to the Soviets in the 1950s. There were WS-117L (for "weapons systems") reconnaissance satellites and MiDAS (Missile Defense Alarm System) missile-detection satellites that allowed America a half-hour advance warning of an incoming Soviet

ICBM attack. There were all sorts of weather satellites (with military applications), and collaborations among NASA, the army, the air force, and the navy that were highly effective. But ways to intercept Soviet missiles and spy on the Kremlin were defensive in nature. Although Kennedy accepted the intelligence findings, he still believed there existed a fierce ongoing battle for global prestige between the United States and the Soviet Union. That's where manned space still mattered. The publicity windfall of a Mercury Seven astronaut in space would be great for America's image abroad. The United States had to lead from strength (or at least the perception of it) to prove to the world that it had the collective will to be the leading space-faring nation.

Nevertheless, the United States clearly was in a far stronger position against the Soviets than Kennedy had alleged, and than much of America feared. From a political perspective, this news was inconvenient for Kennedy—some Republicans faux-congratulated him on ending the missile gap after only eighteen days in the White House. In private, Kennedy would deadpan, "Who ever believed in the missile gap anyway?"

As for space, Kennedy's primary goal was to deny the Soviets any more *Sputnik*-like propaganda victories. But on entering office, he had no clear plan for how to reconfigure NASA to accelerate its progress. For perspective, he turned to Jerome B. Wiesner, director of the Research Laboratory of Electronics at the Massachusetts Institute of Technology (MIT). An electrical engineer who had developed microwave radar and nuclear weapon components during and after World War II, Wiesner became an unpaid advisor to Kennedy following his work on the influential 1958 Gaither Report, lending insight on space, technology, medicine, the environment, and other science-related issues. Shortly after the election, Kennedy asked him to chair an urgent task force to decide NASA's future direction, and to submit his results before Inauguration Day.

The "Report to the President-Elect of the Ad Hoc Committee on Space" (dubbed the Wiesner Report) was handed in on time, on January 10, 1961. In it, Wiesner's task force concluded that Project Mercury was of exaggerated value to the United States and should eventually be discontinued. Convinced that the Soviets had a huge lead on NASA in the man-in-space business and fearful that a hurry-up approach could lead to mission failure and loss of life, the Wiesner Report advised avoiding a "ghastly situation of serious national embarrassment" by utilizing only unmanned probes for future U.S. space exploration. This recommendation against manned space exploration created a seismic wave of dejection within the NASA bureaucracy.

"The Wiesner Report," historian Roger E. Bilstein surmised, "aroused real concern among NASA personnel; there was a definite feeling that the report was neither fair nor carefully prepared." Von Braun was livid over the report's recommendations that the incoming president "stop advertising Mercury as our major objective," and that after the Mercury missions, the manned space program should be discontinued. Why in God's name, von Braun wondered, would the United States allow the Soviets to be the only nation to put men in space and reach for the moon?

Among its recommendations, the Wiesner Report urged replacement of NASA administrator Glennan with a new leader who would focus on the nonmilitary aspect of space exploration. The report was a slap in the face to Glennan himself but also, he believed, to the experts at NASA. Nevertheless, he chose to maintain a gentlemanly silence. When Glennan appeared at a press conference in Chicago, reporters peppered him with questions about the report, but all such attempts, he proudly recorded in his diary, "met with complete failure." What did please Glennan was that Hugh Dryden, his number two and someone in whom he had full confidence, would be staying on as acting NASA administrator until a permanent successor could be found.

Tension at all levels of NASA was causing rifts and bickering, even before the Wiesner Report was officially released. The Mercury Seven worried that they'd just undertaken a grueling yearlong training regimen for nothing. The astronauts couldn't believe NASA was pondering whether humans were needed in space when computers and cutting-edge technology could operate the flights automatically. Adding insult to injury, the Seven were supplanted in the news by a trained ape: a chimpanzee named Ham, who'd been launched into space on January 31, having been trained to pull a lever on cue as a test of an astronaut's (or "astrochimp's") ability to perform tasks in space.

Ham, born in the thick-forested mountains of Cameroon, was trained at NASA's primate program, based at Holloman Aerospace Medical Center in New Mexico (the facility's acronym giving the chimp his name). It was a rather mind-boggling journey for a primate, from the jungles of Africa to the New

Mexican desert to outer space. Ham's sixteen-and-a-half-minute flight aboard a Mercury-Redstone 2 encountered several glitches, including the fact that the spacecraft soared to 157 miles, rather than the intended 115. The mission was termed a success, however, and Ham survived flight impact in pretty good physical shape. While his journey paved the way for a manned flight in coming months, for the time being it only depressed the Mercury Seven to see that a good-natured chimp, one who could communicate through sign language and establish long-lasting friendships, had seemingly proved Wiesner's point that astronauts weren't necessary.

A month after the election, Kennedy had discussed NASA with the one member of his team who was bursting with ideas about space: Lyndon Johnson. Having championed NASA in the Senate during the Eisenhower administration, LBJ was now tasked with finding a suitable successor for Glennan. Kennedy had already delicately sidestepped trial balloons from several quarters, including Johnson himself, that his vice president was available to run the space agency in his spare time. It was one of several balloons floated by the restless veep-elect, who was beginning to feel like not only a caged tiger, but a marginalized one. Another idea involved carving out a role negotiating New Frontier legislation in the Senate—not merely by presiding over the Senate (as is one of the vice president's duties), but by being intimately involved in moving Kennedy's proposals into law. This idea baffled everyone, especially those who had read the U.S. Constitution, and was quickly quashed. Acting to put Johnson's skills to good use, Kennedy made the hard-charging Texan chairman of the National Aeronautics and Space Council (aka the Space Council), where he would have oversight of both military and civilian space activities.

In the wake of the Wiesner Report, the role of NASA administrator was anything but a hot property, and in fact was the last major post to be filled in the Kennedy administration. Many qualified candidates opted to remain in the lucrative private sector rather than accept responsibility for a two-year-old government agency that not only was in transition, but lacked a clear vision or mandate on just where that transition might lead. Johnson struggled to find the right man. In all, seventeen candidates politely declined Kennedy's offer. The eighteenth also wanted to take a pass, but Kennedy wouldn't let him dodge public service.

That candidate was James Webb, a respected fifty-five-year-old North Carolina lawyer and Truman Democrat who knew little about rocketry but a lot about managing big government budgets, being adaptive, and profiting from intersections of the military-industrial complex. Even his detractors knew he was an uncommonly reassuring presence in any situation. Webb was the perfect person to lead America on a path toward what historian Walter A. McDougall has called "technological anticommunism." He had graduated from the University of North Carolina at Chapel Hill in 1928 with an AB degree in education. He served as a pilot in the Marine Corps from 1930 to 1932, then as secretary to U.S. representative Edward Pou. To better position himself for work in government, he enrolled at George Washington University Law School, receiving his JD in 1936.

During the late 1930s and early 1940s, Webb was vice president of the Sperry Gyroscope Company in Brooklyn, New York, supplying flight navigation systems and airborne radar devices to the armed forces and developing lasting friendships with top leaders of the New Deal industrial mobilization order. In early 1944 he was granted permission to leave Sperry to reenlist in the marines, where he became commander of First Marine Air Warning Group (first as captain, then as major). Confidence in Webb's leadership abilities was so great that he was tasked with running the American radar program for the planned invasion of the Japanese mainland, which proved unnecessary after the Hiroshima and Nagasaki bombings.

After the war, President Truman asked Webb to serve first as budget director and then as undersecretary of state for Dean Acheson. In Acheson's opinion, nobody was better at budget issues or big-picture thinking than Webb, though he'd also been responsible for cutting the United States' early ICBM and satellite programs in the spirit of fiscal responsibility. At the end of the Truman administration, Webb left Washington to make a fortune with the oil company Kerr-McGee, in Oklahoma, and became a board member of McDonnell Aircraft.

Seldom does the American system produce such a competent government infighter as Webb. Smart as a whip, liberal in approach, able to see the battlefield of American politics with perspicacity, he was a rare mixture of big-corporate mores, industrial procurement know-how, bipartisan political instincts, good-ol'-boy charm, and budget wizardry, all undergirded by the unimpeachable credentials of a valiant U.S. Marine. Webb had an appealing face, bright blue eyes, broad Southern accent, and a penchant for folksy homilies—

indeed, nothing about him seemed Ivy League; nevertheless, he was often one step ahead of the so-called best-and-the-brightest types that showboated around Washington.

As the Cold War had set in during the Truman years, Webb gained traction in official Washington for clear-mindedness, loyalty, and managerial effectiveness. Gregarious and a natural marketer, he could talk a blue streak with business friends from Raleigh to Reno. If you asked Webb a question, he'd start yapping, answering the original question five minutes later. His chattiness earned him the nickname the "Mouth of the South." While Robert Kennedy mistakenly derided him as a prattling "blabbermouth," just about everybody else in Washington considered him the exact kind of infighter you'd want as a government foxhole ally, or with whom to spend a twelve-hour day and shut the building down late.

In January 1961, the fifty-four-year-old Webb was in Oklahoma City at a dinner honoring his friend Robert S. Kerr, a smooth, old-style oil-and-gas patch power broker who had just taken over as chairman of the U.S. Senate's Space Committee following LBJ's departure to serve as vice president. A petroleum industry millionaire, Kerr was intrigued about the financial aspects of space technology and was a reliable rubber stamp for anything the new vice president wanted done in the Senate. When LBJ had earlier asked Kerr for advice on the best person to helm the U.S. space program, his answer was immediate: James Webb.

At the dinner event, Webb was passed a note asking him to immediately take a call from Wiesner, Kennedy's science advisor. On the phone, Wiesner informed Webb that he was needed immediately in Washington, to meet first with LBJ and then with JFK himself. Uninterested in the NASA job, but feeling hemmed in, Webb dutifully boarded a private plane, which whisked him to the capital that very evening.

Webb profoundly understood the implications of the Wiesner Report, knowing that the fierce battle then under way between Wiesner's skeptical "nay" and Johnson's fervent "yea" could determine the future of the NASA manned space program. When Kennedy met privately with Webb in the Oval Office, he straight-out asked him to head NASA. Webb, like all the other choices, demurred. This time, however, the president refused to take no for an answer, promising Webb that he'd have the power to shape NASA in dramatic ways. NASA, JFK persisted, didn't need a scientist in charge but "someone who understands policy . . . great issues of national and international policy." Kennedy continued: "I want you because you have been involved in policy at the White House level, State Department level." Under pressure, Webb reluctantly agreed. "President Kennedy said, 'I want you for this reason,'" Webb later recalled. "And I've never said no to any President who has asked me to do things."

Many in the space community worried that the Wiesner Report meant Webb would be only a caretaker, slowly backing the United States out of the manned-space business. But they'd misread Kennedy, who, while calling the report "highly informative," chose to just file it away after taking one of its motifs to heart: that failed missions and dead astronauts would be public relations disasters for his administration. Managing that risk meant hiring pragmatic managers, and in that regard, Johnson and Kerr had keen instincts: Webb was the ideal choice for the job, viewing NASA as an extension of the large government projects that had given birth to the atomic bomb, ballistic missiles, and the Polaris submarine. Webb strove from day one to make NASA the "perfect organization," and under his leadership and indomitable salesmanship, it emerged as the most efficient government agency of the 1960s—a well-oiled, high-octane hub where gov-

John F. Kennedy collaborated with James Webb of NASA on prioritizing Project Apollo in the early 1960s.

ernment, industry, and academia worked as a harmonious team. Drawing upon his experiences at the Bureau of Budget and the State Department, Webb became the master of the new field of space age management. Indeed, as Tom Wolfe wrote in *The Right Stuff*, Webb became "one of the ablest and most distinguished of the off-the-ballot politicians, a man who could make bureaucracies run."

Webb took over at NASA on February 14, calming Wiesner-related nerves among the agency's nineteen thousand employees by reassuring them of his commitment to manned space missions. It was music to the ears of Mercury astronaut commander Alan Shepard, who was slated to become the first American in space by mid-March. After conducting a thorough review of the administration's budget and goals, Webb recommended to Kennedy that the NASA allocation be increased by over $300 million in order to fast-track Saturn booster rockets, with the eventual aim of sending Apollo astronauts to go around the moon. This was a bold strategic gambit. Webb understood that the federal budget is not only a practical map of actual expenditures, but also a symbolic one that highlights a White House's priorities. And Webb knew that no matter what JFK said privately about peaceful space exploration, militarization of space was always part of the equation.

According to space historian John Logsdon, it was at that mid-March budget meeting that "John Kennedy, perhaps for the first time . . . had the chance to get a clear picture of the space policy and budget issues requiring his decision." Countering military concerns that setbacks in Mercury testing would reflect badly on the air force (whose Atlas rockets would boost some Mercury modules into orbit), Webb convinced Kennedy that U.S. manned space efforts had to be continued. Tireless Soviet rocketeers such as Sergei Korolev, Mikhail Yangel, Valentin Glushko, and Vladimir Chelomey were working around the clock, preparing to launch a *Vostok 1* cosmonaut in space. With the multidisciplinary game of sending humans beyond Earth's atmosphere already well along, the new president would get pilloried by the public if he abdicated to the Soviets prematurely. Driving his points home, Webb detailed how the space effort was a great global advertisement for American technological prowess and would also benefit the U.S. military establishment—a view shared, he hoped, by Secretary of Defense McNamara, a Harvard Business School MBA and recent Ford Motor Company president who recognized that the manned space program had the potential to produce valuable spin-off technologies and capabilities.

"We feel there is no better means to reinforce our

old alliances and build new ones," Webb told Kennedy. "Looking to the future, it is possible through new technology to bring about whole new areas of international cooperation in meteorological and communication satellite systems. However, the extent to which we are leaders in space science and technology will in large measure determine the extent to which we, as a nation, pioneering on a new frontier, will be in a position to develop the emerging world force as a basis for new concepts and applications in education, communication and transportation, looking toward more viable political, social and economic systems for nations willing to work with us in the years ahead."

Kennedy agreed with the thrust of Webb's wide-angle thinking. He knew, more than any politician of the Cold War era (with the possible exception of LBJ), that the U.S. space race with the Soviets was not only a clash of cultures, economies, and governing systems, but also a challenge to the American democratic way of life. And Americans, he knew, could be motivated by that dare to rise to greatness. Putting an astronaut in space, perhaps even one bound for the moon, was good for the *spirit* of the nation. In Webb, JFK had a tireless advocate who would bring definition to his administration's space policy. Though the new NASA administrator did not know much about astrophysics or rocket payloads, he had a keen grasp of the politics of space that synced with Kennedy's futuristic New Frontier vision. If Webb could run NASA like a well-oiled bureaucratic machine and could maintain friendships with powerful Democratic lawmakers such as Robert Kerr and Albert Thomas, the president would find the right inspirational words to bring the nation along for the great manned space race ride.

On April 12, 1961, Russian cosmonaut Yuri Gagarin became the first human in outer space. President John F. Kennedy was embarrassed that the Soviet feat took place on his White House watch. It helped spur his decision to back a U.S. effort to go to the moon.

11
Yuri Gagarin and Alan Shepard

Everyone has oceans to fly, if they have the heart to do it. Is it reckless? Maybe. But what do dreams know of boundaries?

—AMELIA EARHART

The first hundred days of Kennedy's presidency, while dazzling in style compared with the Eisenhower era, were short on tangible accomplishments, beyond the establishment of the Peace Corps on March 1, 1961. National Security Advisor McGeorge Bundy famously compared the young administration to the Harlem Globetrotters novelty basketball team: a collection of experts capable of anything they chose to do; they only had to decide. Therein lay the difference and the reason that the Kennedy team hadn't "made a basket," as Bundy put it in the April 3 issue of *The New Republic*: they had to decide, with constancy, which policy opportunity to fight for first.

Absent the urgency of a historical moment, we'll never know where Kennedy might have chosen to focus his attentions, because on April 12, 1961, Russian cosmonaut Yuri Gagarin became the first human to journey into outer space. Taking off from Kazakhstan's Baikonur Cosmodrome aboard *Vostok 1*, a two-module spacecraft propelled by a modified version of the mammoth R-7 rocket that had launched *Sputnik*, Gagarin completed a single low orbit with no serious problems, returning to Earth just 108 minutes after liftoff. The bracing *New York Times* headline said it all: "Soviet Orbits Man and Recovers Him; Space Pioneer Reports: 'I Feel Well'; Sent Messages While Circling Earth." With that jarring Soviet success, Kennedy's political honeymoon ended abruptly.

The Soviets had chosen their first cosmonaut wisely. Air force major Gagarin was extremely photogenic, whip smart, fearless, physically fit, and a delightful extrovert with a constant twinkle in his eye. He looked every inch the Soviet hero, with a life story to back it up. Born on a collective farm in the village of Klushino (renamed Gagarin after his death in a 1968 airplane crash) and trained as a foundryman at a steel plant near Moscow, he, along with his family, had been battered by the famine and privation of World War II. After induction into the Soviet Army, he attended aviation school and rose through the ranks of the Soviet Air Force. In 1960, he was among thousands of candidates screened for the cosmonaut program, during which a Soviet Air Force doctor evaluated his personality: "Modest; embarrasses when his humor gets a little too racy; high degree of intellectual development evident in Yuri; fantastic memory; distinguishes himself from his colleagues by his sharp and far-ranging sense of attention to his surroundings; a well-developed imagination; quick reactions; persevering, prepares himself painstakingly for his activities and training exercises, handles celestial mechanics and mathematical formulae with ease as well as excels in higher mathematics; does not feel constrained when he has to defend his point of view if he considers himself right; appears that he understands life better than a lot of his friends." Rising through the selection and training process, he became one of the Sochi Six, an elite, Mercury-like group selected for special manned space training.

The Kremlin well knew the immense geopolitical benefit of beating the United States in technological know-how. Laika the dog had become posthumously famous as the first Earth creature in outer space, symbolizing Soviet technological prowess to countries around the globe. Nikita Khrushchev, a fine propagandist, knew that his first cosmonaut would drive the USSR's prestige even higher, his mission ranking with the voyages of Ferdinand Magellan and Christopher Columbus in the history of human exploration. "The road to the stars is steep and dangerous," Gagarin later admitted. "But we're not afraid. . . . Spaceflights can't be stopped. This isn't the work of one man or even a group of men. It is a historical process which mankind is carrying out in accordance with the natural laws of human development."

As Major Gagarin prepared for his takeoff that April, he was unusually quiet—though not nervous, as judged by his heart rate. The thought that he might be soaring to a fiery death occurred to him but didn't haunt him; how his private life would irrevocably change if he returned to Earth as a space pioneer did. Their imaginations inflamed by state media, the Soviet people would see him as the living embodiment of Communist excellence. Self-aware and self-contained, Gagarin knew fame would be a double-edged sword.

While the *Vostok 1* rocket was similar to that used for *Sputnik*, the space vehicle mounted at its tip was far larger, weighing 5 tons compared with *Sputnik*'s minuscule 184 pounds. It included the cramped capsule into which Gagarin was strapped (or suspended, as he later recalled) and another module containing retrorockets intended to guide the vehicle back to Earth. Fearing that the effects of zero-gravity conditions would cause their cosmonaut to become lightheaded or incoherent, the Soviets had designed *Vostok 1* to be piloted by ground personnel, with Gagarin taking control only in case of emergency. Video and radio communications monitored his mental state throughout the mission, satisfying experts at flight control that his faculties were unimpaired.

Gagarin's 108 minutes in flight were nearly all above the sixty-two-mile (one-hundred-kilometer) mark recognized as the boundary between Earth's atmosphere and space. With all the aerodynamics working as planned, he completed a single orbit of Earth and then headed toward touchdown in a rural area near the Volga River. After reentry the capsule properly deployed its parachute, but still slammed to the ground with such force that its six-thousand-pound bulk bounced several times before coming to rest. Fortunately, Gagarin was not on board: he had ejected at seven kilometers (4.4 miles), parachuting safely into a farmer's field. The fact that the Soviets had not engineered a means for Gagarin to return *inside* his space vehicle was one indication that the Kremlin was rushing to get its man into space first, however rudimentary his mode of return.

For many years, the Soviet Union guarded the secret of Gagarin's parachuting from the capsule. The Fédération Aéronautique International (FAI), the official body recording aerospace records, required that for a record to be awarded the pilot must land with the vehicle. To maintain the Soviet record that Gagarin had landed with *Vostok 1*, the Kremlin perpetuated a lie until the 1990s. In addition, to preserve the exaggeration that Gagarin had circumnavigated the Earth, the Soviets fabricated the locations of the launch and landing; in reality, the orbit was not quite the entire circumference of the Earth.

In April 1961, the U.S. government did not know about Gagarin's fudges and in-flight tribulations. Kennedy dutifully sent a telegram to Soviet premier Khrushchev, congratulating him on the historic launch, but the president was shaken by the fact that the Soviets had beaten America in the manned-space race. He feared, as his naval aide put it, that he was "walking on thin ice" and that the GOP and the press would excoriate him for the Soviet win. At an afternoon news conference, when asked about the Soviet space juggernaut, Kennedy responded with a sprightly, half-distraught air, acknowledging that indeed "we are behind" on manned spaceflight, but refreshingly, he refused to heighten expectations for NASA. "However tired anybody may be—and no one is more tired than I am," he said in a wistful tone, "it is a fact that it is going to take some time."

Kennedy's honest answer lacked the inspirational rhetoric of his "Ask Not" inaugural address, perhaps reflecting the fact that he'd yet to recognize space as the marquee goal of his New Frontier, and as the most visible American way to win the Cold War. On the campaign trail it was easy to mock the Eisenhower administration for the "missile gap." But he was now commander in chief. The sheer financial cost of Mercury and Apollo was daunting. Understandably, he wanted to ponder his space policy options in his first months in the White House without feeling hemmed in. Recalling a meeting held on the very day Gagarin orbited Earth, speechwriter Ted Sorensen later noted that JFK "had no real grasp of the enormous technology involved and remained skeptical about the cost and importance of space missions." The president, however, knew that putting a man in space was the new sine qua non of global prestige. Forced to dwell on Gagarin for even a few seconds, Kennedy turned testy, swatting the conversation away. Yet for the next couple of weeks he filled his black aviation bag with NASA reports for bedtime reading.

Time magazine's Hugh Sidey, perhaps the top White House correspondent in the spring of 1961, with frequent access to Kennedy in the Oval Office, called the president's disposition to Gagarin's accomplishment "disturbing." When the space race was evoked, the conversation petered out. Overall, Sidey admired JFK—unobjectively so, critics carped—but he also knew that Kennedy's long-standing criticism of Eisenhower's go-slow approach to space had hit reality after his own inauguration. Once armed with CIA intelligence that the United States was in fact ahead of the Soviets in the less-newsworthy but tangibly more significant development of ballistic missile technology, Kennedy had adopted much the same cautious position toward space as his predecessor. Rather than focusing on headline-grabbing space launches, Kennedy was looking elsewhere for measurable accomplishments—mainly, fixing an economy burdened with rising unemployment, slumping profits, and depressed stock prices in order to define the New Frontier as being about economic prosperity.

Nonetheless, Gagarin's flight was a public relations disaster that was hard to ignore. Just hours after Webb learned of the Soviet feat, he wrote Keith Glennan, his predecessor, about his nagging concern that the American system lacked a galvanizing ingredient: "My own feeling, in this, and many other matters facing the country at this time, is that our two major organizational concepts through which the power of the nation has been developed—the business corporation and the government agency—are going to have to be reexamined and perhaps some new inventions made." Webb believed there was no talent deficiency per se in the Kennedy administration, NASA, the U.S. military branches, the private sector, or academia—only a lack of a grand collective goal that transcended mere containment of Soviet expansionism.

Kennedy may have been somewhat aloof at that point but, realizing that the United States would have to send a Mercury astronaut into space soon, he asked Webb to have NASA work double shifts. Following this direct order to grind out a solution, engineers at Huntsville and Hampton tested timetables, calculations, and probabilities. Under pressure from the president, NASA narrowed the field for its first manned mission to two candidates. Either John Glenn or Alan Shepard, each training hard and competing in Virginia and Florida, would be the first American in space, and he'd get there *soon*.

When Shepard first learned that Gagarin had rocketed into orbit, he was livid. On March 24, NASA had scrubbed his planned mission due to a technical problem. Shepard, by training and personal disposition, wanted to beat the Soviets, not play second fiddle. It frustrated him that Eisenhower's old science advisor George Kistiakowsky had spooked the president by warning that NASA wasn't yet ready for manned space, and that if a Mercury astronaut were prematurely launched on a Redstone rocket, the attempted suborbital flight would be "the most expensive funeral man has ever had." Chris Kraft, the first NASA flight director, summed up the post-Gagarin frustrations at Cape Canaveral perfectly: "I didn't like it worth a damn," he recalled. "But the only thing to do was get back to work and do our jobs."

Unable to endure any more asleep-at-the-wheel criticism amid the mounting national self-doubt that followed Gagarin's flight, Kennedy faced the fact that the

New Frontier needed a defined space policy. While the world lionized "Ga-Ga" (as Gagarin was affectionately nicknamed), the American public was troubled and astonished by the Soviet feat, and felt diminished by the USSR's space propaganda win. "Of course, we tried to derive the maximum political advantage from the fact that we were first to launch our rockets into space," Khrushchev admitted in his memoirs. "We wanted to exert pressure on the American millions—and also influence the minds of more reasonable politicians—so that the United States would start treating us better."

Life magazine canvassed people from around the world, finding that in all corners of the globe, people knew the score: USSR, 1; United States, 0. An African student in Paris told *Life* that "the Americans talked a lot. Russia kept silent until success came. The results speak for themselves." (When flying over Africa, Gagarin had acknowledged citizens of the continent "trying to break the chains of imperialism.") A German office worker said, "This makes one realize Soviet boasts of ultimate superiority may not be groundless after all." An Egyptian youth summed it up: "The Americans are licked." None of those interviewed by *Life* mentioned the undeniable and very dull fact that U.S. Earth satellites, while lacking a cosmonaut with a movie star's smile, were even then netting essential astrophysical data for future space exploration. Nor did they know that the Saturn 1 rocket that von Braun's team was developing at Huntsville, with its first-stage thrust of 1.5 million pounds, was technologically far superior to the Soviets' guided-rocket capabilities.

At home, even though Kennedy maintained a 70 percent approval rating in the polls, there was an almost tangible sense of an administration slipping backward and losing support. Even in the first hundred days, Kennedy was already eyeing his reelection effort in 1964. For history's sake, he needed to validate his razor-close win over Nixon in 1960. In part, that was a natural continuum for a hard-driving, ambitious politician. But that competitive imperative also sprang from JFK's desire to make the United States the world's sole superpower and guarantor of global peace.

On April 14, Kennedy addressed the issue of NASA with considerably more vigor, convening a meeting of space advisors Webb, Wiesner, Sorensen, and Dryden, as well as David Bell, his administration's budget director. Bell's job was to remind Mercury enthusiasts in the group that every dollar spent on space took a dollar away from domestic programs that affected men, women, and children in real time. Or, alternatively, every space dollar raised taxes on all Americans. There was a slim overlap of what the nation could not afford to miss and what the country could actually afford to do. All presidents deal with the same conundrum all day long, across an array of issues, but in the case of space, the potential costs were gargantuan.

Because *Time* correspondent Hugh Sidey was present for much of the meeting, Kennedy gave something of a theatrical performance that afternoon, as he tilted his chair back precipitously, ran his fingers through his thick hair, and laid out his dilemma. Whenever Gagarin was mentioned, the president would bristle. "Is there any place we can catch them?" he asked, according to a record of the conversation. "What can we do? Can we go around the moon before them? Can we put a man on the moon before them? What about Nova and Rover? When will Saturn be ready? Can we leapfrog?"

Sometimes in history, a single word or concept can trigger a blinding flash that illuminates a presidency or the life of the nation. *Leapfrog*, a word Kennedy first used in his 1960 letter to Princeton University student William Everdell, became that kind of word, taking on a life of its own in NASA culture. Rather than suggesting NASA should skip the methodical steps needed to thoroughly and safely achieve plans already in place, the president was pushing for an audacious goal beyond manned-orbital spaceflight, where the Soviets already owned bragging rights. "If somebody can just tell me how to catch up," JFK implored at the meeting. "Let's find somebody—anybody. I don't care if it's the janitor over there, if he knows." Picking arguments apart, asking probing questions, Kennedy sought allies for his bold contention that the next NASA move had to be very dramatic. But with Lyndon Johnson away in Africa and Asia, the president was deprived of the fervid support his vice president would likely have offered for something as phenomenal as a moonshot.

"It was not much of a discussion," Sidey recalled, noting that the others present sidestepped JFK's leapfrog question by asking for more time to ponder. But Kennedy didn't have more time. The bell of history had rung. It was time to lead.

One day before the meeting, an aide had reminded the president that Franklin Roosevelt had announced in the first year of World War II a production schedule of fifty thousand airplanes annually, a target the corporate manufacturing sector said would be impossible to achieve. But the goal was met—and exceeded. Later, though knowing little himself about the science of nuclear physics, Roosevelt created the Manhattan Project, teaming America's best and brightest scientists under the leadership of J. Robert Oppenheimer to invent the atomic bombs used against Japan.

While the cause of victory in war could inspire the kind of government-business partnerships, patriotic dedication, and at-all-costs inspiration that drives fast, far-reaching technological advances, huge peacetime infrastructure projects could also harness a nation's ability to achieve the seemingly unachievable. In its fast-paced history, the U.S. government had completed the Erie Canal in 1825, the Panama Canal in 1914, and the Hoover Dam in 1935—each of them beginning as an engineering challenge and ending as a symbol of national indomitability and excellence. Conquering these supposedly impossible tasks lifted the nation's spirits, advanced knowledge within an array of professions, and produced immediate innovation-based economic benefits that justified their high federal price tags.

As a governmental expenditure, the multibillion-dollar leapfrog that Kennedy sought didn't fit neatly into either the wartime or infrastructure category. It would not immediately open economic floodgates as rural electrification or hydropower dams might. It could not keep America safe from military attack like the production of fifty thousand warplanes. It might be considered a third category, exploration, but the

U.S. government had rarely been interested in funding epic journeys simply to fill in maps of Earth or sky. It might also be considered a type of big-stick diplomacy in the vein of JFK's idol, Thomas Jefferson, who used the Lewis and Clark Expedition and other missions to the American West as part of an ongoing struggle with Spain west of the Mississippi. "Jefferson's expeditions and much-publicized explorers had been the masks of his conquest," wrote historian Julie Fenster, "the tools of diplomacy that allowed the stakes to rise without forcing a military response."

Aware of historical timing, Kennedy sought a manned-space leapfrog that would transform the goal of expanding human knowledge (like Lewis and Clark) into irrefutable proof of American exceptionalism—humbling the Soviet Union without turning the Cold War into a shooting war. If the mission he sought could, in effect, lead to long-term peace with the Kremlin, it would more than cover whatever cost it incurred. Being in the White House only three months hadn't given Kennedy enough time to reorient NASA policy; but he was poised to do exactly that.

According to Ted Sorensen, the April 14 meeting was where "Kennedy began to really get the feel of what this whole thing might mean to the Presidency and to the United States." Now the mission would truly begin: Finding the right space strategy to once again uncork America's spirit of scientific achievement, engineering ingenuity, and global leadership.

Kennedy's meeting on space took place on a Friday evening, breaking up at dusk. About an hour later, the night plunged directly into disaster when a small fleet set sail from Nicaragua, bound for a swampy inlet on Cuba's southern coast known as Bahia de Cochinos: the Bay of Pigs.

Launched five days after Gagarin's flight, the hastily conceived Bay of Pigs invasion, planned under Eisenhower, was an attempt by "Brigade 2506," comprising fifteen hundred Cuban exiles armed, trained, and funded by the CIA, to topple the Communist government of Fidel Castro, which had itself overthrown the pro-U.S. Batista regime two years before. Overly complex and poorly conceived, with little margin for setbacks, the invasion was doomed from the outset. News of the plans leaked within the Cuban community of southern Florida and was soon transmitted to Castro's government. The counterrevolutionaries came under immediate fire as they began their landing, soon finding themselves outmanned and outgunned by the twenty thousand troops and air support Castro had ordered to the beaches. In and around the Bay of Pigs, Brigade 2506's hopes dwindled after just one day, although the battle lasted officially for three. Hundreds were killed on both sides, including some Americans, and more than a thousand Cuban Americans were taken prisoner. The debacle was a demeaning defeat for the anti-Castro forces, for the CIA, and ultimately for the White House, where Ted Sorensen recalled Kennedy as "anguished and fatigued" and in "the most emotional, self-critical state I had ever seen him."

On Monday night, April 17, Kennedy hosted a gala at the White House, maintaining a cool and graceful demeanor for guests while still seething inside. Soon enough, though, he retreated to the Rose Garden in his white tie and tails, unable to maintain the charade any longer. In the span of one stinging week, the young president had been thoroughly embarrassed both militarily and scientifically. To some extent, it was his own fault: he'd been ill prepared, governing as though through a sideview mirror, and his lack of resolute leadership had been successfully exploited by the Soviet Union and its client state, Cuba.

By itself, the failure of the Bay of Pigs invasion may not have influenced American space policy. However, following on the heels of the Gagarin flight, it made for a nasty one-two punch that damaged Kennedy. The press poured criticism down on him. He was plagued with the inexorable pressure of time. After stumbling into the Cuban mess, he was learning the hard way, in the course of a crucial thirty-day span, that the federal government didn't lead itself. His political survival dictated that he become more involved in and closely attached to the foreign policy issues that defined his sense of the United States and its position in the world. At one White House meeting, his brother Robert, the attorney general, ripped into the CIA. "All you bright fellows," RFK said. "You got the president into this. We've got to do something to show the Russians we're not paper tigers."

The furor over the lopsided Bay of Pigs defeat continued for months, and as Kennedy surely suspected in the Rose Garden on April 17, it would go on to stain his legacy forever. On April 19, when Cuban forces mopped up the last remnants of Brigade 2506, the president scheduled a meeting with Vice President Johnson and Webb. Reflecting on the April 14 meeting about space policy, JFK asked LBJ to investigate the status of NASA's programs and his opinion on the best possible leapfrogging mission. A five-point memo issued by the president for Johnson on April 20 laid out two overriding concerns: settling on the right mission and ensuring that NASA and related agencies were capable of delivering an all-out, do-or-die effort to make that mission happen. That memorandum stands as a clear manifesto of the president's thinking in the aftermath of Gagarin's flight:

> *1.* Do we have a chance of beating the Soviets by putting a laboratory in space, or by a trip around the Moon, or by a rocket to land on the Moon, or by a rocket to go to the Moon and back with a man. Is there any other space program which promises dramatic results in which we could win?
>
> *2.* How much additional would it cost?
>
> *3.* Are we working 24 hours a day on existing programs? If not, why not? If not, will you make recommendations to me as to how work can be speeded up?
>
> *4.* In building large boosters should we put emphasis on nuclear, chemical or liquid fuel, or a combination of these three?
>
> *5.* Are we making maximum effort? Are we achieving necessary results?

If Kennedy's impatience could be read in his April 20 memorandum, Johnson's steadfast ambition could be as easily read in the dateline on his response: April 28, only eight days after receiving the president's requests for answers. Most of Johnson's quickly compiled memorandum told JFK why NASA was important in terms of international relations, making the salient point that "dramatic accomplishments in space are being increasingly identified as a major indicator of world leadership." Johnson had cast a wide net compiling his response, receiving important input from not only Webb, Dryden, and NASA deputy administrator Robert Seamans, but also von Braun, science advisor Wiesner, Defense Secretary Robert McNamara, air force general Bernard Schriever, navy admiral John Hayward, budget director Bell, and three nongovernment leaders: George Brown of engineering and construction firm Brown and Root, Donald Cook of American Electric Power, and Frank Stanton of CBS. Johnson's evaluation contended that the United States had greater resources than the USSR for attaining space supremacy, but that the country lacked the willpower and drive to marshal those resources in a dramatic fashion. But one point in the April 28 memo from Johnson seemed to capture Kennedy's full attention:

> Manned exploration of the moon . . . is not only an achievement with great propaganda value, but it is essential as an objective whether or not we are first in its accomplishment—and we may be able to be first. We cannot leapfrog such accomplishments, as they are essential sources of knowledge and experience for even greater successes in space. We cannot expect the Russians to transfer the benefits of their experiences or the advantages of their capabilities to us. We must do these things ourselves.

Johnson's "Evaluation of Space Program" document was thought provoking. However, Kennedy hadn't asked *if* or *why* America needed a major goal in space; he had asked *which* and *how.* The vice president's memo allowed that circumnavigation of the moon and a manned trip there could be accomplished by 1966 or 1967, but then it slid back to a recommendation that less complicated NASA goals could be attained more rapidly.

Although everybody of importance at NASA and among the president's cabinet and scientific advisors weighed in on what Kennedy should prioritize, it might well be that a letter from von Braun, dated the day after Johnson's memo, made a better and more persuasive argument. A genius at cutting to the chase, von Braun named and ranked four possible goals for the space program, including establishing an orbiting space laboratory ("we do not have a good chance of beating the Soviets"), circumnavigation of the moon, and placement of a radio transmitter on the lunar surface ("a sporting chance"). Von Braun saved his highest grade, "excellent chance," for landing a crew on the moon. "The reason," he explained, "is that a performance jump by a factor 10 over [the Soviets'] present rockets is necessary to accomplish this feat. While today we do not have such a rocket, it is unlikely that the Soviets have it. Therefore, we would not have to enter the race toward this obvious next goal in space exploration against hopeless odds favoring the Soviets."

Von Braun was the "somebody—anybody" whom Kennedy had sought at the April 14 meeting. Ever since they served together helping to select *Time*'s Man of the Year in 1953, they had been political allies of sorts, tied together by a shared repugnance for Eisenhower's low-key response to the Soviet technological advances in space. Von Braun was the ally who could bring the kind of burning vision and take-charge aggressiveness to space exploration that JFK himself was intent on bringing to government. In that respect, he was similar to J. Robert Oppenheimer, who provided scientific leadership to Roosevelt's Manhattan Project. Sensing his opening to build a moon rocket, von Braun gave clear and concise answers to the president's five questions about how the United States could leapfrog the Soviets in space. In the most formidable example of can-doism NASA had available, he stated that the technological journey would be tough, but a moon landing was attainable. To get to the moon, NASA would have to develop a launch vehicle far more powerful than von Braun's eight-engine Saturn; but it *could* be done—and von Braun, having worked on the Nazis' hyperaccelerated V-2 program during the war, knew precisely what could be expected of scientists on a grueling round-the-clock schedule. "We were being rushed—as usual," von Braun recalled, "by Russia's great strides." But, he contended, "with an all-out crash," he believed NASA could accomplish landing men on the moon in 1967–68.

Having received the consensus from Johnson, Webb, Dryden, and von Braun that a moonshot was a difficult but credible option, Kennedy continued to consider his next move as other developments battled for his attention. Internationally, the fallout from the Bay of Pigs fiasco had yet to subside, even as he and his brother were secretly arranging a summit meeting with Khrushchev, to be held in Vienna on June 4. Domestically, the Congress of Racial Equality had begun grabbing headlines with its so-called Freedom Riders campaign, organizing mixed-race groups of young people to travel by bus from Washington, DC, to New Orleans as a way of protesting the nonenforcement of Interstate Commerce Commission rules prohibiting racial discrimination in interstate travel. Like the rest of America, the president watched television coverage with dismay, as the Freedom Riders faced beatings and other violence at the hands of white bigots in Southern states.

Preoccupied with these and other matters, Kennedy contemplated America's lunar future only in fleeting moments as NASA continued to prepare for the more immediate future of a manned Mercury flight. For over a year, Americans had been hearing that such a mission was close, but the success of Gagarin's flight had upped the ante. In response, NASA made the bold but risky decision that it would encourage television and radio coverage of the eventual launch, offering a vivid contrast to the secrecy insisted upon by the Soviets, who were fearful of the optics should their launches fail. JFK himself dreaded the risky launch for the same reason, fearing that any technical malfunction, glitch, or human tragedy broadcast to the world would be blamed on his White House.

A new launch date of May 5 was announced. On May 4, Lieutenant Commander Victor A. Prather, a navy flight surgeon working for NASA, prepared for one final test of the full-pressure Mark IV space suits that had been created for the Mercury astronauts by B.F. Goodrich. The day began hopefully. Prather and another scientist, Malcolm Ross, flew in a Strato-Lab V balloon to a height of 113,740 feet (21.5 miles), a manned-balloon altitude record for decades to come. The flight lasted nearly ten hours and ended as planned, with a smooth landing in the Gulf of Mexico. There, Prather's copilot, Ross, was plucked from the water by a navy helicopter after almost falling back into the sea while trying to grasp the sling lowered for him. When it was Prather's turn, the same problem occurred. The sling had been designed for people in wetsuits or street clothes, not cumbersome twenty-five-pound space suits. With little room to hold on, Prather slipped, fell, and was pulled under as water rushed into the open face guard of his suit. Navy divers couldn't get to him in time, and he drowned—giving NASA, and Kennedy, a reminder of the deadly peril of rushing astronauts into space prematurely.

Even though Prather's flight ended tragically, it had also fulfilled its mission, allowing NASA to send one of the Mercury Seven into suborbital flight the very next day, certain of the Mark IV space suit's efficacy at high altitude.

That first American in space would be Alan Shepard Jr., a physically fit, towheaded navy test pilot from New Hampshire, whose ancestors had come to America aboard the *Mayflower.* Proud of being an eighth-generation New Englander, he had grown up on the family farm and attended a one-room elementary school. Obsessed with model planes, at age twelve he built a full-size glider, strapped himself in as if he were Orville Wright at Kitty Hawk, and took flight, sailing at an altitude of some four feet before crashing.

Shepard grew committed to becoming a top-tier military aviator in the mold of Jimmy Doolittle. When airborne, he had calm judgment and the gift of unflappable concentration. During World War II, he studied at the U.S. Naval Academy in Annapolis, Maryland, earning his bachelor of science degree in 1944. The next year, he began flight training at Corpus Christi, Texas, and Pensacola, Florida, perfecting the art of aircraft carrier aviation. By 1950 he was admitted to the highly competitive Patuxent River Naval Test Pilot

School in Maryland, where he became a military aviation superstar. Going on to test the F-3H Demon, F-8U Crusader, F-4D Skyray, F-11F Tigercat, and F-5D Skylancer, as well as in-flight refueling systems, the flinty Shepard was a standout. Whether it was night-landing on an aircraft carrier in the Atlantic; breaking speed records at Moffett Field, California; or serving as an operations officer for the 193rd Fighter Squadron aboard the carrier USS *Oriskany* in the western Pacific, he was always the right man for the job. Like JFK, Shepard made no excuses for his eye for the ladies, even as he stayed married to a lovely woman whose lot was to make her peace with his flirtations. Five feet eleven inches with blue eyes, he was nicknamed the "Icy Commander" for his detached and intimidating persona. "He was hard to get to know," astronaut Gene Cernan said of Shepard. "But once you cracked the surface, he was your friend for life."

As May 5 dawned, the air hung thick and clammy at Cape Canaveral as the launch of Shepard's Mercury-Redstone rocket, *Freedom 7*, was delayed time and again by weather and mechanical problems. The millions watching on TV grew more anxious every second. Residents of his hometown, East Derry, New Hampshire, could barely sit still. Wearing a close-fitting silver space suit, Shepard was eventually strapped

On May 5, 1961, members of the Kennedy administration gathered in the White House office of the president's secretary Evelyn Lincoln to watch the liftoff of Alan Shepard aboard a Mercury-Redstone rocket. *Left to right:* Vice President Lyndon Johnson, Arthur Schlesinger Jr., Admiral Arleigh Burke, and President John F. Kennedy and First Lady Jacqueline Kennedy were thrilled when Shepard's *Freedom 7* capsule successfully splashed down in the Atlantic Ocean.

into his module atop the seven-story rocket. NASA technicians fastened ventilation hoses to Shepard and fed him pure oxygen. For over four hours, Shepard still managed to exude a dashing aura, telling those calling the shots, "I'm cooler than you are—Why don't you fix your little problem and light this candle?" That phrase, "light this candle," would became associated with Shepard forever.

President Kennedy was in a White House meeting when word arrived that the delays were over and the countdown to liftoff had begun. With tense anticipation and bated breath, he moved into the office of his secretary, Evelyn Lincoln, where a television was tuned to the broadcast from Cape Canaveral. Kennedy's wife, Jackie, joined him there, and at 9:34 a.m. they watched as Shepard's rocket rose from the launchpad without a hitch, sending him hurtling into history.

In all, Shepard spent 14.8 anxious minutes aloft, peaking at an altitude of 115 nautical miles. When it came time during his descent, at 7,000 feet above Earth, his capsule's red-and-white parachute opened. He splashed down in the Atlantic Ocean 302 miles from Cape Canaveral. Still strapped inside the *Freedom* 7 capsule, waiting to be rescued by helicopter and praying not to meet Victor Prather's fate, Shepard proved the American space program was on track. It all seemed strange and magical, yet astoundingly real.

Back at the White House, the president evinced an odd kind of reverie, watching intently while aware of the whispered banter around him. It had been a long twenty-three days since Yuri Gagarin's flight, but suddenly Kennedy was back on top. Only when a White House aide told JFK that Shepard was aboard the helicopter and had been pronounced A-OK by NASA physicians did the president allow himself to say, "It's a success," with a smile. America's post-*Sputnik* space prayers had been answered. Seeing that capsule land in the Atlantic was one of the greatest thrills of Kennedy's life. For his part, Shepard enthused, "Boy, what a ride!" as he was whisked to the USS *Champlain* four miles away.

After doctors checked Shepard's blood pressure, heart rate, breathing, and psychological state, he guzzled down a glass of orange juice and was given a tape recorder to capture his thoughts and feelings. Joking around, he said, "My name is José Jimenez . . . ," a reference to comedian Bill Dana, and then started rambling about every detail of his voyage until he got a shore-to-ship phone call from the president himself.

"Hello, commander," Kennedy said.

"Yes, sir."

"I want to congratulate you very much."

"Thank you very much, Mr. President."

"We watched you on TV, of course," JFK continued. "And we are awfully pleased and proud of what you did."

The key phrase in this brief exchange was "We

watched you on TV." In that, the Kennedys hadn't been alone. Across the country, some forty-five million TV viewers had watched what the *Houston Chronicle* dubbed "the greatest 'suspense drama' in the history of TV." The experience bonded the entire nation, rekindling the collective American spirit like nothing since V-J Day at the close of World War II.

A steadied and reassured Kennedy understood that the public needed Shepard as a reason to cheer for the nation in peacetime. While only a momentary respite from the existential dread of the Cold War and the domestic upheaval of the civil rights era, the coast-to-coast celebrations for Shepard were as real as postwar victory parades; yet, such celebrations were also profoundly different, marking not just victory and relief, but excitement over taking the first steps into an almost unimaginable future. Great presidents (Jefferson, Lincoln, FDR) think about what the world will look like a hundred years into the future. At a time of insecurity, Kennedy finally understood the power of space exploration to unite the nation. "I think [Kennedy] became convinced that space was a symbol of the twentieth century," White House science advisor Wiesner reflected. "It was a decision he made cold-bloodedly. He thought it was good for the country."

After a weekend of debriefing and further medical

tests, Shepard was flown to Washington at the president's invitation and given the red-carpet treatment. His motorcade, which consisted of nothing more than Shepard and his wife in one open car and the other six Mercury astronauts in several cars following, turned into a thunderous spontaneous parade down Pennsylvania Avenue, with a quarter of a million people crowding the streets and sidewalks to celebrate. Washington bureaucrats were overwhelmed by the outpouring of genuine affection for Shepard, with the local newspaper observing that Americans had been hungry for a triumph like Shepard's enthralling leap into space.

At a White House ceremony, Kennedy had the honor of presenting Shepard with NASA's Distinguished Service Medal. In a funny and clumsy moment, JFK accidentally dropped the medal from its velvet box, prompting a visibly displeased Jackie Kennedy to mutter, "Pick it up, Jack!" On cue, the president and Shepard both bent and reached for the medal, almost bumping heads. Kennedy grabbed it first, then, smiling, presented Shepard "this decoration, which has gone from the ground up." The spontaneous line brought the house down. "We had a big laugh out of that," Shepard recalled. Kennedy played host to Shepard for the rest of that day, seeing at close range the astounding power of space to propel the hearts of America. For his part,

Shepard recalled in an oral history that he was "even more thrilled at that moment in talking to [Kennedy] than I had been after the flight."

For a week following Shepard's whirlwind visit to Washington, Americans were still celebrating *Freedom 7* with champagne and slaps on the back. It was a burst of collective confidence, an outpouring of pride that the American century was alive and flourishing. The scar of defeat and shame that Yuri Gagarin had left had vanished. Kennedy became fast friends with Shepard, whom he saw as the masculine personification of the New Frontier. Instead of cowboy hats and six-shooters, the unshrinkable New Frontier heroes wore silvery fabric and rubbery space suits along with round helmets with wide visors. The well-spoken Shepard explained that his bond with Kennedy was based on a shared willingness to take high risks to "meet the challenge" of beating the Soviets in space. In this regard, *Freedom 7* was an update on the *PT-109*. "Thus Alan Shepard became a great national hero in the spring of 1961," historian Steven Watts explained in *JFK and the Masculine Mystique*, "as the embodiment of the Kennedy male spirit."

That spring, Kennedy had truly gotten his baptism by fire in space politics. Gagarin's mission had jolted him into an acceptance of the Mercury program's enormous political value for the New Frontier. Beefing up NASA was now a priority. Von Braun and a first-rate cadre of other experts had given the president the perspective to think clearly about American prospects in space, and Shepard's successful flight had proved the viability of manned spaceflight, putting even the moon in range. The Soviet news agency TASS's rote criticism of Shepard's flight as "very inferior" in every regard to Gagarin's unrivaled mission only reinforced the fact that the United States had dented the Soviet armor. Be that as it may, Kennedy issued a proactive statement, charging that Shepard's flight had provided "incentive to everyone in our nation" concerned with space exploration "to redouble their efforts in the vital field." In a press conference later, Kennedy, basking in the Shepard glow, promised that the federal government would oversee a "substantially larger effort in space."

Just days after Shepard's mission, Kennedy and Johnson assembled a who's who of New Frontiersmen; representing NASA were Webb, Dryden, Seamans, and Silverstein. They met with a half-dozen top Pentagon officials, toward the goal of offering Lyndon Johnson their final recommendations in response to Kennedy's April 20 memorandum. This was the first time NASA and Defense Department officials helped in this type of task force. Everything, from communications satellites and ICBMs to Project Mercury, was discussed. The elephant in the room was whether the United States should commit itself to a lunar landing. After hours of back-and-forth, Robert McNamara instructed Seamans, Deputy Director of Defense John Rubel, and Willis Shapley of the Bureau of the Budget to draft a series of recommendations for Johnson to then submit to Kennedy. Webb, determined to make sure NASA didn't get bigfooted by the Pentagon, insisted that he maintain a hand in drafting the document; McNamara agreed. "In choosing the lunar landing mission as the central feature of its recommended program, the group had no firm intelligence regarding whether the Soviet Union has already embarked on a similar program," historian John Logsdon wrote in *The Decision to Go to the Moon* about the Webb-McNamara Report. "Much in the same way as national defense programs are formulated the group evaluated Soviet capabilities not intentions, and decided the United States could probably beat the USSR to the moon."

The very fact that NASA and the Department of Defense were gladly collaborating was a breakthrough for Kennedy. A consensus had been brokered by McNamara and Webb that admitted that the point of Project Apollo was prestige (proving U.S. technological excellence over the Kremlin's). There really weren't military imperatives for a lunar voyage. However, without the air force's SAMOS (Satellite and Missile Observation System) imagery and electronic intelligence satellites, the air force and the CIA's Corona imagery intelligence satellite, the air force's MIDAS early warning satellite, and the Naval Research Laboratory's GRAB (Galactic Radiation and Background) electronic intelligence satellite, the Kennedy administration wouldn't have had the confidence that an American moonshot was doable anytime soon.

Less than two weeks after the outsized excitement of Shepard's White House visit, Kennedy reserved a long span in his appointment schedule for a much more private meeting with Victor Prather's widow, Virginia, and her two small children, Marla Lee and Victor III. Still tormented by the deaths of the *PT-109* sailors under his command during World War II, Kennedy wasn't constituted to forget NASA's other hero that May, posthumously awarding Prather the navy's Distinguished Flying Cross. Taking Prather's children outside, he encouraged them to play with his own children's toys and swing set. "He talked to them just like their daddy used to," Virginia Prather said.

The president was a father and politician above all else, full of empathy and concerned about the well-being of his countrymen. To him, space exploration was about people, and that is how he could understand it, in his own way. The American spirit had been roused by brave men like Prather and by the Shepard mission, and Kennedy planned to capitalize on this national ardor for winning the space race without delay.

On May 8, 1961, Mercury astronaut Alan Shepard, the first American in space, received a medal at the White House from the hands of President Kennedy in the presence of his wife, Louise.

President Kennedy at the joint session of Congress on May 25, 1961, making his historic American pledge: "I believe that this nation should commit itself to achieving the goal, before this decade is out, of landing a man on the moon and returning him safely to Earth."

12

"Going to the Moon"

Washington, DC, May 25, 1961

At a basic level, the president's Apollo decision was to the United States what the Pharoah's determination to build the pyramids was to Egypt.

—ROGER D. LAUNIUS, *APOLLO'S LEGACY: PERSPECTIVES ON THE SPACE RACE* (2019)

On the afternoon of May 25, 1961, members of Congress assembled in a joint session at the Capitol, eager to hear John F. Kennedy's highly anticipated special address on "Urgent National Needs." The White House had billed the Thursday event as a second State of the Union address, a privilege to which first-year presidents are entitled but rarely exercise. Over four months into the new administration, with a crisis in Cuba and a deteriorating political situation in

Laos, Kennedy was clearly eager for a reset, prompting press speculation that May 25 would be a comeback address, redirecting the president's high-minded optimism and undeniable energy after the double wallop of Yuri Gagarin and the Bay of Pigs. With his "Ask Not" inaugural still fresh in the public's mind, many reporters intimated that something large and unprecedented might be coming—and they were right. "As far as President Kennedy and the space program are concerned, he didn't really get his mind around it until Gagarin went into orbit," Robert Seamans, the deputy administrator of NASA, recalled. "And I guess you can say that President Kennedy also went into orbit."

In the forty-three days between the Gagarin flight of April 12 and the scheduled May 25 speech to U.S. lawmakers, the administration had narrowed American options in space until they pointed squarely at the moon. If Alan Shepard's suborbital flight had fizzled, it's unlikely JFK would have prioritized a lunar voyage—but, as noted, it had been an astounding success. Kennedy had asked his top advisors to consult with a wide array of space scientists and technocrats on every analysis of cost, risks, manpower, alternatives, and administrative responsibility. Lyndon Johnson, the loudest voice calling for a U.S. lunar voyage, had harangued the experts to forget caution and to think big, telling them in a meeting, "You're the people who have to initiate this. Say what you think ought to be done. You may not get all you want but we can't do anything unless you come forward with your proposals."

Kennedy's big question was what a moon landing would cost in money and time. The ballpark answer, twenty to forty billion over a six- to eight-year period, was appalling to him, but he intuited that Congress would lack the nerve to obstruct a moonshot with the public still euphoric over Shepard's triumph. For Kennedy, who had digested the Webb-McNamara recommendation on May 8, going to the moon was starting to win out as the heart and soul of the New Frontier: a combination of heroic journey, national security imperative, scientific windfall, global prestige booster, and potentially transformative technological and economic boon.

Early that May, Kennedy had invited the Mercury Seven astronauts to the Oval Office, along with NASA public affairs officer Paul Haney, and a few others. Playing contrarian, Kennedy probed the astronauts with tough questions. Were manned space missions really essential? Couldn't robots or monkeys perform equally well? How would a system of tracking stations work? At the core of Kennedy's big decision was the precise purpose of a manned voyage to the moon. He kept asking NASA hands: What is the desired effect of manned space? "JFK was obviously prepping himself for questions he would get should he go ahead with his lunar proposal," recalled Paul Haney. "As usual, Alan Shepard and John Glenn supplied most of the answers."

Indeed, many respected American scientists questioned whether manned spaceflights would produce any benefit above what could be obtained through the use of probes, satellites, and other automated devices. Kennedy's NASA supporters dismissed the argument that astronauts were extraneous to the hard science of the trip by countering that even if the moonshot didn't discover anything dramatic about the lunar surface, the collective effort would inevitably reveal a vast amount about human beings, and about the human race. For the rest of his presidency, Kennedy would counter such arguments by pointing out that back in 1903, the Wright brothers were told not to attempt flight at the windy beach just south of Kitty Hawk; however, they disregarded the skeptics, and their first controlled, sustained flight birthed modern aviation.

What Kennedy concluded after weeks of cogitation was that rarely did a leader get the opportunity to oversee an epoch of what historian Daniel Boorstin called "public discovery." Even if a manned Apollo flight to the moon wasn't strictly necessary for collecting scientific data, the very act of trying would revolutionize technology.

From that perspective, though, there was no particular imperative behind the accelerated push for a NASA moonshot. The process of human discovery could have accommodated the steady, workmanlike development of rocketry, which was on track to deliver moon-ready technology by the late 1970s or 1980s. And the moon, having already waited millions of years, seemed in no rush to welcome its first visitors. But politics did not have the same patience. Kennedy knew that by mid-May 1961, the moon had become the ultimate prize in the Cold War rivalry with the USSR over technological superiority and, by extension, global prestige. That lunar sphere in the sky, whose meteor-pocked face had smiled down on every human who'd ever lived and whose phases had given mankind its first calendar, was now a stopwatch counting down to victory or ignominious defeat. With no way to prepare for a moonshot out of the public eye, there was also no possibility of saving face if the other side reached the goal first. If NASA failed, Kennedy failed. If Kennedy failed, America failed.

It is both brazen and disingenuous for a U.S. president to say that a mission cannot fail. As Kennedy knew all too well, space launches certainly could fail. Out at the White Sands Proving Ground, rockets scuttling around the pad and then blowing up were commonplace. From 1957 to 1961, for every rocket launch success at Cape Canaveral, there had been two disasters. The Mercury Seven themselves, on hand to witness the first test of the Atlas rocket that would eventually lift one of them to orbit, watched as it exploded just after liftoff, prompting Shepard to remark, "Well, I'm glad they got that out of the way." There were plenty of pragmatic reasons for Kennedy *not* to embrace the risky and expensive moonshot, but the odds were just close enough to push him toward go.

In truth, nobody knew the Kremlin's true capabilities or intentions. CIA intelligence estimated that the Soviets probably couldn't send a human to the moon within ten years, but beating even that schedule would require government funding and industrial mobilization on a near-wartime scale, committing all involved to a breakneck pace until the mission was accomplished. Such an approach was problematic for NASA engineers; Project Mercury was notoriously behind in its much simpler schedules for astronaut suborbital flights. Bugs and glitches were pervasive in the embryonic manned space program. Kennedy knew, as did Khrushchev, that as in any business, a rush order would cost extra.

In 1961, the U.S. economy was improving from the recession of 1957–58, though not robustly. Inasmuch as

Kennedy had the leeway to propose a major new initiative, antipoverty programs appealed to him the most. In the area of natural resource management, he had an abiding fascination with developing a process for seawater desalination, feeling that making fresh water abundantly available in the Middle East and Africa would improve conditions in those regions, where the United States was vying with the Soviet Union for influence. For its part, Congress had already signaled that it would be most receptive to an infrastructure omnibus bill. Rebuilding bridges, highways, and electric utility grids, as well as other New Deal–style public works programs, would strengthen the economy, first through construction jobs and then in increased capacities for growth. In fact, Democratic members of Congress awaited just such a proposal from the White House, presuming that practically every district in every state would benefit. Rather than being pie in the sky, a lunar voyage promised some of the same economic stimulus.

For NASA administrator James Webb, the real excitement of an accelerated lunar landing program lay in the federal government's ability to pool resources with corporations and academia to get the job done. Existing science, technology, and aerospace hubs such as St. Louis and Los Angeles would be the most immediate beneficiaries, with new additional high-tech centers sure to emerge coast to coast to handle demand. Enormously complicated systems engineering, along with technological integration requirements, would be mastered via NASA coordination. A tidal wave of advanced scientific knowledge would be unleashed. Webb never lost a chance to stress to Kennedy that most of the jobs created by Project Apollo would be at a high educational level, with a correspondingly high pay scale. From Webb's perspective, Apollo was something akin to the Tennessee Valley Authority, the Grand Coulee Dam, the Saint Lawrence Seaway, and the Interstate Highway System, all rolled into one. "What we had in mind," he later reflected, "was to try and build all the elements of a total space competence."

Webb did worry, however, that in its zeal to reap political and economic benefits, the White House wasn't giving adequate weight to the very real technological difficulties inherent in staging the first moonshot. "I'm a relatively cautious person," Webb recalled. "I think when you decide you're going to do something and put the prestige of the United States government behind it, you'd better doggone well be able to do it." But Webb also noted that NASA was ready to execute a moonshot if Kennedy made it a national priority. And a fixed date to toil toward would help organize the all-hands-on-deck effort.

When working on his May 25 speech, the president was also cognizant of the military aspects of space. One of his burning fears was that Soviet satellites soon would be equipped with nuclear bombs, which could then be launched accurately from airspace directly over a target in North America. Under the president's directive, the air force worked on antimissile defense prototypes to do battle against this kind of space weapon, but the program was riddled with flaws across its interlocking network of sensors, inceptors, and kill vehicles. With such a defensive shield far from operable, JFK moved steadily on the diplomatic front toward a ban on all U.S. or Soviet nuclear weaponry in space. Lacking a comprehensive safeguard against America's nuclear adversary, Kennedy instead stressed the need for U.S.–Soviet peaceful exploration of space. The Soviets wanted any such ban to include satellite surveillance, an area in which the American military was heavily invested. While the United Nations was involved initially in the cause of keeping space conflict-free, the obvious fact was that any peaceful future depended on an agreement at the outset between Washington and Moscow.

For the sake of those bilateral negotiations, Kennedy needed to show the world that the U.S. interest in space was cordial and nonviolent, and nothing more. Just as Antarctica had been demilitarized by a 1960 treaty, why couldn't space be, too? While American and Soviet diplomats stumbled in their preliminary moves toward talks, the president repeatedly asked his staff to keep the U.S. Air Force from trumpeting its every space-related triumph. Most notably, its X-15 was continuing to nip at speed and altitude records for a reusable craft. Even more advanced was the Dyna-Soar, another airplane-like spacecraft designed to be capable of entering space, orbiting Earth, and then returning. Assuming it received federal government support, the Dyna-Soar was slated for construction starting in 1963, with initial testing in 1964–65. Neil Armstrong, in fact, had been recruited to the Dyna-Soar corps as a consultant pilot. "Every time the Air Force put up a space shot and any publicity was given to it, [Kennedy] just went through the roof," an acquaintance said later. From the Oval Office, the president would call air force secretary Arthur Sylvester in a rage, asking why he continued to "let those bastards talk."

From the air force's perspective, JFK's stance was just as enraging. When Kennedy took office, the air force had hopes of leveraging a larger role in the administration of space. But air force senior officers soon intuited that the president had embraced von Braun's Huntsville program. And neither of the two Mercury astronauts initially chosen to go into space—Shepard and Glenn—were in the air force. Knowing well how Congress worked, air force generals recognized that they would lose priority without well-crafted publicity. Plain pride was also at stake. Ever since the inception of an American civilian space program in 1958, air force loyalists had been in a constant state of frustration, baffled as to how they could possibly be sidelined from spearheading most space developments. Many air force generals agreed with Kennedy that space needed to be off-limits to weaponry, yet they believed that NASA didn't know how to properly put test pilots into space. They were relentless in lobbying JFK on this point, and on the efficacy of the Dyna-Soar. While NASA's abilities were stretched to the limit just trying to put a "man in a can," as the Mercury astronauts themselves called their short flights, the Dyna-Soar was regarded in engineering circles as a better program, and was making bigger strides. The constant delays in the Mercury program only fed the air force's antipathy toward NASA, and its frustration with Kennedy.

As Kennedy prepared to address Congress about the future of the U.S. space program, a draft of a major study sponsored by the air force was under review in the office of Secretary Sylvester. Called the "Space Plan," it envisioned manned flights, including trips to the moon, and research into medicine and other subjects.

The Space Plan would serve as a valuable document for years, a map of the future of space exploration. Underlying its well-considered predictions, however, lay the contention that leaving the forward edge of space exploration to NASA and maintaining the separation of peaceful and military space programs was "absurd," "unsound," and "arbitrary." In this report, as it did through every other available avenue, the air force continued to campaign against NASA for sole leadership of the manned space program.

Therefore, in thinking through the future prospect of a NASA moonshot on an accelerated schedule, Kennedy had to consider the air force's dissension. This youngest service branch had had powerful friends in Congress, such as Senator Stuart Symington, who had served as Truman's secretary of the air force. It had a built-in determination to win the East-West rivalry between the world's superpowers. And it had the Dyna-Soar, which held enormous promise. But the Dyna-Soar was designed with the potential to carry nuclear bombs, making both it and the air force liabilities in the effort to reach an accord with the Soviets on banning such weapons in space. Secretary of Defense Robert McNamara, in particular, was increasingly suspicious of the air force's motivations in attempting by every means to gain control of the space program, even if it meant that space would become yet another armed theater of Cold War brinksmanship.

When all the considerations were weighed, the most uncontroversial path Kennedy could have taken in the spring of 1961 would have been continuing the Mercury manned space program without radical alteration. Pragmatic incrementalism in rocketry had been the American way throughout the Eisenhower administration, and few in Congress were expecting a bold new direction for NASA just a few months into the new presidency. Even Defense Secretary Robert McNamara, who always supported the Apollo program, testified in 1963 that from a military vantage point there was no "man-on-the-moon" requirement. Faced with more pressing priorities, most presidents would have turned away from the moonshot recommendations in the Webb-McNamara Report as nonessential, but some would have seen in them what Kennedy saw. JFK's idol, Thomas Jefferson, would have blanched at the size of the expenditure (which would make the NASA budget one of the largest in the federal government), but he wouldn't have been able to resist one of the most monumental advances of scientific discovery in human history. For James K. Polk, the very bigness of the project might have proved an irresistible attraction. Theodore Roosevelt would have regarded the moonshot as a fitting errand for a world power and proof of American exceptionalism writ large. Franklin Roosevelt, his boundless imagination matched by his belief in priming the economic pump, would have seen an opportunity for the federal government to spread prosperity in places like central Florida, the Virginia Tidewater region, New Orleans, Houston, and beyond. Among New Deal–style Democrats in 1961, a NASA moonshot's potential to bring both white-collar and blue-collar jobs to must-win congressional districts in the Gulf South, Southwest, and Midwest didn't go unnoticed. With the Deep South's traditional support for the Democratic Party beginning to slip in the face of demographic change and opposition to the civil rights movement, pumping pork-barrel money into Dixieland would be smart politics for Kennedy.

On May 9, Kennedy allowed the Webb-McNamara Report—officially, "Recommendations for Our National Space Program: Changes, Policies, Goals"—to be circulated in Congress. Behaving like a patron, Johnson had been the front man on getting the moonshot recommendations on the president's desk. Had the White House detected screams of outrage from Capitol Hill over its request for an additional $600 million in NASA funding to get Project Apollo going strong, they'd have been able to quietly back away from the moonshot plan before Kennedy's speech. But the Congress of 1961–62 did not, as a body, view the White House as the enemy camp. In that sense, it reflected the nation at the time. Despite the hard-fought, razor-close election of 1960, the majority of Americans looked forward to seeing what the Kennedy years would bring. There was a fair-minded consensus that Kennedy had, after all, been elected to unite the nation, and in the spring of 1961, Congress, to a surprising bipartisan extent, gave the young president room to lead.

With an American moonshot being recommended by all except Wiesner and a few Eisenhower holdovers, LBJ left the United States for a whirlwind trip to Southeast Asia; he wouldn't return until May 24. This allowed Kennedy time to prepare his planned speech to a joint session of Congress on May 25 without leaks or further input from his vice president. Ever since the Shepard success, the president was elbowing LBJ out of the picture, determined to make sure that the Apollo moonshot was interpreted by the press as *his* initiative. The White House had purposely let the *New York Times* know a day in advance that Kennedy would be pushing for a NASA voyage in his upcoming address to Congress.

As the morning of May 25 arrived, Kennedy was busy making last-minute modifications to his speech, which had been drafted by Willis Shapley of the Bureau of the Budget and Theodore Sorensen. Originally, JFK had thought of declaring that America would put a man on the moon by 1967, the fiftieth anniversary of the Bolshevik Revolution. But wisely he went with the "end of the decade," to buy NASA more time. Even on the limo ride from the White House to the Capitol, the president was still editing furiously; having been a member of the House and Senate for fourteen years, he wasn't a fan of long-winded speeches. The talk was crafted to be forty-five minutes long, but by ad-libbing three additional paragraphs, he brought the final length up to forty-seven. Wiesner, though chairman of the White House's Science Advisory Committee, had advised him to avoid the word *science* in his address; the president concurred.

With the speech in his suit coat pocket, the president entered the chamber, shook hands, and strode to the podium. Johnson, in his capacity as Senate president, sat behind JFK at stage right, beside Speaker of the House Sam Rayburn on the left. Nearly everybody in Congress seemed in good springtime spirits. A masterly orator, Kennedy intended to build the rhetoric up slowly, with domestic and foreign policy requests, and then hit lawmakers smack between the eyes with the moonshot gambit. Like a whale after a deep plunge, the president was coming up to spout and breathe life into NASA's Project Apollo.

Just after 12:30 p.m., following a round of polite applause, Kennedy began the address in a strong, confident voice, describing plans to affirm his so-called Freedom Doctrine by financially supporting troubled nations aspiring to democracy, for the global enrichment of humanity. He offered a domestic jobs program and then confirmed his forthcoming summit with Nikita Khrushchev in Vienna. He requested money for foreign aid, offering "our skills and our capital and our food" for countries, among them Vietnam, "upon whom our hopes for resisting the Communist tide . . . ultimately depend."

For the first half hour, Kennedy spoke in a steady voice, soberly discussing increased military spending and disarmament negotiations with the Soviets and receiving commensurately businesslike applause. Listening to the audio recording decades later, one can almost sense that he was holding back a bit, setting up his audience for the jolt of excitement to come. Then, finally, it came: "If we are to win the battle that is now going on around the world between freedom and tyranny," he said with rising urgency, "the dramatic achievements in space which occurred in recent weeks should have made clear to us all, as did the *Sputnik* in 1957, the impact of this adventure on the minds of men everywhere who are attempting to make a determination on which road they should take."

Framing NASA's moonshot as a choice between tyranny or freedom was smart Cold War politics, the surest path to persuade Congress to open its purse. Kennedy knew that GOP conservatives would find it hard to resist a lunar voyage framed as a way to prove democratic capitalism's superiority to single-party communism. He didn't resist the temptation, as Eisenhower had, to equate space accomplishment with excellence in government. Instead, JFK posited that a government that could send a man to the moon must be superior in all else as well. "We go into space," he added, "because, whatever mankind must undertake, free men must fully share." Now was the time, he said, "to take longer strides, time for a great new American enterprise, time for this nation to take a clearly leading role in space achievement, which in many ways may hold the key to our future on earth."

Mentioning the Soviets' successes in building powerful rockets and satellites, Kennedy stressed that the United States had to compete harder. "For while we cannot guarantee that we shall one day be first, we can guarantee that any failure to make this effort will make us last," he said. "We take an additional risk by making it in full view of the world; but as shown by the feat of Astronaut Shepard, this very risk enhances our stature when we are successful." And then the president laid out the grand challenge that would come to define his administration and legacy: "I believe this nation should commit itself to achieving the goal, before the decade is out, of landing a man on the moon and returning him safely to Earth. No single space project in this period will be more impressive to mankind or more important for the long-range exploration of space; and none will be so difficult or expensive to accomplish."

According to Sorensen, Kennedy "sensed . . . that his audience was skeptical, if not hostile" to his moon challenge. Even though this assessment was false, it caused the president to lose confidence. In closing, his voice didn't soar, as it had in his "Ask Not" inaugural speech. His brows were suddenly drawn back a bit, as if questioning his own act. After speaking for more than a half hour with only one inconsequential stutter, he was suddenly tripping over words. There was a note of distraction in JFK's voice as the speech concluded. The audience couldn't help but sense his discomfort. Standing before an assembly of his government, with tens of millions of Americans tuned in on TV and radio, he heard only stony silence and subdued applause rather than the epochal excitement that should have greeted the notion of sending an American hurtling 230,000 miles through nothingness, to another world, by 1970. Jeers might have been preferable. Nonetheless, his brazen moonshot call was among the most courageous statements and greatest gambles ever made by an American president.

Part of a generation that equated space travel with *Buck Rogers* and *Flash Gordon*, Kennedy might have been nervous over his grandiose promise. Sorensen, perhaps in overstatement, claimed the president was inwardly panicking, believing that his challenge was being received with "stunned doubt and disbelief by the members of Congress, both Democratic and Republican." *New York Times* reporter Alvin Shuster wrote that the Senate and House Republican leaders scribbled notes, stared at their hands, and brushed their hair back. When Kennedy finished his speech with more talk of freedom around the world, the audience perked up again, with thunderbolts of applause.

In the limo back to the White House, the president was downcast, worried he had choked, his face full of perplexity. Even though his talk was interrupted eighteen times with applause, he felt the pledge had fallen flat. The Apollo challenge was intended to be his big calling card at the upcoming Vienna summit with Khrushchev. Now he was bewildered. With the virtue of hindsight, he determined that his moonshot plea should have been leaked to other news agencies besides the *New York Times* the previous day, to get Congress and the press energized in advance. Catching lawmakers by surprise, before they had time to internalize what a presidential pledge meant to their careers, had, Kennedy feared, been clumsy. Sorensen commiserated with him.

All this second-guessing was for naught. To JFK's surprise, the next day's *Washington Post* labeled his speech "spectacular." Canvassing an array of congressmen in a "cloakroom consensus," the *Post* concluded that some of Kennedy's smaller, domestic proposals were probably doomed, but that the hefty request for Apollo funding would go through. "He tossed the ball to us," said one member of Congress in the *Post*, "and don't think for one moment that anyone up here is going to drop it." On the Democratic side, Senator Hubert Humphrey of Minnesota led the huzzahs. "The president is ahead of Congress," he gushed, "but not much ahead. And unless I'm mistaken, the country is right up there with him." For many Americans, it seemed as if fate were summoning the nation on a lunar voyage, and a cautious, incremental approach wouldn't take it there. A giant leapfrog seemed both an admirable goal for a superpower and necessary if the Cold War were to be won.

Considering that NASA's total accomplishment in manned spaceflight to date was Shepard's short suborbital flight of three weeks before, Kennedy's decision to shoot for the moon was extra bold. It was also an about-face from the cautious first weeks of his administration. As NASA deputy administrator Robert Seamans admitted, the New Frontiersmen went in just a few months from "doubting the value of any human spaceflight" in the Wiesner Report to calling exploration of the moon "essential." A big part of the president's calculus was that Shepard had proved that space exploration was a TV bonanza flush with undeniable theatrics. If JFK could have a Mercury launch semiregularly, it would be good for his poll ratings. From slowly stated countdowns to roaring liftoffs and victorious treks into space, Kennedy would be the chief political beneficiary. The payoff in these TV broadcasts would be a successful splashdown and recovery followed by the president congratulating *his* astronaut.

Few in the main game of U.S. politics opposed the moonshot gambit outright, although some did grouse about the mission's eye-popping price tag. In a Senate Finance Committee hearing, Republican John J. Williams of Delaware suggested that with additional NASA expenditures, "our debt may reach the moon before we do." Senator Prescott Bush of Connecticut, a Republican, carped that Kennedy's moonshot, coupled with a probable refusal to raise taxes, would "unleash the forces of inflation." The president's own father, Joseph Kennedy, was apoplectic about the plan. "Damn it, I taught Jack better than that!" he lashed out at White House aides. "Oh, we're going to go broke with this nonsense! I told him that I thought it was ridiculous." Former president Eisenhower also fumed that JFK (whom he privately derided as "Little Boy Blue" blowing his horn) had changed the political dynamic and oversold the necessity of a lunar voyage in a speech he deemed "almost hysterical." In early October, spurred on by former NASA administrator Keith Glennan, Ike told faculty at the Naval War College that for Kennedy to "make the so-called race to the moon a major element in our struggle to show that we are superior to the Russians is getting our eyes off the right target." The former president, critical even of the Peace Corps, couldn't comprehend that the calculus behind the moonshot also included the prospect of new technological development that would drive American economic growth and global influence for decades to come.

Hoping to persuade Congress not to fund the moonshot, Eisenhower privately told Republicans at an off-the-record Capitol Hill meeting that "anybody who would spend $40 billion in a race to the moon for national prestige is nuts." But his entreaties fell on deaf ears. Spooked by Soviet successes in space and spurred by Kennedy's aspirational vision, both houses of Congress quickly approved huge increases in NASA's budget. NASA boosters were fond of taunting, "No bucks, no Buck Rogers." The rationale for Kennedy's moonshot was most honestly stated by Senator John Stennis, the chairman of the Armed Services Committee, who days after the president's speech offered this statement for the *Congressional Record*: "Space technology will eventually become the dominant factor in determining our national and military strength. Whoever controls space controls the world."

But Project Apollo didn't come without other costs. With the NASA moonshot now the cornerstone of the New Frontier, the administration's focus shifted away from a major antipoverty program it had been considering for Appalachia, the Deep South, and struggling cities. "It will cost thirty-five billion dollars to put two men on the moon," National Urban League president Whitney Young complained. "It would take ten billion dollars to lift every poor person in this country above the official poverty standard this year. Something is wrong somewhere."

With the exception of Webb, Dryden, and Seamans, few at NASA foresaw the challenge coming their way. Chris Kraft, NASA's first flight director, was writing a report at the Mercury Control Center at Cape Canaveral that May 25 when he turned on CBS News to watch Kennedy's speech. "My head seemed to fill with fog and my heart almost stopped," Kraft recalled. "Did he say what I thought I heard?" Telephoning Webb, a flabbergasted Kraft tried to ascertain what this meant in terms of NASA budgets and research timetables. Kraft also called members of the Mercury team, who were all in a state of happy bewilderment. In his memoir, *Flight: My Life in Mission Control*, Kraft inventoried the prevailing sentiments in his shop: "We've only put Shepard on a suborbital flight . . . an Atlas can't reach the moon . . . we have mountains of work just to do the three-orbit flight . . . the moon . . . we'll need real spacecraft, big ones and a lot better than Mercury . . . men on the moon, has he lost his mind? . . . Have I?" Robert Gilruth, chief of the Space Task Group, leader of the American manned space program, had only one word for Kennedy's arbitrary deadline of 1970: "aghast."

There's an old engineers' saying that their slide-rule tribe typically overestimates what can be accomplished in a year and underestimates what can be accomplished in a decade. Kraft's NASA team certainly hoped this aphorism was true, because Project Mercury was already behind schedule even as the president was committing them to an exponentially more difficult Apollo goal. But if the men and women at Mercury Control in Cape Canaveral were excitedly baffled by Kennedy's speech, von Braun and the other rocketeers at Huntsville's Marshall Space Flight Center were beside themselves with glee. Alabama was regularly in the news for all the wrong reasons. Birmingham, only a hundred miles from Huntsville, was racked by systemic racism, white supremacy, and police antagonism against African Americans. But due to Kennedy's pledge, the focus in northern Alabama shifted virtually overnight to making America proud of the new family of Saturn launch vehicles being developed there for an eventual moonshot.

The race to the moon was on. In practical terms, Kennedy's May 25 speech to Congress put Huntsville at the vortex of the New Frontier. Only sixteen years earlier, von Braun had been working for the Nazis at Peenemünde, aiming to destroy London and Antwerp. Now he was the indispensable partner of a popular young U.S. president determined to use his rockets to go to the moon. If there really was something called the

American dream, von Braun was living it beyond his wildest imagination. Unwaveringly self-confident, an energized von Braun claimed no worries about achieving Kennedy's moonshot goal by late 1969, as long as the federal funding came through. "Of course, the moon [had] a romantic connotation for me as a young guy," von Braun recalled, "but I must confess, that as soon as President Kennedy announced we were going to land there within this decade, I began to identify it more and more with the target in space and time. . . . It was a constant reminder, 'We'll get you before this decade is out.'"

Mercury astronaut Deke Slayton considered May 25, 1961, one of the gold-starred days in American history. No longer did astronauts have to worry about job security. What impressed him most about the moonshot speech was the brash "by the decade's end" challenge. "What Kennedy did with the moon program was to pick a goal people could relate to," he recalled. "It has to be something under ten years; if you give people a thirty-year goal, they won't waste time thinking about it, it's too far away."

The Soviet Union was deeply startled by Kennedy's moonshot speech. Did the United States have the technology for a lunar landing? Had NASA already developed a network of tracking stations? Was the president merely grandstanding? Were his words just propaganda, or was the United States really putting an Apollo moonshot on the front burner? Earlier that year, the Kremlin had issued public declarations about sending cosmonauts to the moon and building a base there, but Sergei Korolev, the top Soviet rocket scientist, was far more committed to launching a huge orbital station and eventually staging manned missions to Mars and Venus. After May 25, Khrushchev denied that America and the USSR were in a space race to the moon, but records unearthed from the KGB archives after the Soviet Union's dissolution showed that Khrushchev had indeed secretly pursued two ultimately unfulfilled moon missions. First, in 1962, spurred on by Kennedy's challenge, Korolev persuaded the Soviet premier to develop an N1 moon rocket. Two years later Korolev again convinced the Kremlin to back a full-bore lunar project. The program known as L3 called for the landing of Soviet cosmonauts on the lunar surface before the Americans. The L3 rocket would be launched into orbit on the N1 rocket, now with a mandated payload capacity of ninety-five tons.

Neither the U.S. moonshot nor the Soviets' own space projects would have been plausible were it not for the humongous size of both countries' economies, which were robust enough to allocate 2 percent of their gross national product on a high-risk lunar voyage. "This and the fact that these nations decided to compete," NASA administrator Thomas Paine explained decades later, "is what would propel us to the moon." On the tech side, advances in liquid-fueled rocketry and digital computing put the tools of spaceflight in the hands of a willing nation.

If presidential greatness is, in the phrase of Harvard professor Richard Neustadt, the "power to persuade," then Kennedy had achieved greatness once again with his May 25 speech. In ten minutes, he'd jump-started what space historian Walter A. McDougall called "the greatest open-ended peacetime commitment by Congress in history," boosting NASA's annual operating budget considerably. For Ted Sorensen, the speech confirmed that the New Frontier was at heart about the "spirit of discovery." Now Kennedy, having put serious chips on that singular lunar number, was about to discover how many large and small hurdles NASA had to overcome for his herculean gamble to pay off.

President Kennedy and Vice President Johnson tour NASA facilities in Huntsville, Alabama. The U.S. government developed ballistic rockets, under the guidance of Wernher von Braun, called Redstone, Jupiter-C, Juno, and the Saturn 1B at the Marshall Space Flight Center there.

13

Searching for Moonlight in Tulsa and Vienna

No commitments have been made, but I believe it is going to be of great importance to develop the intellectual and other resources of the Southwest in connection with the new programs the Government is undertaking.

—JAMES WEBB TO LYNDON JOHNSON, MAY 23, 1961

As fate would have it, the first National Conference on Peaceful Uses of Space had been scheduled for May 26, the day after JFK's speech to Congress, in Tulsa, Oklahoma. Suddenly, the assembled aerospace industry leaders and technologists had a new focus, and a new impetus. Leading aerospace companies wanted a piece of the Apollo action. According to *Business Week*, the meeting hall was "seething with excitement," with the three major television networks scrambling to provide live coverage. Ambitious scientists could be seen circulating among sharp-eyed industry leaders and excited government administrators. With Kennedy's moon pledge, the New Frontier's dining table was set, and every business executive and technologist in Tulsa was clamoring for a prime cut of the main course: congressional appropriations for space hardware.

Tulsa, a city built on the rising fortunes of the oil industry, had been chosen for the aerospace summit due to the influence of Oklahoma's wily Democratic senator, Robert S. Kerr, the chair of the Senate's Committee on Aeronautical and Space Sciences. When it came to promoting NASA and cutting taxes Kerr was Kennedy's most valuable ally on Capitol Hill. Arranging for NASA to partner with the Tulsa Chamber of Commerce to sponsor this conference, Kerr had also induced Kennedy to address the participants via a telephone link from the White House. The president spoke for three minutes, his voice strong yet low-key. Noting that the conference's subject "deals with the very heart of our national policy in space research and explorations, to which I devoted a good deal of my speech yesterday before the Congress," he nevertheless downplayed the challenge the moonshot represented for the nation, instead focusing on the benefits of space research and the responsibility to maintain U.S. leadership in the field. The word *moon* was not uttered once by the president, but there was no mistaking the subtext. Everybody in Tulsa knew what JFK meant.

If Kennedy was subdued in tone, the participants were not. Senator Kerr led the cheerleading, his commitment to NASA having grown a thousandfold in the previous twenty-four hours. He envisioned Sun Belt cities—Houston, San Antonio, Dallas, Oklahoma City, and Tulsa, in particular—becoming NASA boomtowns. "The costs will be tremendous," a buoyed Kerr said of Project Apollo in his keynote address, "but the rewards will be unlimited." *Newsweek* backed Kerr up, speculating that a moonshot would drive record profits at companies such as North American Aviation, Space Technology Laboratories Inc. of TRW, and Lockheed Missiles and Space Company.

Other speakers at Tulsa spoke of the various benefits to computer science and aerospace engineering. For Apollo to put an astronaut on the moon, new program management and systems integration approaches would have to transpire in the technocratic NASA culture. Dr. Abe Silverstein, NASA's director of spaceflight, looked to ancient history in order to put the future into perspective. "Man has progressed," he told the conference, "from the inside of the cave by accepting challenges." Silverstein made clear to the conference that NASA was looking to buy technology in six distinctive areas: launch vehicles, command vehicles (which would also serve as return capsules), command capsule propulsion units, lunar landers and ascent stages (which return astronauts from the lunar surface to lunar orbit), communication and tracking networks, and ground infrastructure. The most critical components of the Apollo effort, Silverstein made clear, were the command/return vehicle and rockets designed to launch payloads into low Earth orbit.

Max Faget, principal developer of the Mercury capsule, was fixated in Tulsa on the idea that the new generation of space capsules shouldn't be as claustrophobic as the tiny Mercury capsule. Envisioning a new spacecraft that was roomy and open, he decided that the base diameter for the Apollo capsule should be at least fourteen feet (approximately two and a half times Mercury's diameter). Meanwhile, Apollo manager Robert Gilruth, hungry for corporate bids and proposals, promised to send 1,350 invitations to various representatives of government and industry for a conference to be held July 18–26 in Washington, DC.

Von Braun, speaking with gravitas as director of

the Marshall Space Flight Center, spoke directly to the many space hardware suppliers in the audience. His Tulsa speech was technical in nature, describing the Saturn I rocket, which would launch that October. Von Braun had originally built the Saturn I to loft Pegasus satellites into orbit. But his rocket was, he said, an all-purpose military booster ideal for Kennedy's moon challenge. Before long he sought to replace the Saturn I with a new derivative Saturn IB, which offered a more powerful upper stage and cutting-edge instrumentation.

After a full day of well-thought-out speeches by what United Press International described as "one of the greatest concentrations of space experts," Webb took the stage to deliver an upbeat spiel that not only summed up the mood at the Tulsa conference but also raised the stakes on Kennedy's initiative. While the president had said that America *could* put a man on the moon if the nation made a gangbuster effort, Webb contended that the country *had to* go to the moon. Offering the heaviest artillery of logic and imperative thinking about JFK's lunar initiative to date, Webb suggested that for Congress to reject Project Apollo funding would be a gross dereliction of duty. "We have the scientists, we have the technology, we have the resources and we have the power and knowhow to do the job," Webb told the audience. "Not to do it would jeopardize the nation's future."

Webb, full of gumption, would soon prove the indomitable linchpin of the Apollo effort, working with his two major deputies, Dryden and Seamans, to pull together the multiple strands and streams of American space research and align them into a single efficient and mighty effort. There was the Space Task Group, which was responsible for managing Project Mercury and would also assume the burden of overseeing the technical aspects of Project Apollo. In Pasadena, Caltech's Jet Propulsion Laboratory was already scientifically studying and mapping the moon. Responsibility for launching rockets was being handled by NASA's Office of Launch Vehicle Programs, Marshall Space Flight Center, and the U.S. Air Force. Ultimately, the list of contributors to the Apollo effort would include some twenty thousand companies and more than four hundred thousand individual citizens—practically if not literally stretching to the moon and back.

Somewhat surprisingly, President Kennedy sent Edward R. Murrow to represent the White House at the Tulsa conference. The former CBS News broadcaster had recently joined the administration as head of the U.S. Information Agency (USIA). Murrow had taken the job on the stipulation that he'd be afforded a high degree of access to White House decision making, telling JFK, "If you want me in the landings, I'd better be there for the takeoffs." At first glance, the president and Murrow seemed like an ideal pair to sell the moonshot to American taxpayers. Murrow had been raised in difficult economic circumstances in the state of Washington, where a first-rate public education left him with a nimble mind and remarkable communication skills. While Kennedy steadily matured during the course of his political career, almost willing himself into projecting self-assurance, Murrow seemed to have been born with an overflow of confidence. During World War II he had become famous reporting from London's streets with a CBS Radio microphone during the harrowing Battle of Britain, describing in photo-clear language the carnage he was seeing. When Murrow said a building was on fire, you could almost smell the smoke. Throughout the war, he recruited a crackerjack team of foreign correspondents who collectively became known as the Murrow Boys.

Back home after the war, Murrow's popularity skyrocketed as he expanded his media reach into the new medium of television. Although he was a product of commercial broadcasting, with all its dependence on advertisements and sponsors, he maintained his high standards and integrity in reporting the news, telling the truth as he saw it. On evening programs now hallowed in TV history, Murrow pushed past the corporate stance of cowardly neutrality and lashed out, most famously, at the reprehensible tactics of Joseph McCarthy, which led to the Wisconsin senator's being condemned by lawmakers in the mid-1950s.

Murrow's first major public appearance as USIA director came at the National Press Club in Washington the day before Kennedy's moonshot speech. Discussing the propaganda war with the Soviet Union, he surprised the audience by speaking out against racial segregation in the South, including that in the nation's capital. Murrow lamented the fact that America's backward Jim Crow segregation policies forced accredited diplomats from African countries such as Ghana and Nigeria to live in a restricted way. "Landlords will not rent to them," Murrow said, "schools refuse their children, stores will not let them try on clothes, beaches ban their families." Emphasizing both the duties of our shared humanity and the damage such prejudice can cause in international relations, Murrow noted that the images of police violence against Freedom Riders in Jackson, Mississippi, that very week, broadcast to the world by newspapers and television, cost America "as much influence as anything the Soviets might do." Clearly stating his own priorities just one day before

THE PROJECTILE PASSING THE MOON.

The earliest fictional account of a visit to the moon was written circa AD 170 by Lucian of Samosata. People no doubt had been dreaming of going to the moon for thousands of years before then. Approximately 1,700 years later, Jules Verne sparked the imagination of millions with his novel *From the Earth to the Moon* (1865). Though Verne was a Frenchman, his book told the story of an American effort to build a moon vehicle, with Florida as the locale of the launch site.

ABOVE: In the high-tension atmosphere of the Cold War, *Time* magazine's choice of Soviet premier Nikita Khrushchev as its 1957 Man of the Year proved unpopular with readers. The editors made a good case, though, for Krushchev's dominance in "a year of retreat and disarray for the West." The most stunning symbol of Soviet aspiration was *Sputnik*, the first manmade satellite. *From* Time, *January 6, 1958 © Time Inc. Used under license.*

RIGHT: In 1958, scientists at the Jet Propulsion Laboratory in Pasadena, California, plotted the course of satellites by hand. Receiving tracking signals from earthbound stations, they triangulated them on a curved aluminum surface to chart progress.

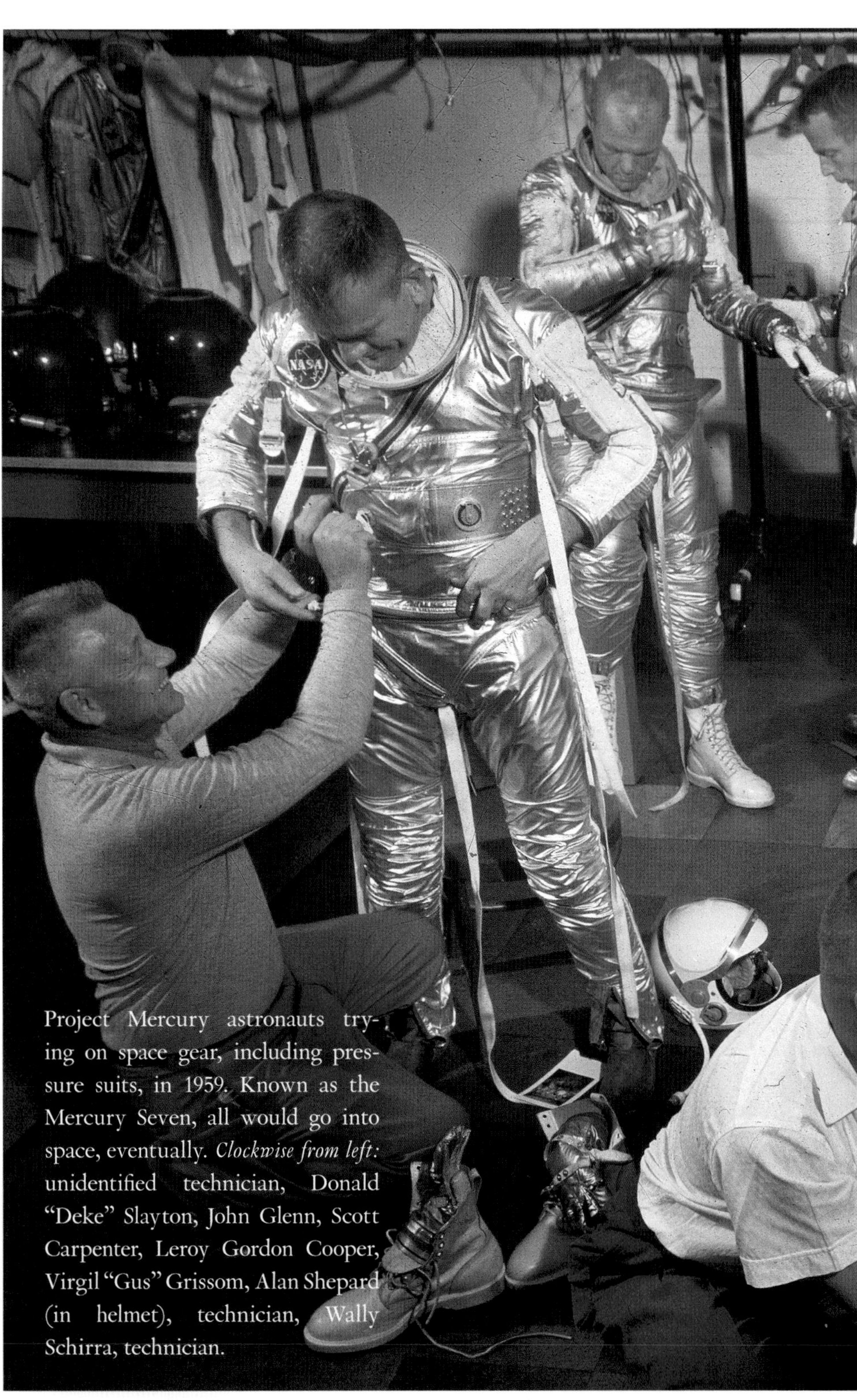
Project Mercury astronauts trying on space gear, including pressure suits, in 1959. Known as the Mercury Seven, all would go into space, eventually. *Clockwise from left:* unidentified technician, Donald "Deke" Slayton, John Glenn, Scott Carpenter, Leroy Gordon Cooper, Virgil "Gus" Grissom, Alan Shepard (in helmet), technician, Wally Schirra, technician.

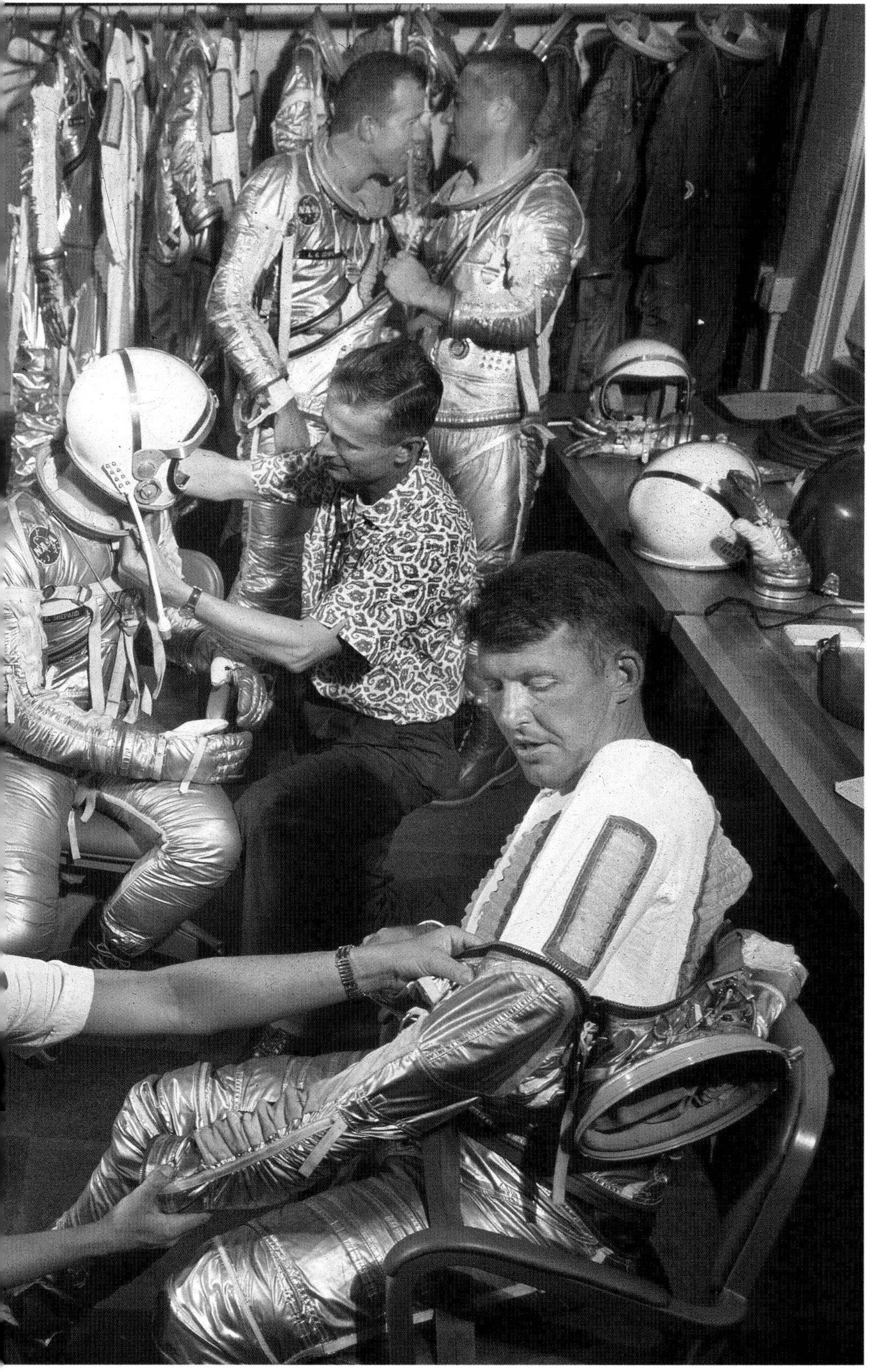

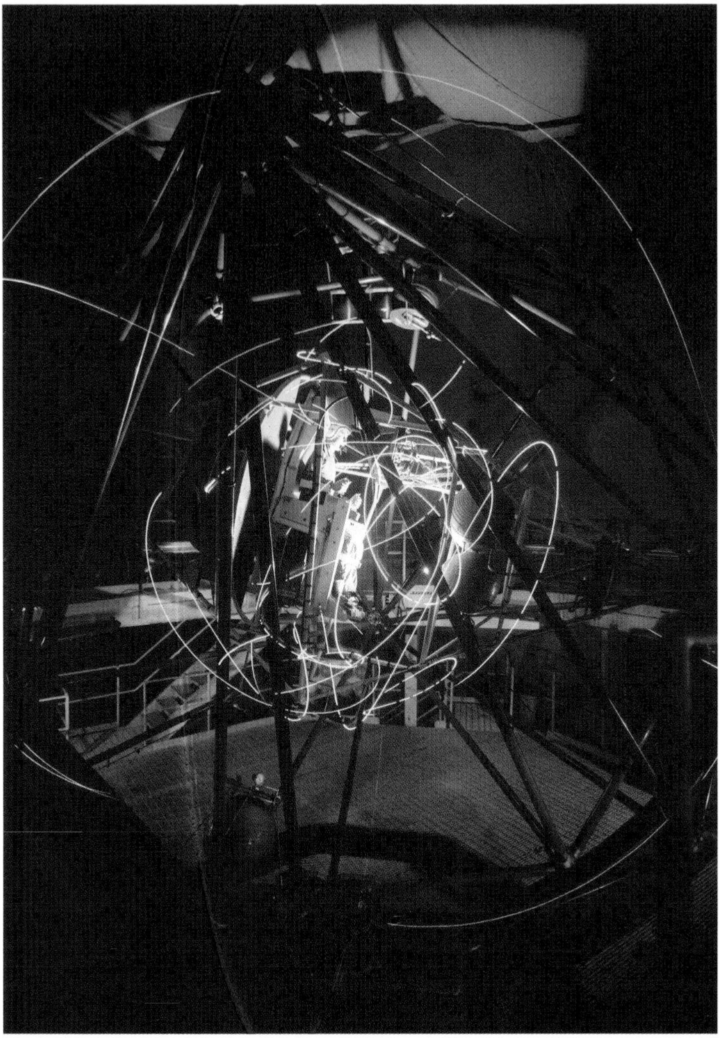

ABOVE: Like the other Mercury astronauts, Gus Grissom (*above*) trained on a multiple-axis machine specially built to replicate and even exceed the amount of spinning actually encountered in space travel. The "gimbal rig" tumbled in all directions, testing the recruit's capacity to function under such conditions as well as his ability to regain control of the machine.

RIGHT: The launch of the *Freedom 7* mission at Cape Canaveral, Florida, on May 5, 1961, was viewed on television by forty-five million people. Alan Shepard, the sole occupant of the capsule, became the first American in space. That fact made his short flight, lasting only fifteen minutes, a watershed in the Cold War.

UNITED

On February 23, 1962, President Kennedy visited Cape Canaveral to honor John Glenn, who three days earlier had become the first American to orbit the earth. He was photographed receiving a personal tour of Glenn's *Friendship 7* space capsule. Kennedy and Glenn would become warm friends, outside

ABOVE: In June 1963, when NASA made the decision to end the Mercury program, the seven original astronauts, who had all been with the program since its inception in April 1959, were photographed together. *From left to right:* Gordon Cooper, Wally Schirra, Alan Shepard, Gus Grissom, John Glenn, Deke Slayton, and Scott Carpenter.

RIGHT: Engineers working through the night in 1963 at NASA's assembly center near Edwards Air Force Base in Kern County, California.

FACING PAGE: Gordon Cooper, being strapped into the capsule *Faith 7.* On May 15, 1963, he piloted the final Mercury spaceflight, orbiting the Earth twenty-two times and staying in space for more than thirty-four hours.

On September 11, 1962, President Kennedy attended a briefing at Cape Canaveral as part of a two-day whirlwind tour of space facilities throughout the South. The briefing was staged at the Launch Complex 34 blockhouse, a windowless control room that was large enough for 134 employees during a rocket launch. Seated with Kennedy are *(left to right)* NASA administrator James Webb, Lyndon Johnson, Launch Operations Center director Kurt Debus, an unidentified officer, and Secretary of Defense Robert McNamara.

On Saturday, November 16, 1963, President John F. Kennedy traveled to Cape Canaveral, where he met with German-born Dr. Wernher von Braun, the leading U.S. Army rocket engineer of the space era, to view the massive Saturn booster rocket, standing next to a model of one.

On July 16, 1969, almost six years after John F. Kennedy's death, the thirty-fifth U.S. president's dream of an American voyage to the moon came to fruition. Apollo 11 brought together the talents of many thousands of workers, but carried only three on the mission: Neil Armstrong, Buzz Aldrin, and Michael Collins. The Saturn V rocket, pictured above, propelled the Apollo 11 astronauts into space. Four days later, on July 20, Armstrong and Aldrin took the most daring step of all, climbing into the tiny *Eagle* lunar module and detatching from the command module, in which Collins remained. The spider-like *Eagle* then made its way to the surface of the moon.

ABOVE: "The *Eagle* has landed," Neil Armstrong reported to Houston's Apollo Mission Control on July 20, 1969. After six hours of further preparations and mandatory rest, Armstrong climbed down a ladder and set foot on the moon, saying, "That's one small step for [a] man, one giant leap for mankind." Armstrong later took this iconic photo of Buzz Aldrin posing with the American flag planted on the Sea of Tranquility. While Old Glory still stands on the moon, ultraviolet (UV) radiation has bleached the colors white.

After more than twenty-one hours on the moon, the *Eagle* successfully lifted off and approached the command module. The complex docking procedure was completed, reuniting the three astronauts for the journey home to Earth. Only after their safe arrival on July 24 could the Apollo 11 mission be fully celebrated as the epochal accomplishment that it was.

Kennedy's moonshot announcement, Murrow asked, "Is it possible that we concern ourselves too much with outer space and far places, and too little with inner space and near places?"

The Associated Press account of Murrow's speech was carried in newspapers large and small. Out of an address of more than an hour on a range of topics, it highlighted his juxtaposition of America's focus on space while civil rights inequities and systemic poverty existed even in the nation's own capital.

Traveling to Tulsa, Murrow continued to be candid, preaching civil rights in front of space geeks. In a segment of the conference devoted to the use of satellites for global communications and weather forecasting, most of the speakers enthused about the potential benefits for humanity, if only the Americans and Soviets could agree on a peaceful satellite policy. But Murrow sounded a warning note, arguing that everything that seemed open and egalitarian about satellite communication also threatened to enable the spread of what he melodramatically called "filth" to households worldwide.

Shocking many attendees, Murrow continued to represent the argument that NASA was a distraction from the real societal problems America faced in 1961, though he also reluctantly admitted that Apollo offered the brightest hope for enhancing national pride. Murrow believed part of the risk of Kennedy's space strategy was the uncertainty of whether the New Frontier America of outer space could really coexist with America's troubled "inner spaces." That, too, would be learned before the decade was out.

Just how committed Murrow was to civil rights became apparent in the coming weeks, when he challenged Webb over NASA's having no African American astronauts. Webb told Murrow that astronauts needed to have very specific aeronautical qualifications, and there were no black test pilots. Taking the matter up with Kennedy directly, Murrow urged the White House to insist on diversity. "The first colored man to enter outer space, will, in the eyes of the world, be the first man to have ever done so," Murrow wrote, cognizant that a vast majority of the world's population was non-Caucasian. "I see no reason why our efforts in outer space should reflect with such fidelity the discrimination that exists on this minor planet."

Despite genuine criticisms from the right (over cost) and left (over the diversion of resources from social justice and antipoverty initiatives), congressional leaders found it easy to move Kennedy's moonshot idea toward a vote in July. The one expected Democratic adversary, the thorniest burr in the New Frontier saddle, was Senator William Proxmire. Complaining about the unnecessary emphasis on speed in NASA's beat-the-Soviets goals, the Wisconsin senator claimed that the space race was warping the U.S. educational system by siphoning off scientific talent from America's universities and by eating up funds that could be used to educate a new generation of scientists. He dismissed Kennedy's idea that the space program itself was supposed to inspire students to pursue such careers; the increase in space activity during the 1950s and the torrents of publicity surrounding it had led in the opposite direction. "The Russians are now graduating some 125,000 engineers and scientists a year," Proxmire said, "compared to our 45,000. That is in contrast with the situation 10 years ago when we were graduating about 55,000—more scientists then than we are now—and the Russians were graduating about 36,000. These statistics are of deep concern. What we have to do is concentrate on scientific education rather than on these spectacular leaps to the Moon as a first priority."

The world of science also contained critics of Kennedy's accelerated moon mission. *U.S. News and World Report* printed an interview with Dr. Hans Thirring, former head of the Institute for Theoretical Physics, in Vienna, who compared the idea of a manned moon voyage to attaining perpetual motion, moving the Rocky

Mountains to another location, or turning an animal into a different species—"not an utter physical impossibility" but presented with "economic impossibilities." From a purely empirical perspective Thirring felt Kennedy's timeline for landing an astronaut on the moon infeasible. "I am quite sure it will not be done within the next 10 years," he said. "And I think it very likely not to happen within the next 30 or 40 years."

Thirring advocated for sending an unmanned vehicle to the moon first, and other notable scientists agreed with him. Dr. Robert Boyd, a British physicist who had worked closely with NASA since 1959, thought that a human in space was "really rather a nuisance." Holding that the massive expenditure planned for the moonshot could be better spent on medical research, Boyd concluded that "I'm not saying that it's unwise, for example, of the United States to do manned space flights. This may, in fact, be the best political thing they can do in the circumstances in which they find themselves. But just taking humanity as a whole and the question of what we would do if we were all sane men, I think we wouldn't be spending money sending man into space. . . . Personally I am rather sorry that, frequently, science is dragged in as the justification for what I really regard as a political exercise."

The arguments of Murrow, Proxmire, Thirring, and

Boyd had merit. They pointed to the most daunting risk undertaken by Kennedy: by backing the moonshot with $20 to $40 billion, he seemed to be turning his back on poverty, civil rights, education, environmental conservation, and medical research. A similarly massive government expenditure in any of these areas would have made a huge near-term difference in people's lives. By prioritizing the moon voyage, Kennedy was gambling on an even larger long-term boost, but he also had to accept the very real possibility of coming up short. If the Apollo program ended in embarrassment or tragedy, all the paeans to the nobility of having made one's best effort wouldn't stack up against the likelihood of what might have been accomplished if he had embraced another field as the heart and soul of the New Frontier, and with the nation's full financial backing. This was the dark side of the moon program, the persistent voice whispering at the societal good that could have been accomplished by devoting more than $20 billion to, say, finding a cure for cancer or building high-speed trains from coast to coast.

But the attraction of the moon went beyond what Dr. Boyd had called "a political exercise" for JFK. It appealed to the president *as a Kennedy*, as someone who had absorbed his parents' ultracompetitive attitude as well as the underdog outlook inborn from their Irish-Catholic heritage. Worries that the United States was lagging behind the USSR in space didn't intimidate Kennedy; more likely, they had the opposite effect. Traveling Europe and the Pacific as a young man, JFK had seen firsthand the inspirational role America played in people's everyday lives. Opinions of its culture might vary, but American exceptionalism was widely embraced, and was too valuable a commodity to lose. The president's wartime role as a PT boat captain, part of a cadre chosen for quick thinking and leadership, gave him a natural bond with the NASA astronauts, who had been handpicked for those same qualities and more. Even his association with Hollywood, via friends such as Frank Sinatra and Dean Martin, gave him an understanding of the need for patriotic heroes, manufactured or not. To Kennedy's mind, Alan Shepard was the New Frontier's John Wayne, an exemplar of American bravery and can-doism in the name of national greatness.

As the icing on the cake, Kennedy brought to the moonshot decision the experiences of his own political career. For fifteen years, he'd experienced Americans' fascination with his youth, radiance, potential talent, and physical vitality. He represented America's future without even uttering a word. And it was for these reasons that voters who met JFK were, even if they fought against it, instantly captivated by him. The most vivid experiences of his life—both his public life and the one he hid from the world (his poor health, philandering, Joe, the grislier details of the *PT-109* incident)—carried him to the decision to go to the moon. His philosophy of courage was that life is short, bold steps forward are immortal, so *act*. Apollo, he understood, transcended party politics and regional differences. It was a story unto itself. Why not create a generation of space heroes? Why not use NASA as a venture to jump-start American technology? Even the eventual price tag of $20 to $40 billion, a staggering amount in 1961, was almost natural for Kennedy, who had seen his father, Joe Sr., spend vast fortunes to get what he wanted on Wall Street or in Hollywood, never with cause for regret.

Faith that von Braun and his team could develop the proper Saturn rocket for a moon launch was widespread in Kennedy administration national security circles. But the flight mode—how to land astronauts on the moon and bring them back alive—was still fiercely debated. "What was difficult for us were so many unknowns with getting there," recalled Chris Kraft, then NASA's sole flight director. "It sort of made us all question whether it was possible or not."

Luckily, NASA wasn't working in a vacuum. Earlier in the spring, a NASA committee had begun pondering the best general design for a possible lunar landing, and major U.S. aerospace firms had been doing cost analysis configuration control, and probability reports. A trio of prominent companies (Martin, Convair, and General Electric) had already started designing possible three-person Apollo capsules, hoping to secure a massive government contract if the mission was approved. Other companies were developing the onboard and ground-control systems that would be needed, including computers, navigation, flight control, thermal protection, and life support. Without a decision on the flight mode, however, nothing von Braun or others built would get off the ground in Cape Canaveral.

Within the space world, debate centered on the precise rocket staging, trajectory, and rendezvous needed to bring Apollo astronauts to the moon and back. NASA was weighing three primary options. Topping the list was direct ascent (DA), in which an enormous rocket would lift off from Earth and plow for the moon, landing and then blasting off for home once the job was done. Although this concept required the smallest number of orbital maneuvers, the big disadvantage was that the proposed Nova rocket would have to be substantially larger than any rocket that von Braun was working to build. Perfecting the rocket for a lunar mission would have proved a difficult enough challenge, but physically the rocket was deemed too gargantuan to produce at NASA's new Michoud facility in Louisiana and too enormous to test at what eventually became the John C. Stennis Space Center near Bay St. Louis, Mississippi.

The second alternative was the Earth orbit rendezvous (EOR) approach. For EOR, multiple boosters would launch spacecraft in rapid succession and they would be joined in Earth's orbit into a full spacecraft. That ship would then fly to the moon, land, discard a module, and return home in a similar fashion to the DA method. Von Braun's Huntsville team promoted this solution because it mitigated the necessity of contracting a huge Nova-class vehicle while they also built more Saturn rockets. The catch, however, was the multistage complexity of the orbital assembly.

The distant third idea was the lunar orbit rendezvous (LOR), a concept championed by Dr. John C. Houbolt of NASA's Langley Research Center. While the Jet Propulsion Laboratory had proposed a modified rendezvous profile where refueling would occur on the lunar surface, the primary LOR profile was to conduct any refueling in Earth orbit (not on the moon). Under Houbolt's plan, this type of sortie would use a powerful three-stage rocket to launch an assembly of three spacecraft into orbit: a modular command module (CM); a service module (SM), containing main propulsion, fuel cells, and attitude-control systems; and a lunar module (LM), a newfangled contraption to reach the moon's surface. In lunar orbit, two astronauts would use the LM to land while one crewman remained with the combined command service module (CSM) in a "parking orbit" around the moon. Once scientific experiments on the moon's surface were complete, the LM would ascend back to lunar orbit and rendezvous with the CSM, which would then fly the three-man crew back to Earth. The big advantage to this LOR approach over Nova was the single-vehicle launch. Furthermore, from a technical standpoint, the LOR payload had a much smaller mass, which made it possible to develop a rocket that would not be as colossal as Nova. The downside to LOR was that the astronauts had no means of escape should any rendezvous maneuver flounder. NASA understandably worried about the public relations abomination of dead astronauts floating around in a stable lunar orbit or if they got marooned on the moon.

As of 1961, Houbolt's idea of docking two spacecraft in orbit was purely theoretical. NASA engineer Laurence K. Loftin Jr. summed up opposition to the LOR concept perfectly: "We thought it was too risky," he said. "Remember in 1961 we hadn't even orbited Glenn yet. We certainly had done no rendezvous yet. And to put this poor bastard out there, separate him in a module, let him go down to the surface and fire him back up and expect him to rendezvous. He didn't get a second chance; it had to be dead right the first time. I mean that seemed like a bit much."

The businesslike Loftin spoke for most NASA executives when he said the LOR was an engineering pipe dream—fine in the drawing room, but a probable disaster if implemented. "Houston's first reaction to Houbolt's suggestion was rather negative," von Braun confirmed. "While conceding that in principle it should be possible to save launch weight by leaving part of the fuel required for the return flight in lunar orbit, rather than soft-landing it on the moon and carrying it out of the lunar gravitational field again, Houston felt that Houbolt's equipment and weight assumptions for the lunar module had been highly unrealistic. For instance, his original LM concept did not have a pressurized cabin for the ascent stage and the LM guidance and control system for the tricky descent to the lunar surface was considered an inadequate rig 'consisting of a plumb bob and a reticle.'"

For Glennan, Eisenhower's NASA administrator,

the fact that the agency was still debating how to bring an astronaut back alive from the lunar surface meant the game was over before it began, giving him reason to label Kennedy's speeded-up moonshot as high-stakes folly. Poisoning the well on Capitol Hill against Apollo became Glennan's hobby in retirement. Regularly, he'd tell NASA leaders and old cohorts that the president and von Braun were on ego trips. Lobbying Webb, Glennan said, "No, Jim, I cannot bring myself to believe that we will gain lasting 'prestige' by a shot we may make six to eight years from now. I don't think we should play the game according to the rules laid down by our adversary." Webb, rich in recent accomplishment and enjoying JFK's full backing for his agenda, dismissed Glennan's dissent as sour grapes. For Webb, going to the moon on a "crash" basis didn't leave room for intellectual musings, Monday-morning quarterbacking, or go-slow Cassandras. His standard defense of Apollo was that while Kennedy's moonshot was extremely audacious, so, too, had been the Panama Canal, D-day, and the Manhattan Project.

Twice in 1961, Houbolt, a tough-minded infighter, was almost fired for skipping proper channels and writing directly to Seamans about his LOR proposal. By the end of the year, Webb and von Braun leaned toward Houbolt's lunar orbit rendezvous plan as the best option. What won both him and Kennedy over was that LOR required only one Saturn rocket. What he didn't know until late in his presidency was that the official name of the jumbo Apollo moon rocket was Saturn V. In the next few months, Kennedy and Webb came to believe that von Braun's Saturn I turned Saturn IB was his ticket to the moon. And if the United States truly wanted to land astronauts on the moon, LOR was the ticket to ride. Although Houbolt's plan occasioned fierce, ongoing debate, by the summer of 1962 it became the decided strategy for the Apollo mission.

Even while the flight mode controversy raged, nearly everyone could agree on one thing: that von Braun's eventual Saturn rocket would invariably have to be one of the most ingenious technological innovations of the twentieth century. Von Braun's Saturn V moon rocket didn't fly until 1967. At the time of Kennedy's speech to Congress, von Braun was still building the Saturn I, the first of which flew in October 1961.

Just hours after Kennedy finished speaking to Congress, test pilot Joe Walker flew an X-15 rocket plane from Edwards Air Force Base in California, traveling five times the speed of sound. This was a new record for a vehicle operated by a human occupant (versus the rockets that carried Shepard and Gagarin, who had little or no control, respectively, over their flights). Almost without exception, the next day's newspapers ran the summary of the president's address next to the more immediately exciting news of the X-15 and its zooming pilot. To the air force, Walker's record-breaking flight was further proof that the reusable X-15 was more important for the future of space exploration than von Braun's single-use Saturn rockets. To Kennedy, who'd just staked his presidential reputation on the goal of a rocket-powered moon mission, the timing of Walker's X-15 flight could not have been pleasing.

In the aftermath of his May 25 speech, Kennedy turned his attention to his upcoming summit with Khrushchev in Austria, where sensitive topics such as the status of divided Berlin would likely overshadow his push for a U.S. moonshot. Still, his speech at least assured the world that the American commitment to space was second to none. Publicly, the Soviets feigned good wishes to the United States regarding its highly implausible lunar initiative, and some scientists working internationally suspected that Moscow was secretly delighted by it—and had even been maneuvering America into just such a plan. These scientists theorized that a laser-like concentration on the moon would distract American focus, funding, and brainpower from more practical military projects, such as aircraft carriers and ICBMs.

At the two leaders' introductory luncheon on June 3, the day before the Vienna summit officially opened, Kennedy brought up the safe topic of Gagarin's flight of two months before. Bursting with pride, Khrushchev described the flight, with a friendly tidbit about all of the trepidation the Soviet leaders had that the beloved cosmonaut would lose his mind from the effects of space. Instead, he noted, there was no problem, and Gagarin even sang folk songs among the stars. Inevitably, the two men began talking about the Americans' grand design for a trip to the moon. "With respect to the possibility of cooperation in launching a man to the Moon," noted the State Department's official memo about the conversation, "Mr. Khrushchev said that he was cautious because of the military aspect of such flights."

Just what Khrushchev meant by that was not recorded, but the notion succeeded in putting Kennedy on the defensive. He responded with an "inquiry" to Khrushchev "whether the US and USSR should go to the Moon together." The world would have toasted these Cold War rivals collaborating on a moonshot, but Khrushchev demurred. "At first he said no," according to the account of their conversation, "but then said 'all right, why not?'" though his first answer was clearly his last. Asked later about the discussion by his son Sergei, Khrushchev admitted that "if we cooperate, it will mean opening up our rocket program to them. We have only two hundred missiles, but they think we have many more." Khrushchev worried that Kennedy might launch a first ICBM strike if the disparity were revealed. "So when they say we have something to hide . . . ?" Sergei pressed.

"It is just the opposite," his father said with a chuckle. "We have nothing to hide. We have nothing. And we must hide it."

For the next two years, the president would periodically send up a trial balloon for the idea of a joint U.S.–Soviet moon venture, then back off and reiterate the national security necessity of being first to the moon. These inconsistencies, which space historian William Kay sorted into the categories of "competition" and "cooperation," place the space race firmly in the context of the Cold War. Kay contends that it was in fact a two-pronged strategy designed to keep the United States on the positive side of world opinion by appearing (as in Vienna) to value peace above all else. Kennedy's words were about peaceful coexistence in space while his actions were aimed at America's winning the race. With

Cold War competition running hot over Berlin, Cuba, and Southeast Asia, Kennedy could and did extend the bauble of outer space collegiality and collaboration as a means of ameliorating the very real aggressions that existed between Washington and Moscow—a peaceful gesture that sacrificed nothing on the broader scale, and couldn't be interpreted as a sign of weakness.

The rest of the Vienna summit did not go as well as the luncheon. Khrushchev was domineering and unyielding on most subjects, especially about wanting the NATO nations to leave Berlin. When JFK tried to frame the division of the city in moral terms, he left the impression that the United States would take no offensive action in response to East Germany's threats to cut West Berlin off from the West. This rhetorical error didn't go over well. "Mr. Kennedy's reaction was not unlike that of many Western officials who have negotiated with the Russians for the first time," wrote James Reston of the *New York Times.* "He approached the conversations thinking he knew what to expect. But nevertheless, he was astonished by the rigidity and toughness of the Soviet leader." Privately, the president mumbled to Reston, "He just beat the hell out of me. I've got a terrible problem if he thinks I'm inexperienced and have no guts." That assessment, blunt and honest, was shared by other Washington insiders. Khrushchev had belittled Kennedy in Vienna by threatening to crush America in Berlin.

The summit had been a bust, and America's geopolitical problems quickly got worse. In Berlin, giant construction machines were getting gassed up and wall-building materials were stockpiled along the border between East and West. In his May 25 speech, Kennedy had spoken of failure as a matter of missing an arbitrary deadline. During the grim hours after meeting Khrushchev in Austria, the president saw more clearly what failure could be. In this tense Cold War environment, a shooting war seemed just one poorly chosen word away. It turned out, however improbably, that Kennedy's moonshot pledge had given the United States the upper hand over the USSR in the psychological game of one-upmanship. If the Americans were playing defense in Berlin, they were, by contrast, on the offense in the fields of manned space-reconnaissance aviation, satellites, ICBMs, and moonshots.

John F. Kennedy and Premier Nikita Khrushchev at the Vienna summit on June 4, 1961.

A U.S. Marine Corps helicopter lifts astronaut Virgil "Gus" Grissom from the Atlantic Ocean after his *Liberty Bell* 7 capsule lands, following a suborbital flight. President Kennedy would telephone Grissom aboard the USS *Randolph* to congratulate him on surviving the ordeal.

14

Moon Momentum with Television and Gus Grissom

The truth was that the fellows had now become the personal symbols not only of America's Cold War struggle with the Soviets but also of Kennedy's own political comeback. They had become *the* pioneers of the New Frontier, recycled version.

—TOM WOLFE, *THE RIGHT STUFF* (1979)

In the spring of 1961, on a limousine ride from the White House to the National Association of Broadcasters meeting with Johnson, Shepard, and FCC chairman Newton Minow, President Kennedy cracked a telling joke at the vice president's expense: "You know, Lyndon," he said with sly delight, "nobody knows that the vice president is Chairman of the Space Council. But if [Shepard's] flight had been a flop, I guarantee you that everybody would have known you were the chairman." Laughter filled the car, but LBJ wasn't amused. Minow, the quick-draw wit, deadpanned, "Mr. President, if the flight would have been a flop, the vice president would have been the next astronaut."

Humor's best when laced with truth, and Johnson knew Kennedy really would have made him the sacrificial lamb if *Freedom 7* had ended in disaster. The mocking tone of the president's tease, with its implication of superiority, stung LBJ. Back in 1957, when JFK was deriding failed Vanguard launches, Johnson had been the driving force in the Senate behind the establishment of a civilian-run NASA. While Johnson had met with ex-Peenemünders numerous times to discuss the nuances of rocketry, Kennedy had focused on the alleged missile gap. From *Sputnik* onward, Johnson had championed NASA as a potential bonanza of tech-driven wealth for numerous states. "Space was the platform from which the social revolution of the 1960s was launched," LBJ explained in his memoir, *The Vantage Point*. "If we could send a man to the moon, we knew we should be able to send a poor boy to school and to provide decent medical care for the aged."

So LBJ was paradoxically pleased yet galled that

JFK had assumed the mantle of "President Moonshot," stealing all his space policy thunder. In a debate on the minutiae of space exploration, Johnson probably would have outshone Kennedy. But the vice president never got such an opportunity, and he learned to hide his low-burning resentment, serving as a loyal New Frontier soldier for the good of the nation. Playing second fiddle, often out of the White House loop, a frustrated Johnson used leverage over NASA administrator James Webb and Senator Robert Kerr to help make sure that NASA's proposed Manned Spacecraft Center would be built in Houston. "Many friends of Lyndon Johnson and [Congressman] Albert Thomas made a hell of a lot of money when Houston was selected," astronauts Alan Shepard and Deke Slayton wrote in their dual memoir. "Those already wealthy smiled as their coffers bulged with new funds."

For all Johnson's political acumen, he never cottoned to the fact that horse-trading and arm-twisting had taken a backseat in politics to being telegenic and delivering a good sound bite. By a long shot, Kennedy was the best communicator of his generation, able to conquer the moon and stars with uplifting oratory and a photogenic smile—skills in which Johnson was sorely lacking. Off camera, Johnson had a captivating way of speaking, peppering his commentary with off-color anecdotes and hoarse laughter. But when the camera ticked on, he seemed tongue-tied, uncomfortable in his own skin, too gravel-road Texan by half.

Kennedy was the apotheosis of a revolution started by Franklin Roosevelt, using media—in this case, TV—to move the presidency closer to the American people than ever before. In the early days of the republic, citizens could just drop in on Presidents Jefferson, Monroe, or Tyler at the Executive Mansion. As the country grew, the president became more isolated, particularly after the assassinations of Lincoln and Garfield. Between 1933 and 1945, though, President Roosevelt became a staple presence in American homes via thirty "fireside chat" radio broadcasts, which he used to explain major New Deal and World War II policy decisions in a conversational rather than oratorical style, as if talking one on one with every American.

By the early 1950s, as TV gained cultural power, presidential aspirants suddenly had to be telegenic. Dwight Eisenhower treated TV as a fad, a radio with images, but nevertheless sought to master it by appointing actor-director Robert Montgomery as a White House consultant, taking the Hollywood veteran's advice on how to relax for the camera. Under Montgomery's guidance, Ike undoubtedly improved between

1953 to 1961, but he was limited by looks, delivery, and temperament. Had Hollywood cast Ike in movies, it would have been as a Midwestern banker or the principal of a Pennsylvania military school.

Kennedy, on the other hand, didn't need coaching. Television was familiar turf for him, and he had an inexhaustible fascination with the medium, especially after broadcasts became one of the most potent weapons in his presidential campaign. TV had been a boon for him in the Senate as well: in February 1957, when the McClellan Committee hearings probing corruption in labor unions were broadcast live, Americans got their first sustained exposure to Kennedy, young, calm, and handsome beside his staid and wizened colleagues. The most telegenic politician of his era, or perhaps any era, Kennedy formed an instant partnership with TV that would last the rest of his life.

San Francisco sociologist Don Mahan never knew Kennedy personally, but "somehow I had come to know this man as a fellow human being, as a friend, as trite as that may sound. . . . I do not believe that this situation is at all unique with me." The medium, Mahan explained, seemed to amplify JFK's stature. "Maybe I watch more newscasts and special news reports than the average television viewer," he noted, "but even the average viewer has to be aware of John Kennedy as he was around so much. I am sure that he was entirely conscious of the potential of the medium."

Every part of the way Kennedy comported himself on TV was riveting and irresistibly kinetic, with a naturalness and energy that cut right through the dull, ordinary hum of late-1950s political discourse. Kennedy brought verisimilitude to the small screen, showing a levelheaded persona shaped by all the influences and events of his life: the hard-driving family, the Ivy League education, his wartime heroism, and his Catholic idealism. Playing to his father's sense of assurance and taking lessons in schoolboy popularity from his brother Joe Jr., JFK had learned to flip from introvert to extrovert when the occasion demanded. By all accounts, his two siblings who died young, Joe Jr. and Kathleen ("Kick"), were better equipped by nature to shine in social settings, but Jack learned to keep up, ultimately transforming from sickly and standoffish youngster to athlete and life of the party—and learning that he enjoyed it. While his early days in Congress suggested he still might have been happiest alone with a pile of books and newspapers, his evolving sense of self finally coalesced into the cool persona Americans came to know through their living room TVs in the early 1960s: charming, accomplished, and slightly detached, embodying good taste, erudition, up-tempo pragmatism, and apparently effortless grace.

Arriving at the White House with confidence born of his perceived wins against Richard Nixon in their televised debates, Kennedy immediately tapped into one of the basic truths of TV's ravenous appetite: a measure of spontaneity is worth the risk of gaffes. Starting on January 25, 1961, he began holding regular televised news conferences, bringing unscripted and unrehearsed banter to the usually sober world of broadcast news. The effect was seismic, and sometimes a cause for worry, too, among his die-hard supporters. Would he make a mistake? Bungle a policy idea? Or simply wear out his welcome in American homes? Pierre Salinger, the White House press secretary, had originated the idea of televising the conferences, and according to him, "Kennedy felt very strongly that he should go ahead."

Kennedy was smart to trust Salinger's instincts. The press conferences, typically scheduled for late afternoon, were staged on average every sixteen days. The American people had never before witnessed their president fielding questions like a shortstop, displaying not only his knowledge (or admitted ignorance) of a vast array of issues, but his diplomacy in choosing precisely how to respond. It was much like the Mercury Seven manned space launches from Cape Canaveral: you knew the time to tune in but not what would ultimately happen. Within any one press conference, JFK would whiff or make outright factual mistakes, but he would also display flashes of pluck under verbal fire, along with irresistible wit. Some questions pleased him, a few rankled—and in each case, the audience could sense his genuine feelings. Overall, even the president's detractors had to admit that he handled the majority of the press's questions with well-chosen words, a raised eyebrow, or a half smile that communicated clearly what he thought about a particular issue. What was more, he kept his answers brief, avoiding that enemy of all good TV: unconstructed rambling.

On average, the press conference broadcasts drew about eighteen million viewers, which Salinger considered fantastic. Some observers caught on to the potential of the televised news conference immediately. Washington correspondent Virginia Kelly, for one, pointed out in the Long Beach *Independent Press-Telegram* that Kennedy's extemporaneous speaking, so rare compared with most previous presidents, would inevitably "reveal his innate characteristics clearly." She noted that his voice, along with his "personal mannerisms, facial expressions, gestures and all the intangibles of personality," impacted the points he made. She evoked a comment he made about Soviet activities in the Congo: "in cold print," she said, his words seemed dark and "stern," conveying "a solemn warning"; on TV, however, the tone of his voice made the statement seem "firm but temperate." For that reason, Kelly predicted that the press and public would seek out both versions of the president's news conferences. The words mattered, but as of 1961, so did their delivery.

James Reston of the *New York Times* recoiled from the TV rendition, calling the televised news conference "the goofiest idea since the hula-hoop." Among other things, Reston felt that the president's exposure would overwhelm that of his opponents, disrupting the political system. The editors of the *Shreveport Times* agreed with that stance and went further, insisting that "the real factor in all of this is the dignity and prestige of the office of the presidency of the United States, the bringing it down to the point of a periodical show." The chance for Americans to see their president doing his job regularly was exactly what the *Los Angeles Times* found admirable about the broadcasts, even if they were "undisguised exercises of the Presidential power." The editors pinpointed the mood of a nation fascinated with its young president: "Whatever the doubts his advisers and well-wishers—or even those who hoped he would come a cropper—had about the televised press conference, they know now that television must have been invented for the use of this President."

On a parallel track, TV might have been invented for the use of space explorers. Throughout history, earthly explorers to unknown locales had returned with copious sketches and maps in order to give their sponsors a sense of their new discoveries. Space, however, made it much more of a challenge to answer the perennial question "What's it like there?" Not even the most advanced space scientists could answer that question before the Soviets launched *Sputnik* in 1957 and began to learn the answers firsthand. Some of the earliest Soviet satellites sent photographic and even television imagery back to Earth, though not in real time. The Americans soon followed, committing to live broadcasts from Cape Canaveral.

For the three major American TV networks, NASA space events offered hours of gripping drama, charismatic personalities, constant action, and stunning visuals. As an added bonus, there were no fees like those they typically paid to broadcast sports and entertainment events—NASA supplied the feeds from space (first audio, later video) for free—though covering blastoffs, touchdowns, and other terrestrial aspects of the flights remained expensive (to lessen the burden, the networks soon formed a pool and cooperated on the NASA video feeds). Another innovation was the taping of NASA's launch rehearsals, which gave the networks an on-hand supply of riveting action footage.

The use of shared footage put more pressure on the running commentary that would distinguish each broadcaster's programming. ABC News entrusted its space coverage to the serious-toned Jules Bergman, an eminent journalist from New York City who had been academically trained in scientific reporting. NBC News relied in the late 1950s on Roy Neal, a reporter-producer educated at the University of Pennsylvania. With a longtime interest in rocketry, Neal had been the network's specialist in the subject since the beginning of the decade; he was also a leader in the effort to arrange pool coverage among the networks. Frank McGee, who learned aerospace jargon technology with enthusiasm, became the Peacock Network's space anchor once Neal moved into the ranks of broadcast producers. CBS News, for its part, initially entrusted space coverage to Charles von Fremd, a Yale graduate in his thirties. Von Fremd was ambitious, but not as ambitious as his colleague Walter Cronkite, who started covering Mercury missions with Alan Shepard's *Freedom 7* flight. Ever since Cronkite had written about the U.S. Army Air Corps in World War II for United Press International, befriending flyboys stationed in Great Britain and accompanying a bombing mission over Germany, he'd seized military aviation and space exploration as his special beat.

NASA space exploration was a major story of the early 1960s, but covering it came with a caveat: the anchor at the launches had to have a confident grasp of complex concepts and keep hundreds of facts close at hand. The reward might be tempting (high ratings and association with the space program), but the risk of a career-ending blunder on the airwaves loomed even higher with sharp reporters. Cronkite conquered this difficulty first by making a serious study of space and rocketry. More cleverly, he filled notebooks with well-organized details of the topics in question and was an expert at consulting his notes without the camera catching him at it.

Unlike Murrow and the many CBS News correspondents he inspired, Cronkite wasn't known for courting controversy; instead, he favored an objective reporting of the day's events. Writing later about Cape Canaveral as it was in the 1960s, he noted that "while the eyes of the rest of our population might have been downcast as the nation dealt with a succession of problems—civil rights, assassinations, Vietnam—it seemed that everyone at the Cape was looking up, up into the skies that invited their conquering touch." Few could discern Cronkite's personal opinions on most topics, but it was clear that NASA's missions had particular appeal to him. In fact, even after being named anchor of *CBS Evening News* in 1962, he made the unconventional decision to continue hosting the network's live coverage of space launches.

The time lines of NASA development and TV's infiltration of American life coincided almost precisely. This convergence meant that fifty million people could see their tax dollars at work, watching the moonshot draw closer to success with each ensuing launch. No exploration in human history had been so democratic. In exchange for a massive expenditure, Americans as individuals became stakeholders in the grand drama, partners in the adventure along with the astronauts and engineers, and the president himself. Kennedy, for his part, acted a dual role: as the man whose vision had lit the fuse beneath Apollo and, simultaneously, as one of the millions of Americans glued to the TV coverage of the program. Americans watched their president on TV viewing TV broadcasts of space launches that they themselves had paid for. The combination of Kennedy, television, and space had the desired effect. As the leading French space scientist (and proud Marxist-

Leninist) Roger Bonnet later recalled, he'd initially been a cheerleader for the Soviets to win the manned-space race, then changed his mind when, under Kennedy, the Americans adopted "an open policy" of space information that made Shepard's flight and the Mercury program feel like they belonged to the world.

An April 1960 poll revealed that across Europe, a majority of citizens believed the Soviet Union was the undisputed leader in space, after years of "competition without war." Echoing this finding, a contemporaneous statistic from the U.S. Information Agency reported that only 7 percent of French citizens polled believed the United States would overtake the USSR in the coming decades. Then came Kennedy's May 25 speech, which framed manned spaceflight as the barometer for America's capacity to mobilize resources and win the Cold War battle "between freedom and tyranny . . . the battle for men's minds . . . the minds of men everywhere who are attempting to make a determination of which road they should take." There was a democratic openness to NASA's space launches. Under Webb's leadership the space agency didn't shield its activities until success was guaranteed, as the Kremlin did. Like Kennedy, NASA astronauts stood up before the cameras and laid it all on the line. By contrast, in the Soviet Union, where obsessive secrecy was the governing ethos, even the phone book was classified.

Building on Kennedy-injected momentum, NASA was preparing for the second televised Mercury launch in the third week of July. With an appropriations vote on the president's moon initiative pending in Congress within weeks, the need for a smooth, successful mission was paramount. In the pilot's seat would be Virgil Grissom, a curmudgeonly pillar of tenacity, unpretentiousness, and midwestern work ethic. The *Liberty Bell 7* capsule in which he would take his projected fifteen-minute flight was more sophisticated than Shepard's *Freedom 7*, and among the new astronaut's missions would be personally testing the spacecraft's reentry.

Grissom hailed from Mitchell, Indiana, where as a boy he'd delivered newspapers and picked cherries and apples for twenty-five cents per tree. With his spare cash, young Virgil took flight lessons at the local airport, dreaming of becoming a military pilot. During World War II he joined the U.S. Army Air Corps, improving his piloting skills in Texas and Florida. After the war, Grissom used the GI Bill to enroll at Purdue University, where he earned a BS degree in mechanical engineering. While at Purdue, he acquired the nickname "Gus" during a card game, when someone misread "Gris" on the scorecard.

Married and with a young child, Gus Grissom reenlisted in the air force after college and was assigned to Presque Isle Air Force Base in Maine, becoming a member of the Seventy-Fifth Fighter Interceptor Squadron. Officers were impressed with his superlative airmanship and how he made an airplane tame and obedient, and he was soon sent to Korea to fly F-86 Sabres, usually in a "finger four" formation, in which the lead plane fired at enemy aircraft while the rest provided cover. "I usually flew wing position in combat," Grissom recalled, "to protect the flanks of other pilots and [to] keep an eye open for any MiGs that might be coming across." All told, he flew one hundred low-flying combat missions, cutting North Korean bridges and roads and engaging Russian-built MiG-15s in aerial combat without even once sustaining a hit on his own plane.

With a chest full of medals, including the Distinguished Flying Cross, Grissom returned to America to become a flight instructor at Bryan Air Force Base, Texas. The general impression of Grissom from his superiors was that he had unflappable grit, a natural feel for flight, and an easy camaraderie. Fellow Mercury Seven astronaut Gordon Cooper described the five-foot-seven Grissom perfectly as "a little bear of a man and a country boy, but when it came to flying he was steady and no nonsense." After successful stints at the Air Force Institute of Technology at Ohio's Wright-Patterson Air Force Base and at the U.S. Air Force Test Pilot School at Edwards Air Force Base in California, he was a natural pick for the Mercury program.

There was, however, a hiccup, or perhaps a sneeze. While undergoing a series of medical examinations as part of the program, it was revealed that Grissom suffered from hay fever, which took him out of the running to be the first American in space, clearing the way for Shepard. Grissom, devoid of envy, patiently waited his turn until July 21, when, hay fever be damned, the decorated air force test pilot suited up to become the second American astronaut to reach suborbital space. At last the air force had one of its own headed into the galaxy.

Just after seven that morning, Grissom was hustled into the capsule atop a towering eighty-three-foot-tall Redstone by his backup, John Glenn. The U.S. Weather Bureau meteorologist thought the conditions were optimal. Glenn gave Grissom a go-get-'em pat good-bye, and then seventy hatch bolts were torqued down. The launch went off without a hitch. Soaring to an altitude of 118 miles, Grissom traveled 302 miles in fifteen minutes and thirty-seven seconds. With more time aloft than his predecessor in space, Grissom had time to peer through his capsule's window and radio back reports of stars as pure as diamonds against an ethereal blue universe. Back at Cape Canaveral, Shepard was serving as capsule communicator, and the two astronauts traded brief awestruck commentary about the beauty of the Earth before heavy cloud cover hindered Grissom's view. The amazement with which Gus tried to convey his space experience was endearing: "I can see the coast," he radioed back, "but I can't identify anything."

Reentry went smoothly for *Liberty Bell 7*, with the craft traveling thirty miles per hour faster than had Shepard's *Freedom 7*. But once the capsule splashed down in the Atlantic, a technical malfunction caused the side door to blow open, sending seawater gushing in. The same type of disaster had killed Victor Prather. Grissom felt the capsule begin to sink. Without waiting for the two navy helicopters assigned to retrieve him, he scrambled out. Minutes of life-or-death drama ensued. When the choppers arrived, Grissom began trying to help the first one attach a cable to the capsule, which was also supposed to be retrieved, but ocean water started entering his space suit through the open collar. Panicking, Grissom waved for help, but the airmen on the helicopter misconstrued his message. Fortunately, the crew of the second helicopter recognized his predicament and quickly lowered a rescue hoop, hoisting him to safety. Abandoning the effort to lift the water-filled *Liberty Bell 7*, both helicopters returned to a nearby navy ship. When asked by a journalist how he felt, Grissom replied, "Well, I was scared a good portion of the time; I guess that is a pretty good indication."

TV viewers didn't see any of the difficulties, because live video of the retrieval wasn't available. As the news anchors struggled to sort out what was happening, CBS and NBC each ran the "rehearsal" footage, shot days before, which showed a picture-perfect operation, with two helicopters flying low over open water, one of them carrying the *Liberty Bell 7*. On NBC, the video feed then reverted to moving images of mission control, also prerecorded. Within a couple of minutes, an announcer from New York interrupted the coverage to say, "A moment ago, you saw some pictures of a helicopter lifting a space capsule out of the water. Those were films made of the trial runs, the practice and training for this program, and were not actual scenes of Grissom being lifted out of the water at this time." The live coverage then continued, but there was an obvious delay, with no confirmation of Grissom's safe return. So it was that the NBC announcer from New York interrupted a second time to clarify that the film of the technicians in mission control had also been shot in advance.

Uneasy viewers were confused and suspicious. What they saw was at odds with what they were hearing. Eventually, live reports from the navy ship told the full story and confirmed that Grissom was unharmed. The confusion, however, was widely criticized, with wags calling it "the great Cape Canaveral tape caper." Recognizing that they'd betrayed their obligation to present news truthfully and endangered public trust in their NASA programming, the two guilty networks were contrite. "It won't happen again," promised Richard Salant, president of the CBS News division. "We know we can't juxtapose fact and fiction."

That morning, John F. Kennedy had been in his bedroom at the White House when aides informed him that coverage of the *Liberty Bell 7* mission was about to begin. Turning on the TV set, he watched with nervousness, amazement, and, finally, joy. Once Grissom was safely aboard the USS *Randolph*, Kennedy placed a telephone call to the astronaut of the moment. Still drenched, Grissom chatted with the president. Photographer Dean Conger of *National Geographic* was on duty and documented the historic call with his camera. "That was the last thing he wanted right then," Conger recalled. "[Grissom] still had his space suit on . . . probably full of sea water."

From Kennedy's perspective, the *Liberty Bell 7* voyage was astounding, the excitement irresistible, and yet, by then, the Mercury suborbital program was already something of a relic. Grissom's flight was significant but, as a baby step, only slightly more ambitious than Shepard's launch of ten weeks before. Grissom himself, sloughing off having almost drowned on reentry, "was angry about being blamed for his space craft having sunk," astronaut Wally Schirra recalled. "Gus was a tiger."

Just hours after Kennedy's telephone call to Grissom aboard the *Randolph*, he signed a bill to enlarge NASA's manned space programs, especially to achieve an Apollo lunar voyage. Praising Grissom, he authorized NASA to spend more money in the coming year; the world paying close attention to space. "Once again we have demonstrated the technological excellence of the country," Kennedy said. "As our space program continues . . . it will continue to be this nation's policy to use space for the advancements of mankind and to make free release of all scientific and technological results."

Over the summer, anyone involved in NASA, along with companies such as North American Aviation (Los Angeles), Grumman (Bethpage, New York), and McDonnell Aircraft (St. Louis), was learning to translate Kennedy's leapfrog strategy into giant strides forward, wheeling and dealing for federal contracts. Under JFK, the United States, for better or worse, truly was becoming the *Republic of Technology*. "Ever since Kennedy declared his intent to go to the moon, all hell has broken loose here," von Braun wrote to his father. "At the moment, we are working on plans which put in the shade everything we have done before and against which even our [early-generation] 'Saturn' pales."

For a time during the summer, von Braun was under consideration to direct Project Apollo. Certainly, he had the engineering know-how, government experience, and leadership skills to do the job. In fact, many Americans, then and since, believed that he did run the project, a testament perhaps to his incessant promotion of space exploration on television, in publications, and at personal appearances. But according to Michael Neufeld, von Braun's biographer, his chance of assuming the Apollo directorship may have fallen victim to the opinion of NASA veterans resentful of his "big-spending and self-promoting ways." Von

Braun remained in Huntsville, heading the Marshall Space Flight Center, exultant in his excitement to build a moon rocket. Director or no, von Braun remained, with John Kennedy, a most effective salesman to put a moon launch over with the American public.

The specter of a lunar voyage caused a disruption to von Braun's rocket program, where the enormous Saturn C-1 rocket was then being readied for initial testing in the fall. Even though it was the biggest rocket ever constructed in the United States, von Braun knew, as he'd suggested to his father, that NASA's new directive made the C-1 immediately obsolete. What would replace it, though, was dependent on which of the three competing flight mode options was ultimately selected: direct ascent, Earth orbit rendezvous, or lunar orbit rendezvous. That decision, in turn, would determine what kind of vehicles would ride his rockets into space, and *that* decision would have to be taken into consideration in determining the final rocket design. All von Braun knew that summer was that whatever option was selected, the launch rocket would need to be far larger and more powerful than the C-1.

Although von Braun had originally objected to the lunar orbit rendezvous plan championed by John Houbolt, he began falling into line as the method gained traction. On November 7, 1962, that flight mode indeed became NASA's official selection. To facilitate the system, NASA chose the C-5, an upgrade from von Braun's C-1 rocket. Designed in a grand collaboration between NASA and the private sector, the rechristened Saturn V would have three stages, stand 363 feet tall, and weigh 3,100 tons. To lift that hulking weight, the first stage had five huge engines that could generate millions of pounds of thrust by burning liquid oxygen. Together, they would send about 100,000 pounds (45 metric tons) of spacecraft all the way to the moon. The Saturn V, the longest rocket built, used Boeing, North American Aviation, and Douglas Aircraft as primary contractors.

Because the Saturn V rocket was so large and complicated to construct, NASA divided the task among contractors. The first stage would be built by Boeing at Michoud Operations in New Orleans, an enormous indoor facility along the Mississippi River–Gulf Outlet. In October 1961, NASA also announced the establishment of the Mississippi Test Facility to test large Saturn boosters for Apollo. The second and third stages were assigned to North American Aviation and Douglas Aircraft Company, respectively. A NASA contractor built a specially enlarged airplane, known as the "Super Guppy," to carry the third stage to whichever testing site was chosen to trial-run Apollo rockets. The larger first and second stages would be transplanted by barge to Cape Canaveral for the assembly phase. IBM also was added to the first-wave lead contractor list for Saturn V.

Over the summer and fall of 1961, however, before those facilities became fully operational, NASA had to confront the stark fact that it wasn't technologically prepared to execute a moonshot. Beyond the sheer power difference between the converted ballistic missiles used to launch Mercury and the Saturn V being planned for Apollo, there was a massive difference in technical complexity. "As planning for Apollo began," recalled NASA deputy administrator Robert Seamans, "we identified more than 10,000 separate tasks that had to be accomplished to put a man on the moon. Each task had its particular objectives, its manpower needs, its time schedule and its complex interrelationship with many other tasks."

To structure its development and mastery of space travel's technical intricacies, NASA adopted systems engineering (step by step). The one-man crews of Project Mercury were already proving that an astronaut could function in Earth's orbit and return alive. In the next phase, known as Project Gemini, two-man crews would perfect the techniques of rendezvousing and docking in space. Additionally, Gemini would prove that humans could survive in space for up to two weeks, during which time astronauts would venture outside the capsule for humans' first attempts at space-walking. The Apollo phase would begin once the successes and research data accumulated by Mercury and Gemini had been digested. Using three-man crews, Apollo would begin testing its orbiter and lander in Earth and lunar orbits. Only when all these steps were perfected would an Apollo mission attempt to carry out the moon landing.

According to Webb, President Kennedy expected the collective effort required for Gemini and Apollo to include as many states, governments, universities, and corporations as possible, all playing a transformative role at multiple levels and locales. In later years, Webb told a surprising story about JFK's mandate, describing a visit from "a sophisticated senior official of a large corporation" holding many aerospace contracts. Webb recalled, "He hit me right between the eyes with the question: 'In the award of contracts are you going to follow 100 percent the reports of your technical experts, or are there going to be political influences in these awards?'"

"My answer was just as direct," Webb wrote. "'In choosing contractors and supervising our industrial partners,'" he told the executive, "'we are going to take into account every factor that we should take into account as responsible government officials.' This meant that NASA officials would be required to meet President Kennedy's basic guideline—that we would not limit our decisions to technical factors but would work with American industry in the knowledge that we were together dealing with factors basic to 'broad national and international policy.'"

From Webb's managerial perspective, the largesse of Kennedy's moonshot plan was akin to Eisenhower's building interstate highways, a massive public works undertaking that continued throughout the 1960s and provided enormous economic stimulus across the nation. That priority recalls a sardonic comment made by Shepard. Asked what he'd thought about in the moments before his historic Mercury flight, he replied, "The fact that every part of this ship was built by the lowest bidder."

In fact, it was not that simple. Webb could have made Shepard feel much better, or worse, with the knowledge that each part was actually made by the company or university or government arm that fit into an overall scheme for U.S. progress in aeronautics, missile technology, computer science, and military aviation. Cost and technical quality were factors in the decision, but they weren't the only ones. Even if a single factory or a single city could have made every part required for the Apollo missions inexpensively and exquisitely, Webb believed the president wouldn't have approved. Kennedy wasn't interested in merely producing a rocket and lunar module to beat the Soviets. He was into building an era, the Space Age, as his enduring legacy. Part of JFK's vision was to bring the American South and Southwest into the world of cutting-edge technology.

Of those 10,000 tasks that had to be accomplished in order to put a man on the moon, 9,998 could wait, but two were extremely pressing: choosing a site for the Apollo launchpad and choosing a location for NASA's new manned space project headquarters. Since Cape Canaveral fell within the ten-mile buffer zone between the launchpad and human habitation that was required under NASA guidelines, it was initially assumed that it could not host the Apollo launches. New sites under consideration included the Gulf Coast of Texas, isolated areas of Hawaii, Christmas Island in the Pacific Ocean, and various military bases in desert regions of New Mexico, California, and Nevada. But none of those alternatives had quite the geographical advantages of Cape Canaveral, which among other things boasted proximity to Port Canaveral, a deep-water harbor that could easily accommodate large ships bringing materials to NASA. On August 24, NASA announced that it had found a workaround that would allow it to employ Canaveral for the Apollo launches: buying eighty thousand acres of orchard land on Merritt Island and points between the cape and the Florida mainland, at a cost of $750 per acre, or $60 million all told.

The transformation of this part of the eastern Florida coastline had been slow during the early years of satellite launches, but the floodgates opened once NASA announced this decision. Traffic surged, schools were built, and tract housing sprang up to accommodate the twenty thousand workers who would soon be manning the space facility. "The American test site," Walter Cronkite remembered, "was set up on a remote, snake-infested swamp called Cape Canaveral on the Florida coast east of Orlando. As the test site grew, so did the nearby villages of Cocoa Beach and Titusville, until they replicated every boomtown in every bad movie ever made—cheap hotels, bars, girlie joints, their wares proclaimed in gaudy neon. This was the environment into which reporters lucky enough to be assigned the space beat plunged."

When David Brinkley of NBC News visited Cape Canaveral not long afterward, he showed that his interest lay in the landscape of Earth, not the moon. In his wry way, he investigated the mentality that had changed a quiet stretch of beach land into the amoral boomtown described by Cronkite. While on the air, Cronkite concentrated on what was good about Cape Canaveral, but Brinkley used his airtime to interview those at the hub of the NASA wheel and to lament that "while the jazzy and new were booming in Cocoa Beach, the old and the quiet were left to die." The difference in the two reactions to "Koo-Koo Beach," as it was newly dubbed by the locals, reflected Brinkley's more complicated, humanitarian outlook and Cronkite's focus on the big story.

Amid slapdash human development, Secretary of the Interior Stewart Udall was giving thought to protecting the area's natural habitat. On his recommendation, Kennedy soon green-lighted what soon became the Merritt Island National Wildlife Refuge on a thirty-five-mile barrier island in the buffer zone next to the NASA launch areas. Officially an overlay of the Cape Canaveral Launch Center, and subject to closure when NASA activities demanded, the refuge is home to more than five hundred different species of wildlife (including alligators, scrub jays, and manatees) and more than a thousand varieties of plants. It's also part of a major bird migration corridor, with seven distinct habitats for nesting and feeding, proving that aviaries and aeronautics can coexist.

Merritt was one of four natural areas along America's coastlines that Kennedy created around this time as part of a push for seashore conservation. The Kennedy years, in fact, gave birth to the modern ecological movement after the president embraced the findings of Rachel Carson's visionary book *Silent Spring* (1962), which called attention to the effects of pesticides and other human contaminants on natural ecosystems. NASA went on to play an important role in raising consciousness among Americans about becoming better environmental stewards. Photographs of Earth taken from space (including the famous "Blue Marble" photo showing our planet illuminated for the first time) fostered a sense of global community. NASA's weather satellites improved hurricane and blizzard forecasting and also collected evidence that air pollution was causing a greenhouse effect, raising the temperature of the planet. Other scientific evidence collected by NASA during the Kennedy years proved that Mars and Venus were dead planets, and in the 1960s, people began to worry that this could someday happen to Earth as well.

The other major decision was where to build Apollo's headquarters. Back in 1958, the town of Greenbelt, Maryland, had been chosen as the site of a spaceflight center that would later be named for Robert Goddard.

As NASA's first flight operations campus, it transformed the surrounding area, drawing hundreds of researchers and engineers. Many municipalities envied or resented Greenbelt's good fortune. When NASA announced that another flight center would be needed for the Apollo program, city development boosters, chambers of commerce, members of Congress, and Kennedy's own vice president began lobbying for their hometowns.

If the story of NASA's economic effect on suburban Maryland and the Florida coast was transparent, the mechanism that brought the Manned Spacecraft Center to Houston was the opposite: a brew of back rooms, boardrooms, barbeques, and Texas politics bubbling thick. The story began when the Humble Oil and Refining Company, the Texas-based precursor to ExxonMobil, began seeking development opportunities for thousands of acres of scrubland it had acquired between Houston and the Gulf Coast. In a deal with Rice University, the company agreed to donate a portion of that acreage to the university, contingent upon its seeking a contract to build a federal research facility on the site.

That's when George R. Brown, said to be the richest person in the Lone Star State, got involved. Brown had made much of his fortune as a principal with the international road-and-dam-building firm Brown and

Root. He'd attended Rice for a few years as a young man, served on the university's board of trustees for twenty-five years (fifteen of them as its president), and in 1961 he set his sights on a Rice-NASA partnership, a deal that would heighten the university's prestige while also opening construction opportunities for his firm. If Brown could pull off bringing the Manned Spacecraft Center to Houston, the economic windfall would rank in Texas history with the discovery of oil at Spindletop in 1903 and the building of the Houston Ship Channel in 1914. Overnight, Houston would once again become a boomtown.

As an exercise in rock-hard politics, the power behind the Houston decision rested squarely on two men: Congressman Albert Thomas and Vice President Lyndon Johnson. Thomas, George Brown's roommate at Rice, had represented Texas's Eighth District (which included Houston and, by extension, Brown and Root) in the House of Representatives since 1937. Johnson also had close ties to the top players at the construction firm, having both granted and received myriad favors over their long relationship. As chair of the Space Council, he'd already made sure Brown was consulted by April 1961. In the course of the council's meetings, Brown learned of the proposed $60 million budget for a new space center. Months ahead of others, Houston's power brokers were in talks with prominent politicians about a course of action to lure NASA's manned space flagship facility.

Through Brown's offices, Humble Oil committed to donating to Rice University about sixteen hundred acres of the Clear Lake area in southern Houston. The school, in kind, agreed to donate a thousand of those acres to NASA, while selling the rest at a cost above market rate to the space agency for the Manned Spacecraft Center. In other words, if the plan came to fruition, Humble would avoid the appearance of a bribe to NASA and Rice would receive $1.8 million for land it had never really owned. Rice had been tuition-free since its founding in 1912 (though it would begin charging fees in the mid-1960s), so new revenue was a perennial problem.

While Brown worked his considerable salesmanship on James Webb in June and July, Massachusetts, Missouri, Louisiana, and California put together their own strong packages in bids for the space center. Each state had excellent sites that met the prime criteria: they had tracts of available land, water routes for shipping, and universities vying to work side by side with NASA. They didn't, however, have the hard-driving Albert Thomas, who exerted control over space budgets as chair of the House Subcommittee on Independent Offices Appropriations. Jack Valenti, Thomas's frequent campaign manager, recalled that the congressman ruled his committee "like a divine right monarch." Thomas had been known for constraining budgets for the government's space activities throughout the 1950s. In the midst of the effort to pass Kennedy's expanded space expenditure bill, he made remarks referring to the search for an ideal site for the new manned space center. "The key to the selection," Chairman Thomas said, "seems to lie in Congressional approval of the vastly increased budget for space asked by this administration." That salient point was true.

So it was that Kennedy incurred resentment among the Massachusetts boosters, who couldn't understand why he didn't promote his home commonwealth for the Manned Spacecraft Center. Massachusetts had much to recommend it—plentiful acreage, access to the Atlantic sea-lanes, the rocket legacy of Robert Goddard at Clark University, and proximity to top-tier research colleges such as Harvard, MIT, Tufts, and Boston University—but what it lacked was Thomas and his ability to wield power over congressional budgets. With Johnson doing the legwork, the president's request for funding for a moonshot passed through Congress without interference from Thomas; and by August, Houston's two leading newspapers, the *Chronicle* and the *Post*, began running articles intimating that the city would soon have the Manned Spacecraft Center.

On September 19, 1961, Webb officially announced that Houston had indeed won the prize. NASA's Manned Spacecraft Center would function as the headquarters for the Apollo missions and future human spaceflight programs. An economic development maven, Thomas had tried to lure the Atomic Energy Commission to Houston in the 1950s and failed. But now, thanks to Kennedy and Johnson, NASA was coming to his district to roost.

Shedding its old "Bayou City" moniker, Houston also became known as "Space City, U.S.A.," as waves of astromania swept the town. Browsing through a Houston Yellow Pages in the immediate years after Webb's announcement, one sees local pride blindingly obvious in the names of new businesses: Space City Bar-B-Q, Apollo Broadcasting Company, Space Age Laminating Company, Astro Babysitters Agency, and on and on. At the Apollo Restaurant and Lounge, the space burger on the menu was "out-of-this-world." Even the names of Houston's new professional sports franchises, the Astros (baseball) in 1962 and the Rockets (basketball) in 1971, had NASA connotations.

Congressman William Cramer of Florida and Senator Benjamin Smith of Massachusetts both slammed

Kennedy for bowing to political heat from the powerful Texans, but the deed was done. In the coming months, the Space Task Group transferred from NASA Langley Research Center in Hampton, Virginia, to Houston, where its members worked in makeshift quarters throughout the city while awaiting completion of the new federal laboratory. Adjacent to the very spot where the Manned Spacecraft Center would be built around the Clear Lake region, Humble Oil spearheaded the formation of the Friendswood Development Company, to create a planned community, known as Clear Lake City, on fifteen thousand acres connected to the NASA compound. Over the years that followed, the company developed homes, apartment complexes, office buildings, restaurants, shopping and recreation centers, and all the necessary amenities of a modern town, drawing many NASA transplants. By the end of the decade, the area's population exceeded forty-five thousand.

Through the rest of his public career, Webb was vulnerable to criticism that his agency had bowed to the Texans' political pressure. He shrugged this off with his proof-is-in-the-pudding attitude: Houston pulled off the Apollo program. For Brown, though, the victory was just another proud episode in a long career of federally funded infrastructure deals. That

Brown and his crowd crossed ethical and, perhaps, legal lines was a matter of debate for decades. And yet, once the contracts had been signed, those same men typically delivered a first-rate job. That was the case with NASA and Houston: it ended up being a perfectly calibrated match. Everyone from President Kennedy on down received what he wanted in the decision—except, of course, the disgruntled boosters from states like Massachusetts whose bids were rejected. But their time would come. After all, the space program still had 9,998 other tasks to spread around the nation.

Astronaut Gus Grissom climbs into his *Liberty Bell 7* spacecraft on July 21, 1961. The Mercury-Redstone 4 rocket successfully launched the *Liberty Bell 7* that morning. This was the second in a series of successful U.S. manned suborbital flights.

Ohio-reared astronaut John Glenn poses on January 20, 1962, during a training session before his February 20, 1962, NASA spaceflight aboard the Mercury capsule *Friendship 7*, in which he would become the first American to orbit the Earth. After visiting the Seattle World's Fair in 1962 Glenn's capsule found a permanent home at the Smithsonian Institution in Washington, D.C.

15
Godspeed, John Glenn

Let's do the John Glenn Twist! Yeah! Oh!
Round and Round and a-round
Three times around the world he goes,
Up in space, Orbitin' in space
And the whole wide world knows. Knows.

—RIO DE FRANCISCO,
"THE JOHN GLENN TWIST," 1962

Within sixteen days of Gus Grissom's triumphant flight on July 21, 1961, Kennedy was faced with the unwelcome obligation of congratulating Khrushchev on yet another manned launch, one that represented a striking advance in space exploration. The president was staying with his wife and two children at their home on Cape Cod for a long-planned weekend of sailing and relaxing along the sun-drenched beaches. Due to Soviet military maneuvers in Berlin and CIA-intercepted rumors about new Soviet space activities, it turned into a working vacation. Rather than invite old friends to Hyannis Port, JFK asked Adlai Stevenson to be his guest. Now the U.S. ambassador to the United Nations, the two-time Democratic presidential nominee stayed in one of the white clapboard guesthouses in the Kennedy compound.

It was a maddening weekend. Kennedy's forebodings about the international situation were correct. Late on Saturday night, August 5, U.S. intelligence informed him that the countdown had begun for a Soviet rocket launch. Before the exact nature of this latest Vostok mission was fully known, the president was made aware by the CIA that it was due to lift off at about 2:00 a.m. (Eastern Daylight Time) on Sunday, August 6. Going to bed early, JFK expected the Soviet space mission to be over by the time he woke up. So great was the technological leap of the spaceflight, however, that it was still under way and going strong the next morning. Kennedy pondered his options for a fitting American response. Stevenson recalled the president as being composed and alert, but stewing—constantly running his fingers through his tousled hair in frustration with the rising stakes for the space game between the United States and the Soviets, and disheartened that his Cape Cod downtime had been snatched away.

Gus Grissom's flight had been suborbital, lasting barely fifteen minutes. The full-orbiting *Vostok 2*, by contrast, broke every record on the books. Just four months earlier, Gagarin had circled Earth once and made history. One can imagine hearts at NASA sinking as *Vostok 2* cruised past that mark, then past the three orbits that many Soviet scientists believed to be the absolute limit of human endurance in space, then past five, then ten. Eventually, *Vostok 2* made an astonishing 17.5 orbits of Earth in just over a day's time.

The cosmonaut piloting the *Vostok 2* was Gherman Titov, an expert skier and gymnast who, at twenty-five, was the youngest person ever to fly in space. Always pushing the "edge of the envelope" as a test pilot he was also the first person ever to nap in space, a feat that seemed extraordinary to a generation still wide-eyed by the thought of spaceflight. Titov actually slept for thirty minutes during the flight. With life-support equipment and radio and television devices monitoring his condition, his mission proved that astronauts or cosmonauts no longer had to operate on an anxiety-ridden red alert; they could relax, live, work, and sleep in space, suffering little more than the space version of motion sickness. In fact, sometime during his thirteenth orbit, Titov, after a fitful start to slumber, became so comfortable that he overslept his nap. This fact brought many smiles at the secret Star City, outside Moscow, where Soviet cosmonauts trained, lived with their families, and benefited from village school facilities and a shopping district.

Piloting *Vostok 2* personally—unlike the previous "man-in-a-can" flights, controlled from Earth—Titov still had time to snap photographs from his cockpit. He also used a Konvas-Avtomat movie camera to film Earth for ten glorious minutes before reentry. Ejecting once his capsule had pierced the atmosphere on return, he parachuted to a landing near Krasny Kut, Saratov Oblast, six hundred miles southeast of Moscow. Having fulfilled the Soviet dream, an unqualified winner, he was then driven three miles to where his capsule had made a hard-impact landing, to recover his film and journal.

Titov became a hero as excitement electrified the USSR, his name uttered with reverence across the land. In the United States, by contrast, the cosmonaut was perceived as just another dastardly Communist spoiler. Members of Kennedy's inner circle, especially Arthur M. Schlesinger Jr. and Ted Sorensen, privately fumed to Webb that NASA should have given their boss advance notice of the Titov mission. What astonished NASA officials was that *Vostok 2* had achieved goals they had earmarked for their sixth and last Mercury mission, still four flights down the list. The gap between the U.S. and Soviet space programs had grown, not shrunk.

Vostok 2 was a serious concern for Kennedy that weekend at Hyannis Port, but it wasn't the only one. In light of a heavy migration of people from East Germany to the capitalist-democratic enclave of West Berlin, the Soviet Union was making increasingly firm demands that the isolated city be reunited with the Communist nation that fully surrounded it. Sweeping aside NATO objections regarding Berlin as ludicrous, Khrushchev increased ground troops in East Germany, boosted Kremlin military spending by a third, accelerated the Soviet space program, and ratcheted up his rhetoric on Soviet nuclear superiority. In Moscow, the Twenty-Second Congress of the Communist Party of the USSR had unleashed inflammatory ultimatums at America.

Over that very weekend, while Kennedy awaited reports at his Cape Cod home, negotiators from Britain, France, and the United States were engaged in high-level talks in Paris, determined to fashion a coordinated response to Khrushchev's insistence on a unified Communist Berlin. Refusing to buckle under Kremlin pressure and determined to protect Western interests in the divided city, Kennedy made a highly visible show of American military preparations in West Germany. Aides worried that the Berlin confrontation might turn the Cold War into a shooting war in Europe, perhaps even a nuclear war. However, even during the crisis, Kennedy continued pursuing U.S.-Soviet cooperation, seeking a nuclear test ban treaty, engagement in space, and a global agreement on the use of communication satellites, which was opposed by the USSR.

While the many levels of U.S.-Soviet relations intensified the significance of Kennedy's response to the space news, Titov's flight and the Berlin crisis converged in public discourse. At the same time as the *Pittsburgh Press* fretted over the Berlin crisis under the headline "East Germans Fleeing Reds One a Minute," astronomers at the city's Allegheny Observatory used high-powered telescopes to catch vivid glimpses of *Vostok 2* passing overhead, appearing as "a very bright star." In Charleston, South Carolina, one newsman reported that *Vostok 2* "looked about the size of a marble." Other commentators were more damning. "You can guess which country appears to be struggling," said Sir Bernard Lovell, one of Britain's leading astronomers, "and it is certainly not Russia." His counterpart in France, Professor Alexandre Ananov of the Astronautic Society, agreed. "I fear," he said, "the lag can never be made up." And one Taiwanese official opined, "It is obvious that Khrushchev was going to use the space flight as a weapon intended to intimidate the West on the Berlin issue."

Choosing to play it cool, a steadfast Kennedy refused to acknowledge publicly that Khrushchev was using space exploration for geopolitical intimidation in Europe. His eventual response, delivered secondhand via Adlai Stevenson on Sunday afternoon, consisted only of a bland statement of "admiration" for *Vostok 2*, attached to the polite hope that Titov was in good health. Hours passed, then days, and Kennedy wouldn't elaborate further on the mission. Behind the scenes, he officially instructed NASA to readjust its schedule and speed up the next Mercury launch.

Khrushchev, however, had plenty to say. Striking while the iron was hot, the Soviet premier delivered a blistering ninety-minute monologue broadcast live to his own nation, but intended for the entire world. In it, he angrily blamed Kennedy and the "degenerate" American political system for taking an aggressive stance regarding West Berlin. Hoping to intimidate JFK, Khrushchev gave an extended description of the nuclear war that would ensue if the United States were not more careful with its routine Cold War "threats." If a war started over Berlin, Khrushchev promised with visceral antipathy, then the USSR would "strike a crushing blow" against the American homeland. It seemed clear that, from the Kremlin's perspective, Titov's triumph was more about the Cold War politics of divided Berlin than about space exploration. Close to home, former secretary of state Dean Acheson saw Khrushchev's démarche less as a comment on Berlin or outer space than as an attempt to shatter America's will to resist, while upending U.S. power and global influence by forcing a backdown in Europe and Asia.

With or without *Vostok 2* and Kremlin bullying, Kennedy stood fast to his commitment that West Berlin would remain free and independent of East Germany. The Potsdam Conference of 1945—in which three of the wartime Allies (the United States, the USSR, and Great Britain) agreed that people could move freely in any sector of Berlin—was in the U.S. president's favor, and most UN member countries backed America's position. East Germany, though, was alarmed by the exodus of professionals, intellectuals, and skilled laborers migrating from East to West, which by August 1961 had reached an average of two thousand per day, principally through West Berlin. Fearing the costs to its economy, on August 13, East Germany began erecting a twenty-seven-mile-long barbed-wire fence dividing socialist East Berlin from the democratic western part of the city. Dubbed an "anti-fascist protection rampart" by the East and the "Berlin Wall" by the West, it went up unannounced, and over the weeks that followed, it was reinforced by concrete, guard towers, and other fortifications. This alarming, provocative development outraged Kennedy, but since West Berlin still remained in Western hands, he didn't overreact. Within a few weeks, the crisis had quieted down. "It's not a very nice solution," the president said, "but the wall is a hell of a lot better than a war."

That was true enough. But Khrushchev wasn't done with his heavy-fisted attempt to intimidate the United States. For the Berlin crisis of August 1961 led immediately into yet another confrontation with the Soviet Union, this time over nuclear testing, an issue that was tied directly to the space race. Kennedy had learned firsthand how the Soviet space program had been used by Khrushchev as a diplomatic and near-military tool at the height of the Berlin crisis. No longer could manned space be compartmentalized as a benign scientific adventure in the cosmos; it was the Cold War.

At NASA, the methodical process of putting a human on the moon within ten years had received a prod from

Vostok 2, because nothing motivates a bureaucracy like wounded pride. Reorganizing to better manage the agency's various missions, James Webb established four program offices: Advanced Research and Technology, Space Science, Applications, and Manned Space Flight.

Among Webb's priorities in the aftermath of *Vostok 2* was a refinement of the space hardware NASA had contracted, to better match its evolving mission. Among the early casualties were von Braun's Mercury-Redstone rockets. According to astronaut Deke Slayton, *Vostok 2* "permanently kill[ed] Mercury-Redstone 5" in favor of larger Atlas (air force) rockets.

The Mercury-Redstone Launch Vehicle had been designed specifically for suborbital flights such as those accomplished by Shepard and Grissom. It was America's first manned space booster. Administrators had originally planned for at least two more such missions before putting an astronaut into full orbit, but on August 18, NASA announced that the data collected from Shepard's and Grissom's suborbital missions had been carefully analyzed and more than sufficed. No further Redstone tests would be needed. The federal space agency was speeding up Mercury's plan to send an astronaut, reportedly, to orbit Earth on a U.S. Air Force rocket.

Generally speaking, "speeding up" a technical process isn't advisable, especially when human lives are at stake. Acceleration could mean a spaceship torn apart in flight or a disastrous communication failure. But *Vostok 2* had forced Kennedy's hand, and NASA's. Although quickening production added human and technical risks, fears of a Soviet hammer-and-sickle flag planted on the moon loomed large, leaving the Kennedy administration few other choices. In Huntsville, the challenge sparked a new wave of determination to quickly develop Saturn rockets. Much federal funding was at stake. "The next flight by an astronaut would be on the Atlas," Deke Slayton wrote later. "We hoped by the end of the year."

The U.S. Air Force's Atlas rocket, which provided five times the thrust of the army's Redstone, was considered a "man-ready" vehicle, able to take a person into space, but it had been prone to disintegration during unmanned testing. With construction not yet begun on von Braun's new three-stage Saturn design, and deployment still years away, acceleration of the Mercury program rested on whether a reliably safe Atlas was ready. In September, an unmanned Mercury capsule was fired off on top of an Atlas as a dress rehearsal for a manned voyage. After taking off without a hitch, the capsule parachuted to a perfect landing in the Atlantic near Bermuda and was retrieved by the U.S. destroyer *Decatur.*

With Houston chosen as NASA's manned-space hub, Webb now needed to choose the right technical manager for Apollo. The two most obvious candidates, von Braun and Silverstein, director of NASA's Office of Manned Space Flight Programs, distrusted each other. (Silverstein, of Jewish heritage, couldn't stomach even being physically near von Braun, with his Nazi past.) For a number of reasons, Webb didn't consider either of the brilliant engineers quite right for the management job. He tried recruiting Captain Levering Smith, deputy head of the U.S. Navy's Polaris program, but Smith chose to stay with the nuclear navy at the new rank of vice admiral. Eventually, Webb chose an executive in the private sector, D. Brainerd Holmes of RCA, for the job. At the time, Holmes had been busy overseeing the design and implementation of the Ballistic Missile Early Warning System, setting up enormous high-tech installations in Arctic Alaska, Greenland, and Scotland. Taking a large salary cut, Brainerd left RCA and joined NASA on November 1, as director of the Office of Manned Space Flight. Just five weeks later, Holmes established a third NASA project, Gemini, as a training program. Featuring two-man crews, Project Gemini would bridge the gap between the one-man Project Mercury missions and the three-man Apollo launches.

In late November, NASA blasted an Atlas rocket from Cape Canaveral. It reached full Earth orbit for the first time, giving its crewman, a chimpanzee named Enos, the distinction of becoming only the third hominid to reach so deeply into space, after Gagarin and Titov. At a famous press conference, Kennedy deadpanned: "The chimpanzee who is flying in space took off at 10:08. He reports that everything is perfect and working well." Although scheduled for three orbits, the mission was ended after two revolutions due to technical malfunctions, and Enos returned safely to Earth—a triumph, but the wrong kind of triumph. In the wake of Titov's mind-boggling flight and East Germany's construction of the Berlin Wall, Kennedy worried that the national fascination with the moonshot was eroding as NASA continued its methodical program of unmanned missions. A chimpanzee, cute as he might be, was no substitute for another Shepard or Grissom.

Sensing that JFK's moonshot was, for the time being, more aspirational than newsworthy, the White House press corps largely avoided the subject, but sometimes the wrong kind of news would creep in. As a case in point, Dr. Al Hibbs, chief of space sciences at the Jet

Propulsion Laboratory in Pasadena, a NASA affiliate, told the press on November 21 that the United States had "less than a 50-50 chance" of beating the Soviets to the moon. Coming from one of the country's leading aerospace technology institutions, Hibbs's gloomy assessment was quoted in newspapers all over America.

Hibbs wasn't alone. Other space experts were also skeptical, including Dr. Walter Dornberger, von Braun's former boss at Peenemünde, who was then chief scientist for partner Bell Aerospace. On November 30, he told a space symposium in Louisiana that the Soviets were far ahead in developing ICBMs and artificial satellites. Blaming this deficit on U.S. indifference from 1945 until the *Sputnik* launch in 1957 toward manufacturing ballistic missiles, Dornberger cast doubt on America's ability to catch up. "The Russians will be able to intercept and shoot down our satellites within a few years," Dornberger predicted. "We will not be able to shoot down their satellites." Whether or not his analysis was entirely accurate, he frighteningly spoke as though war in space were inevitable, another roiling theme in light of *Vostok 2*'s show of continued Soviet innovation.

If late 1961 was a challenge for NASA and America's space morale, the bright spot was the marine who was waiting to take the national stage. Forty-year-old Ohioan John Glenn was the oldest of the Mercury Seven astronauts, yet he often looked the youngest. With his boyish face, bright smile, penetrating blue eyes, and general air of collegial optimism, he was impossible not to like. Slated to take the third suborbital Mercury mission, he was bumped up to a full orbital flight when the program jumped ahead in August, and he spent the last months of the year training with his backup, Scott Carpenter of Colorado.

While Glenn waited for his big moment, Kennedy was focused on other aspects of space exploration: opening orbital space to communications satellites and closing it to nuclear arms. Just four years after *Sputnik*, U.S. companies were vying to exploit the potential commercial viability of satellites. Two American physicists at Johns Hopkins University's Applied Physics Laboratory had also made dramatic strides in monitoring satellite radio transmissions, thereby establishing the conceptual foundation for global positioning systems (the beginning of today's GPS). Underdeveloped regions of Latin America, Asia, and Africa could be brought into modernity, it was thought, via cutting-edge satellite technology. "Never before has a major scientific venture involved such mutual dependence" between industry and government, wrote George J. Feldman of Massachusetts, chief counsel for the Committee on Astronautics and Space Exploration. JFK was emerging as the poster president of the same military-industrial complex Dwight Eisenhower had warned about in his farewell address. A new era of space-related inventions and entrepreneurial innovations was poised to begin. But this communications revolution required America to secure new "open skies" treaties with other countries, notably the USSR, a nation that equated satellite communications with spying.

On December 5 at the United Nations, the U.S. delegation submitted a resolution on the peaceful uses of space. Heavily promoted by Ambassador Adlai Stevenson, it called for a registry of satellites, cooperation on weather satellites, and extension of communication via satellites to all nations as soon as possible. The Soviet Union responded negatively. Kennedy expected nothing less, but he still needed such an agreement.

Two days later, Glenn's mission, named by him *Friendship 7* (in homage to his fellow Mercury cohorts), was suddenly delayed. This caught nearly everyone at NASA by surprise, especially Glenn. The reasons for pauses and postponements ranged widely. One was the undeniable fact that NASA's "hurry-up plans," in the words of Associated Press reporting, were too rushed amid too much Cold War tension. Another was the fear that a *Friendship 7* disaster just before Christmas would cripple NASA permanently, drying up its congressional funding. Public doubts about the moonshot would grow. The equally tense negotiations over the UN satellite resolution were another consideration. From Kennedy's point of view, the double-whammy PR disaster that would ensue were Glenn to be incinerated on liftoff and his satellite treaty to fail at the United Nations made the December launch too risky. Timing was everything in politics and diplomacy, and JFK simply wasn't ready to gamble his long-term peaceful plans for space on a made-for-TV "keeping up with the Soviets" launch. And then there was the official NASA reason for delays: mechanical problems. Deeming the president's caution reasonable, Glenn accepted the postponement without public complaint, even though inside he was desperate for the launch to happen.

Khrushchev played on Kennedy's equivocations to his geopolitical advantage. The Soviet premier gave another of his bombastic televised speeches, which seemed to be directed mainly at the American president, pounding his fist repeatedly as he bragged about the superiority of his country's nuclear arsenal. The Soviets were known to have renewed atmospheric nuclear testing, which had prompted the United States to conduct its own atmospheric exercises on a similar scale. Khrushchev's speech seemed to portend an end to all JFK's hopes for serious disarmament, or at least for an effective test-ban treaty.

One aspect of Khrushchev's provocative speech stabbed at another of Kennedy's priorities. Underscoring Dr. Dornberger's impression that the Soviets' main goal was to dominate the battlefield of space, Khrushchev boasted more than once about his nation's ability to arm rockets of the kind that had already put Soviet cosmonauts into orbit. If that capability became a reality, then the demilitarized idea of a moon voyage would be absurdly naïve.

The speech was meant to intimidate the United States but it came across as primarily hyperbolic blather. Nevertheless, within two days, the USSR made the truly surprising announcement that it would support the "peaceful uses of space" resolution. The agreement gave Kennedy a major win, blemished temporarily over the Christmas holiday when Defense Secretary Robert McNamara announced that he was authorizing research into equipping U.S. rockets with nuclear weapons. After the press pounced, McNamara backed away from the controversial claim.

On December 7, 1961, the twentieth anniversary of Pearl Harbor, Dr. Robert Gilruth of NASA revealed to Kennedy the plan to develop a two-person spacecraft as the necessary bridge between Mercury and the moonshot. Alex Nagg from NASA headquarters in Washington dubbed the effort Project Gemini, after the constellation that included the twin stars Castor and Pollux, a sign of the zodiac controlled by Mercury.

On December 22, McDonnell Aircraft of St. Louis was given the government contract to develop Gemini capsules, following up on their success building more than twenty Mercury capsules since 1959. One stipulation was that the first Gemini capsule had to be manufactured within fifteen months. After that, a new capsule would be delivered to NASA every sixty days. "Gemini's a Corvette," Gus Grissom recalled of the upgrade. "Mercury was a Volkswagen." Meanwhile, the Titan II, to be assembled by Martin Marietta of Baltimore, was selected as Gemini's launch vehicle; it was a modified ICBM developed by the U.S. Air Force. The company also had test stand facilities in Littleton, Colorado. On January 15, 1962, Gilruth explained NASA's three new "Project" organizations to the public: one organization for Mercury, one for Gemini, and one for Apollo—all civilian, of course, though that public image belied the agency's deep collaboration with the military. And new Earth landing systems for both normal atmospheric entry and various abort contingencies were being developed at considerable cost to avoid the mishaps of *Liberty Bell 7*.

Like most other Americans, John Kennedy started 1962 concerned about the space race. In his State of the Union address, he acknowledged that the United States might not win: "This nation belongs among the first to explore [the moon]. And among the first, if not the first, we shall be." It wasn't the strongest of promises. But Kennedy ably gave himself cover just in case the Soviets pulled off a moonshot. "Our aim," the president continued, "is not simply to be first on the Moon, any more than Charles Lindbergh's real aim was to be first to Paris. His aim was to develop the techniques and the authority of this country and other countries in the field of the air and the atmosphere." That point is arguable; the Orteig Prize of twenty-five thousand dollars had loomed large for Lindbergh, and its sponsor specifically stipulated that it would go to the aviator who was first, not "among" the first, to fly nonstop between New York and Paris. Nonetheless, Lindbergh was not the only "among the first" to fly nonstop across the Atlantic Ocean, the British duo of John Alcock and Arthur Brown, most notably, having accomplished that feat in 1919. Having grown up in that era, JFK knew that Lindbergh's fame was not diminished in the least by Alcock and Brown.

Like a doctor speaking to a patient's family before a serious operation, Kennedy was trying to lower his audience's high expectations. A genius at reading public sentiment, he understood that American citizens were aware that the United States might lose the race to the moon, and he wanted to assure them that was all right. If that were to occur, Kennedy would try to minimize the Soviet achievement, as Eisenhower had done after *Sputnik*. Regardless of the Soviets, NASA was still going to the moon, because first or not, going to the moon was a noble venture. Kennedy then charged onward in the speech to the more upbeat topic of the strides that had been made toward "peace in space" and the myriad of wonders that satellite technology would enable.

Kennedy was far more specific about space exploration in his first one-on-one conversation with John Glenn while the astronaut was waiting for his *Friendship 7* mission to be cleared following problems with the Atlas rocket fuel tanks, which led to postponements on January 16 and then again on January 20. Glenn was, in fact, surprised to receive a summons to the White House in early 1962. He and Kennedy had a casual chat, "one human being to another—as one 'guy' to another, if you will," the astronaut recalled a couple of years later of his February 5 meeting. "He just wanted to talk about what was planned on the flight and I went into some of the details of what we expected to experience. . . . He brought up whether we felt very personally every possible thing had been done to ensure our safety and I told him that when we first came into the program one of the things we were told, by [Space Task Group chief Robert] Gilruth, was that we had veto control over anything that was to occur on the project. That at any time we, as experienced test pilots, saw something going on that we didn't like or there was an area that we thought needed more testing or anything that we weren't satisfied with, to let him know. . . . The President thought that was an excellent way to conduct such a project."

Impressed that Glenn had won the Distinguished Flying Cross five times and flown fifty-nine missions in the South Pacific, sometimes with baseball great Ted Williams as wingman, Kennedy asked so many questions (about g-forces, rocket control, and NASA planning) that the astronaut offered to come back with models of the Mercury-Atlas 6 and Mercury capsule Number 9 he'd soon be riding. The president agreed, and Glenn returned to the White House days later for an even longer discussion. The constancy of Glenn's code of honor was something Kennedy was deeply impressed with. It reminded him of his deceased brother, Joseph Kennedy Jr. "John tries to behave," a friend of Glenn told a *Life* reporter, "as if every impressionable youngster in the country were watching him every moment of the day."

Realizing that Glenn was a man of honest faith, persistence, and the instinct for courage, Kennedy initiated a warm friendship with his next Mercury astronaut, one that transcended the obvious fact that both were banking heavily on the success of *Friendship 7*. Meanwhile, Glenn, who had flown hundreds of daunting navy missions in training props, warplanes, and experimental jets over the previous twenty years, realized that he needed to address the possibility that he might die during his Mercury spaceflight. For the first time, he spoke to each of the members of his family about this dreaded possilibity, hoping to ensure that none of them had, or would have, any regrets.

Friendship 7 was an airtight, watertight, and soundproof marvel of compact engineering, containing more than ten thousand components, seven miles of wiring, and shielding able to protect its pilot from both Arctic cold and three-thousand-degree heat. Built by

McDonnell Aircraft in St. Louis, the capsule boasted a recessed fiberglass couch (specially contoured to fit Glenn's body); a gleaming instrument panel with more than one hundred dials, switches, and lights; a wide-angle window and periscope; and a parachute, recovery gear, and emergency exit system.

On February 20, 1962, Glenn awoke at 1:30 a.m. for his fourteenth scheduled attempt at launch. For once the sky was serene and unclouded, the Atlantic calm, and nobody could think of a reason not to make a go. The final countdown for *Friendship 7* began a little after nine that morning, with the launchpad bathed in sunshine. Word spread that in a little more than eighty minutes the marine astronaut would lift off on the 125-ton Atlas-Mercury 6, en route to circling the planet. It would be only a minor exaggeration to say that the least nervous person in the country was Glenn himself, lying on his back stoically, his pent-up frustration over the long delays had transformed into well-focused hope and faith as the clock ticked down. "Don't be scared," Glenn telephoned his wife, Annie, who was at their home in Arlington, Virginia, their two children at her side. "Remember I'm just going down to the corner store to get a pack of gum."

At 9:47 a.m. Eastern time, the Atlas rocket fired 367,000 pounds of thrust and rose with an initial slowness that some observers thought majestic and others thought worrisomely hesitant. That Atlas vehicle had a relatively low thrust-to-weight ratio. As the propellant burned, the vehicle became lighter and accelerated faster. Scott Carpenter, serving as astronaut capsule communicator for the mission, wished for his friend "Godspeed, John Glenn." Even though Glenn's earphones didn't pick up Carpenter's parting message, the salutation was caught on tape, went viral within NASA culture, and Walter Cronkite of CBS News turned it into the catchphrase of the entire Mercury flight. Tom O'Malley, General Dynamics' test director, prayerfully added, "[M]ay the good Lord ride with you all the way."

In living rooms from coast to coast, America vibrated with anticipation as it watched Glenn's Atlas rocket soar skyward, with coverage on every TV network. At that moment, commuters in New York City left their trains empty while they stayed glued to televisions in Grand Central Terminal. Millions of schoolchildren around the country watched as well, on sets borrowed and brought in just for that day. In Dover, Ohio, businesses locked their doors so that employees could watch the historic launch on TV or listen to the play-by-play on the radio. In Trenton, New Jersey, a bank robber got away with almost nine thousand dollars that Wednesday morning and then stopped for a quick drink at a bar, where he ended up staying to watch Glenn's flight on television. The police caught up with him there. In Grand Rapids, Michigan, a judge and jury were hearing testimony in a case involving a stolen television when the judge suggested they turn the TV on to watch the launch—which they did. Up the road in Detroit, operators at Michigan Bell Telephone reported to their bosses that there might have been an equipment failure, because their usual heavy load of calls had completely dried up in the moments before liftoff.

Eyes and ears all across America were fixed on Glenn. In Salt Lake City, folks brought transistor radios into restaurants and listened to the nail-biting coverage. At a coffee shop fortunate enough to have a television, an employee reported packed crowds who "watched as though they were spellbound." In spring training for their inaugural season, the New York Mets were practicing leading off first base when Casey Stengel, never the most easygoing manager in baseball, stopped the drills so his ballplayers could watch the launch. On site at Cape Canaveral, tears flowed as *Friendship 7* moved past its initial hesitancy and headed toward the heavens. Even the most cynical reporter called out, "Go, baby, go!"

More than forty million American homes had tuned in to the Glenn mission. Traveling seventeen thousand miles an hour, three times faster than Shepard had, Glenn described the African coast, gorgeous rainbow radiance, and blazing blue bands glowing around Earth. "Wonderful as man-made art may be," Glenn would write of his space odyssey, "it cannot compare in my mind to the sunsets and sunrises, God's masterpieces." On CBS, the invariably enthusiastic Cronkite anchored for ten straight hours, with the banner "Man in Orbit" splashed across the screen beneath him. With no live cameras aboard the spacecraft (as there would be for journeys later in the decade), Cronkite and the other space anchors painted a picture with words for their TV audience, marveling at Glenn's stoic composure, cooped up in a claustrophobic capsule with just over twenty-four hours of breathable air. Ensconced much more comfortably at the CBS News Cape Canaveral facility, Cronkite became the maestro of the historical moment, whose dissemination through TV and radio, according to the *New York Times*, "united the nation and the world . . . in a common sharing of the excitement, tension and drama" around the flight.

However, all was not going well with Glenn's mission. Mercury Control in Florida was jarred by a telemetry signal "segment 51" from *Friendship 7*, indicating that the heavy-duty heat shield was probably loose. Glenn was told about the malfunction by astronaut Gordon Cooper over the radio. Without one erg of emotion, Glenn acknowledged the technical problem but stayed calm and carried on. Back at Cape Canaveral, the collective nervousness was palpable, and technicians began to discuss emergency plans. The capsule's designer, Max Faget, was consulted on the vessel's aerodynamics. Lieutenant Colonel John "Shorty" Powers, the Project Mercury information officer, told the TV networks that if the heat shield became unhinged, Glenn's life would be in danger. Even if the shield were only slightly loose, that could complicate the planned ocean splashdown. Fear mounted at NASA and around the world that *Friendship 7* might incinerate, that the aerodynamic effects of reentry could tear the capsule wide open. "I knew that if the shield was falling apart," Glenn later wrote, "I would feel the heat pulse first at my back, and I waited for it."

Before long, Glenn lost radio contact with Mercury Control due to a technical malfunction. Nobody knew the status of *Friendship 7*. It was the first blackout in a manned spaceflight, a ghastly four-and-a-half-minute silence amplifying the incredible distance between Glenn and the world. The tension was so great that, for millions of viewers, time froze. When radio contact was restored and Glenn again spoke to Mercury Control, a collective sigh of relief could be heard at NASA. Their astronaut still had a fighting chance of survival. The descent back to Earth proved a wild and bumpy ride, but the primary parachute opened as planned when *Friendship 7* reached 10,800 feet. At 2:42 p.m. Eastern time, the capsule splashed down in the Atlantic, about forty miles off course. For a long spell, Glenn sat in the bobbing capsule, sweating profusely—he'd been trained not to open the hatch prematurely. "How do you feel?" Shepard asked from Mission Control. "Oh, pretty good," Glenn said. "What is your general condition? Are you pretty well?" Shepard continued to inquire. "My condition is good," said Glenn. "But that was a real fireball, boy. I had great chunks of retro-pack breaking off all the way through."

Drenched in sweat, Glenn shed his harness, clutched his survival kit, and prepared to make an emergency exit if necessary. He didn't have to. In short order, the destroyer USS *Noa* arrived. Glenn opened the hatch door, made a fast exit, and once aboard the navy ship, was taken immediately to medics on the aircraft carrier USS *Randolph.* The whole operation was a giant win for NASA. Although dehydrated and five pounds lighter than when he stepped into the spacecraft, Glenn was in good shape. Kennedy called him with words of praise and pride. No longer was America "second best" in space.

Once the astronaut was retrieved, the world delighted with an outpouring of love and excitement that easily surpassed what had greeted the previous Mercury astronauts on their return. Jubilant space mania was in the air. An Ohioan with the scruples of a Boy Scout had orbited Earth three times in just under five hours—once every 88.29 minutes, a total of 81,000 miles, at a speed of 17,545 miles per hour and an altitude of 160 miles. Glenn, the "Clean Marine," became the most famous heroic explorer of the American century since Charles Lindbergh. "The best moment," Cronkite wrote in a syndicated column, "was when Shorty Powers announced from Mercury Control: 'We have a hale and hearty astronaut.'"

Fierce patriotism rose like a sudden swamp fog around America. Morale zoomed from worst to first and got people talking about the moonshot again. Astronauts Shepard and Slayton summed up the post–*Friendship 7* euphoria best: "The distance to the moon was starting to lessen." Because everything seemed to have changed on one memorable American morning, a California journalist suggested that February 20 be a national holiday ever afterward. "Orbit Day? Space Day? Glenn Day?" Bob Wells of the Long Beach *Inde-*

pendent asked readers to weigh in, and a dozen decent ideas came forth.

Kennedy flew to Cape Canaveral three days after the *Friendship* 7 flight, to receive a briefing from Glenn and pose for a series of photo ops, where he treated the astronaut as if he were a brother. Bursting with pride, Glenn brought his wife, Annie, and their two children to meet his friend the president. Over the days that followed, at Cape Canaveral and then back in Washington, astronaut and president spent many hours together, delivering public relations victories both for JFK's New Frontier and for NASA. Glenn was not a wit on par with Kennedy, but he could more than hold his own on history and politics. Jackie Kennedy noted that both men projected an aura of "cool" self-control. The president was riveted as Glenn told him about seeing three sunsets and three dawns in just his four-hour, fifty-six-minute flight.

As the two families spent more time together over that summer of 1962, the First Lady became, as her friend Cyrus Sulzberger recalled, "vastly impressed by John Glenn, the astronaut. She says he is the most controlled person on earth. Even Jack, she said, who is highly self-controlled and has the ability to relax easily and to sleep as and when he wishes, to shrug off the problems of the world, seems fidgety and loose compared to Glenn. Glenn is the most dominating man she ever met."

Joining the chorus singing Glenn's praises was von Braun. He was thrilled that now the United States had put up three astronauts (Shepard, Grissom, and Glenn) compared to only two Soviets (Gagarin and Titov). At a press conference from Florida, the rocketeer said that with Glenn's Mercury flight, America had taken a "Bunyan step" forward toward the moon. Soon, a genuine friendship was forged between the rocket scientist and the charismatic astronaut. When Glenn went on a goodwill mission to West Germany and Switzerland a few years later, he wrote a postcard in German to von Braun that read, "Here I am in Lucerne and you are in Huntsville. What a switch!"

A week after Glenn's flight, Kennedy again met with the space hero at the West Palm Beach airport to fly back to Washington together on Air Force One. In the coming months, Kennedy commissioned the French-born industrial designer Raymond Loewy—whose logos for Lucky Strike cigarette packages and Ritz cracker boxes were popular—to revamp Air Force One. Following JFK's orders, the words United States of America were emblazoned on the blue-white fuselage. Jackie had wanted her five-year-old daughter, Caroline, to meet Glenn, so they came to the Air Force

One tarmac for a quick meet-and-greet. "We were on the plane, and the president boarded, and behind him came Jackie with little Caroline, holding her by the hand. Jackie said, 'Caroline, this is the astronaut who went around the Earth in the spaceship,'" Glenn recalled. "'This is Colonel Glenn.'" Caroline stared at Glenn with a look of confusion in her eyes. She started looking all around Air Force One, a deep sadness engulfed her face. Fighting back tears, her voice trembling, she asked, "But where's the monkey?"

JFK found himself saying nice things about Ham and Enos to please Caroline. Once Air Force One lifted off, Kennedy proofread the speech Glenn planned to deliver to a joint session of Congress. It was solid. "I still get a hard to define feeling inside when the flag goes by," Glenn later told the assembled lawmakers. "I know you do too." Perhaps someday, Glenn hoped, Old Glory would be planted on the moon. Later that day, at a White House Rose Garden ceremony, Kennedy connected his own lifelong affinity for sailing in the Atlantic to NASA's exploration of the cosmos. "We have a long way to go in this space race," the president said with Glenn by his side. "We started late, but this is the new ocean, and I believe the United States must sail on it and be in a position second to none."

Glenn returned the respect of the Kennedys, becoming a valued member of their inner circle. He and Bobby Kennedy, whose knowledge of space was quite limited, grew especially close. But Jack was the one who encouraged Glenn to consider a career switch to electoral politics. Asked about his political affiliation, Glenn replied that he was registered as an independent, sparking a concerted and ultimately successful effort by the Kennedy brothers to woo Glenn to the Democrats. Glenn had become such a valued New Frontier icon, given a ticker-tape parade in New York City reminiscent of Lindbergh's in 1927, that NASA official Charles Bolden reported that JFK didn't dare "risk putting him back in space again."

From the start, Glenn recognized that the president had a rare intelligence and a lively interest in even some minute details of the space program. Glenn was happy to oblige, giving Kennedy a friendly education in the hard science of space and providing a firsthand understanding of what going there had been like. Two years later, Glenn recalled JFK's ability to remember details from a conversation they'd had in the weeks preceding the flight: "When I came back, he recalled quite a number of these things I had said in [our] preflight meeting on the 5th of February. Most of the things that we had expected in space flight were encountered and

Vice President Lyndon Johnson watches as Lieutenant Colonel John Glenn shakes hands with John F. Kennedy, on February 23, 1962, the day the president presented the astronaut with NASA's Distinguished Service Medal at Cape Canaveral, Florida. Glenn received the award after orbiting the planet three times in his little capsule, *Friendship 7*.

he recalled these all very accurately. He evidently had remembered all the things we talked about that day."

As Glenn came to know Kennedy better, in weekends on Cape Cod or at White House dinners, he tried to answer a fundamental question: Just what was the president's personal interest in space exploration? JFK was the driving spirit putting America into the business of space, yet his motivations remained something of a mystery to Glenn. Was it just smart politics? Or did Kennedy truly enjoy the prospect of a manned voyage to the moon? Some observers contended that the president was interested only in the Cold War and that if there had been no Soviet Union, there would have been no moonshot. Had a more benign nation such as Denmark, Australia, or Brazil marshaled its resources to lead the way into space, would JFK have backed a $25 billion American effort to surpass it?

Ultimately, John Glenn believed that Jack Kennedy indeed would have. "His attitude toward the whole project changed a little bit as time went on," Glenn recalled in a 1964 oral history interview. "I think early in the program, from statements I have read and from personal remarks when we were together, that he saw it originally as more of a competitive thing with the Russians. That we couldn't let them best us in this scientific field. Period. This, of course, is one phase of the program. However, those of us in the program have felt that the program is completely worthwhile even if there was no such place as Russia, just on the basis of being an exploration and research capability. I think his statements and his feelings on this came more around to the latter as time went on." Glenn believed that Kennedy at heart was an avatar of "public discovery," pushing for science and exploration both to advance human knowledge and for the economic value they'd bring. "His vision set an inspiring example, and I saw that the Kennedy charisma could move millions to contribute to something I thought was vital," Glenn wrote of JFK, "a democracy of energized participation in which people shared their talents with the nation and kept it improving and evolving."

The success of Glenn's mission created a popular culture boom around all things space. His ticker-tape parade in New York City was like a Macy's Thanksgiving Day parade, with Glenn standing in for all the helium balloons. Fashion designers in Soho such as André Courrèges, Paco Rabanne, and Pierre Cardin went astro-chic in advertisements in *Vanity Fair* and *Harper's Bazaar*. Architect John Lautner built his iconic Chemosphere in the Hollywood Hills, which hovered like a flying saucer above the valley, while Eero Saarinen built his elegant Gateway Arch in St. Louis (constructed between 1963 and 1965) with a futuristic feel. Seeing UFOs in your own backyard became the strange rage. Edward Craven Walker designed the Astro lamp (aka lava lamp) to give owners an outer space experience in their own bedrooms. The injection-molded and stackable Polyside chair was modeled after satellites, while kitchen appliances were marketed as Space Age gadgetry. Prototypes of rocket automobiles were designed, while TV shows such as *Lost in Space* and *The Jetsons* became wildly popular. In the world of music, the hit song became "Moon River," composed by Henry Mancini with lyrics by Johnny Mercer; it won the 1962 Grammy Award for Record of the Year and Song of the Year. Handsome TV host Andy Williams made "Moon River" the theme of his popular show. When Bobby Kennedy and his wife, Ethel, went to California for a brief holiday they picked John Glenn up at the airport with Andy Williams in tow. "It was so wonderful," Ethel Kennedy recalled. "We were all in a car together, driving around, singing 'Moon River.' It was a grand kick."

To observers at home, *Friendship 7* seemed to have catapulted America into the lead in space, though in truth Glenn's three orbits paled beside Gherman Titov's seventeen laps and the overall greater complexity of the *Vostok 2* mission. Nevertheless, NASA's manned space program was clearly gaining momentum. *Friendship 7* was the triumph of Kennedy's "soft power" approach to convincing the world that American technology was more advanced than Soviet. When Kennedy presented Glenn the NASA Distinguished Service Medal on February 23 for his successful mission, the astronaut graciously averred that he was just a "figurehead" for the New Frontier's "big, tremendous effort."

Notes of congratulations poured into the White House that late February, from schoolchildren, teachers, fellow politicians, and average citizens, and Kennedy happily answered a random selection. A number of people wanted to use Glenn for Cold War propaganda advantage over the Soviets. "May I humbly offer a suggestion to your Excellency?" Carl H. Peterson wrote Kennedy. "Would it not be a splendid idea to appoint our famous Astronaut Col. John Glenn to be an Ambassador of Good Will to all Nations of the World that will want to know what is in outer space which Col. Glenn can explain so well? This could inadvertently make the road for Khrushchev rockier."

In fact, Glenn's flight seemed to have shocked Khrushchev into an uncharacteristic silence. Eventually, he sent the obligatory congratulatory note to the White House, expressing stilted praise for Glenn's courage before echoing Kennedy's own suggestion about space cooperation at their meeting the previous June. "If our countries pooled their efforts—scientific, technical and material—to master the universe," he wrote, "this would be . . . acclaimed by all peoples who would like to see scientific achievements benefit man and not be used for 'cold war' purposes and the arms race."

Not wanting to suggest that Kennedy was using the

Glenn launch "for 'cold war' purposes and the arms race"—though he clearly was, as was Khrushchev—the White House lost no time in responding publicly. Within hours of receiving the premier's dispatch, the president was standing at a press conference, duly celebrating Glenn's mission but giving even more time in his introductory remarks to the Soviet leader's suggestion and, even more so, his own history of calling for U.S.-Soviet cooperation in space. "I am replying to his message today," JFK told the assembled reporters and his television audience, "and I regard it as most encouraging." With the arrival of Khrushchev's message of goodwill, the science of space disappeared and the two leaders' tit-for-tat game returned—the one in which the need to be seen winning the propaganda war was almost as important as actually winning the Cold War.

That spring, the Kennedy administration sponsored *Friendship* 7 on a second mission. The craft was loaded onto a giant C-124 cargo plane and sent on a thirty-city "Around the World with *Friendship 7*" tour to promote U.S. space achievements; one wag called it "the Fourth Orbit." At the Science Museum in London, throngs of spectators showed up on opening day, so many that thousands of would-be visitors had to be turned away. Around the world, newspaper photos showed *Freedom*

7 next to a trumpeting elephant in Ceylon (now Sri Lanka), a mariachi band in Mexico, and a children's choir in Nigeria. People often waited for three or four hours just to see the pride of McDonnell Aircraft and NASA. In Tokyo, half a million people queued up to photograph themselves next to the capsule when it was on display at a downtown department store. In India, over 1.5 million people lined the streets of Bombay (now Mumbai), hoping for a glimpse of the space capsule. Every balcony, window, and roof was jammed with enthralled spectators.

Like Glenn himself, *Friendship 7* was a golden advertisement for Kennedy's New Frontier agenda, and a fine public relations counterstatement to the Soviets' trumpeting of Gagarin one year before, not to mention the commemorative stamps, coins, postcards, and other mementos that had celebrated Soviet space superiority ever since *Sputnik*. As Glenn wrote McGeorge Bundy, Kennedy's national security advisor, the "fourth orbit" wasn't merely a propaganda extravaganza but "a well-thought-out scientific [education] program that could eventually benefit all peoples of the world as the scientific exploration it is."

The incredibly popular Glenn visited the Seattle World's Fair on May 10, which showcased the landmark Space Needle. Mobs of people followed the astronaut around, craning their necks for a glimpse of his reddish crew-cut head, hoping for an autograph or photo. Accompanied by von Braun, Glenn traveled on the futuristic Monorail and rode the elevator to the top of the Space Needle, which had been built especially for the exposition. The highlight was when Glenn attended a NASA conference at the Seattle Opera House that day. Gherman Titov had already appeared at the World's Fair and claimed he "saw neither angels or gods" in space. Asked for a comment on that assessment, Glenn shrugged it off, claiming his orbit had reaffirmed his bedrock Christian values and democratic beliefs.

Surrounding Glenn at the opera house were members of the "100,000 Foot Club," a group of test pilots who had attained that altitude in balloons, rockets, and other aircraft. But Glenn, who gave a better-than-decent presentation, was the man everyone wanted a piece of. The *Seattle Post-Intelligencer* joked that the NASA press conference might as well have been Glenn and "eight guys named Joe." What nobody realized was that the quiet Purdue University graduate and Korean War ace sitting next to the panel chairman was the man who would eclipse even Glenn's fame and achieve Kennedy's moonshot challenge by the end of the decade: Neil Armstrong.

Colonel John Glenn at a ticker-tape parade in New York City.

Kennedy tours Marshall Space Flight Center with von Braun.

16

Scott Carpenter, *Telstar*, and Presidential Space Touring

Space represents the modern frontier for extending humanity's research into the unknown. Our commitment to manned programs must remain strong even in the face of adversity and tragedy. This is our history and the legacy of all who fly.

—JOHN GLENN

Had John F. Kennedy lived to a fitting old age, he might have looked back on the late winter and spring of 1962 as the sweet spot of his presidency. In early April, his approval rating stood at 79 percent, as reported by pollster George Gallup. The number of Americans who reported disapproving of him was just 12 percent, with 9 percent expressing no opinion. Even more startling than the overall high rating was the fact that most respondents couldn't point to *anything* that Kennedy had done wrong. "He has yet to provoke any solid bloc of opposition to his programs or actions," Gallup noted on April 2. Only seven months out from the midterm elections, the Republicans still hadn't found a major issue to damage Kennedy and his party. The only public gripe, according to the poll, was that he was absent from the White House too often, notably at his family's oceanfront homes on Cape Cod and Palm Beach.

If America's aerospace industry shareholders had been polled, Kennedy would probably have scored 100 percent favorable. Instead of dispensing the entire plum of Project Apollo development to a single contractor, the administration spread the financial allocations among hundreds of happy companies. It was the New Frontier's infrastructure stimulus approach applied to a lunar voyage, space exploration, computer science, and aeronautical activities in general, making it possible for NASA "to acquire billions' worth of exotic new hardware and specialized features without overrunning the initial cost estimates and without even the slightest hint of any procurement scandal in that vast empire."

As noted, due to the Saturn V rocket's massive size

and complexity its three stages would be produced by three separate companies. Boeing won the contract to build the first stage (SI-IC, with five F-1 engines) at Michoud Operations in New Orleans, Louisiana, a 1.8-million-square-foot manufacturing facility that had built Patton and Sherman tanks for the Korean War. Located beside Bayou Sauvage and offering ship access to the Gulf of Mexico, Michoud adopted the slogan "from muskrats to moon ships." Just weeks after choosing Michoud, NASA determined that the Saturn V test site would be on the Pearl River, in Hancock County, in southwestern Mississippi, only a half-hour drive from Michoud.

Saturn V's second stage would be manufactured by North American Aviation in Seal Beach, California, and its third stage would be built by Douglas Aircraft at Huntington Beach, California. Rocketdyne, a subsidiary of North America located near Los Angeles, supplied engines for the Saturn V: five of the F-1 type for the first stage; five of the J-2 design, using a different fuel mixture, for the second stage; and one J-2 for the third stage, which sent the future Apollo astronauts to the moon. Two hundred times heavier when assembled than von Braun's V-2, the three-stage Saturn V rocket would stand 363 feet taller than the Statue of Liberty. Numerous other auxiliary companies—including Collins Radio, and Minneapolis-Honeywell—also received lucrative NASA contracts for the Saturn V.

Besides John Glenn's successful Mercury mission, there had been another event in early 1962 that contributed mightily to JFK's record level of popularity. On February 14, all three networks broadcast *A Tour of the White House with Mrs. John F. Kennedy*, an hour-long special in which Jacqueline Kennedy unveiled the just-renovated Executive Mansion. Solidifying the unique rapport the Kennedy family had with the American public, the First Lady made the $2 million restoration seem like a grand triumph, representing the nation's history and esteemed place in the world. Her eyes bright and vivid, and displaying style, grace, and poise as she thoughtfully and politely guided viewers from room to room, the First Lady won over slews of new admirers from around the world. On the night of the broadcast, three out of four TVs in America were tuned in.

Underneath all the New Frontier post-Glenn optimism and the First Lady's refreshing elegance, the protracted crisis in Berlin still simmered. Although the possibility of a U.S.-Soviet war over the city had grown more remote, it remained Europe's scariest Cold War hotspot. Every day of his presidency, Kennedy worried about the fate of West Berliners, and he was determined to keep NATO fortified. The U.S. State Department grappled on a daily basis with East Berliners engaged in escape attempts. On the broadest scale, fear of nuclear war was widespread. Convinced the Soviets would only respect American toughness, JFK endorsed a Pentagon recommendation to deploy von Braun–designed Jupiter nuclear-armed missiles in Turkey and Italy. All the while, the White House maintained a tactical back channel of diplomatic communication with the Kremlin. Out of the public glare, the Cold War rivals were at least covertly talking about cooperation in outer space and nuclear test bans, even as both pressed ahead with ICBM development and advanced reconnaissance systems that could be construed as "weaponizing space."

Throughout 1962, NASA continued to be the focus of federal space-related initiatives and congressional appropriations. The army was the most cooperative branch of the armed services in ceding the space initiative to NASA, acquiescing to NASA's takeover of former army jewels such as Caltech's Jet Propulsion Laboratory and Huntsville's Marshall Space Flight Center. The navy likewise agreed to leave the lion's share of space research activities to NASA, though it stubbornly retained control over the development of an array of satellites, especially those designed for mapping and positioning.

That left the air force, which was still unmollified about a civilian-run NASA. Its sticking point, as ever, was that *space* was *air*, and if the air force dominated the air, as its name affirmed, then it should dominate space as well. While the navy and army could move aside for NASA without losing much stature, the youngest military branch might easily have renamed itself the "space force" in order to drive its point home. Instead, it renamed the field in question, promoting the word *aerospace* to help justify expanding its aeronautical mission beyond the atmosphere. Determined not to be bigfooted by NASA, the air force hotly pursued Defense Department grants for major space probe and launch initiatives, especially for the continued development of the Dyna-Soar reusable rocket masterminded by ex-Peenemünder Walter Dornberger. According to the Kennedy administration's guiding edict, space programs in any of the military branches had to be nonaggressive in nature. Peaceful space exploration had to be the modus operandi of space research—even if the potential for a more offensive military use was close at hand.

Hoping for diplomatic negotiations that would prevent the militarization of space, Kennedy grasped at the trial balloon Khrushchev had sent up after Glenn's *Friendship* 7 flight. On March 7, 1962, he wrote a three-page letter to the Soviet premier suggesting that the two superpowers "could render no greater service to mankind through our space programs than by the joint establishment of an early operational weather satellite system." His feelers to Khrushchev mentioned future collaborations, such as the exchange of space-related equipment, scientific data, and aerospace personnel. Toward the end of his letter, Kennedy once again offered to share the glory, and the expense, of manned space exploration. "Some possibilities are not yet precisely identifiable," Kennedy wrote, "but should become clear as the space programs of our two countries proceed. In the case of others[,] it may be possible to start planning together now. For example, we might cooperate in unmanned exploration of the lunar surface, or we might commence now the mutual definition of steps to be taken in sequence for an exhaustive scientific investigation of the planets Mars or Venus, including consideration of the possible utility of manned flight in such programs."

Kennedy didn't specifically mention the demilitarization of space in this letter, but it was inherently on the negotiating table. Behind the scenes, the White House had made it well known to the USSR and Great

Britain that the president was hungry for a nuclear test ban treaty. Furthermore, the State Department dangled before Khrushchev the possibility of joint U.S.-Soviet weather satellite systems, cooperation on mapping Earth's magnetic fields from high altitudes, and shared data on space medicine being innovated at the U.S. Air Force School of Aerospace Medicine, in Texas. Even the possibility of a futuristic joint U.S.-Soviet manned flight to Mars or Venus was open to negotiation. On the peace offense, the surprisingly openhanded Kennedy boasted of NASA's steady technological advancements, knowing full well that the Kremlin loathed and feared transparency. America continued broadcasting its Cape Canaveral satellite and manned space launches on live TV, while the Soviets maintained ironclad secrecy. The U.S. government's message was crystal-clear: JFK's New Frontier stood for freedom and openness. By contrast, the Kremlin was all about hidden totalitarian agendas. After John Glenn's mission, a fair-minded judge could assert that the United States was beating the USSR in the race for global prestige.

On the same day that Kennedy sent his friendly proposal to Khrushchev, he was asked at a news conference about the possibility of joint U.S.-Soviet exploration of outer space. The president referred to his March 7 letter and promised to release its contents. What Kennedy wanted the world to appreciate was that the United States thought of space as a peace initiative—but in reality, this was only half true.

Khrushchev, fearful of being duped by the American leader, was less attached to the humanistic goal of reserving space for scientific exploration. Nevertheless, the Soviet desire for peace exemplified by the nuclear slogan "Let the atom be a worker, not a soldier" also applied to their space rockets, where excitement centered on the hope that they would blaze a trail of scientific research, not be destroyers of cities. But behind the scenes, the Soviets were willing to weaponize space, having begun high-altitude testing of nuclear bombs during the last half of 1961, extending a program of surface and air-dropped atmospheric tests that had resulted in 2,014 detonations between 1949 and 1962.

Back in 1958, when Kennedy was running for reelection to the U.S. Senate, the Eisenhower administration had also detonated a series of nuclear warheads in space. At that time, the Soviets had taken the high road and chosen to impose a moratorium on themselves—which Khrushchev broke in late August 1961, under pressure from hard-line militarists. The Soviets' first launch during Kennedy's presidency exploded a nuclear warhead within Earth's atmosphere. A pair of tests that followed in October broke through and detonated in outer space. After that, the Soviet tests abruptly stopped. In the aftermath, Kennedy was determined not to allow the Soviets a technological advantage, yet he remained intent on brokering a ban on atmospheric nuclear testing, fearing another tit-for-tat dynamic that could lead to disaster. Careful never to describe the cosmos as a Cold War battlefield, Kennedy told *Time*'s Hugh Sidey, "Ever since the longbow, when man has developed new weapons and stockpiled them, somebody has come along and used them. I don't know how we escape it with nuclear weapons."

When in December 1961 Khrushchev described how easily the Soviets could substitute nuclear explosives for cosmonauts in orbiting *Vostok* rockets, Kennedy grew alarmed. Space started approximately sixty-two miles above Earth—declared such by the founding director of the Jet Propulsion Laboratory, Theodore von Kármán—and Kennedy wanted it maintained as a world peace demarcation line. Defense Secretary McNamara immediately authorized a leading Southern California think tank, the Aerospace Corporation, to work on a U.S. antiballistic missile defense system. By no coincidence, this company, founded only two years before in the midst of the Dyna-Soar's development, was closely aligned with the air force. Within a few weeks, however, the president rethought the idea, rescinded the brief to Aerospace Corporation, and refocused on deescalating the Cold War in space—a wiser long-term policy objective, he believed, than diverting money to build a multibillion-dollar cocoon over American airspace. Nearly a quarter century later, President Ronald Reagan would propose a similar program, the Strategic Defense Initiative (popularly known as "Star Wars"), which would cost taxpayers at least $30 billion before being scrapped due to technical failures, budgetary constraints, and concerns that it could complicate arms-limitation talks. Under President Bill Clinton, however, SDI was reformed into the Ballistic Defense Organization. (The extent of U.S. defensive coverage today is classified. But, according to what is in the public domain, programs already in place provide detection and tracking capabilities, and ground- and sea-based interceptors can launch kinetic vehicles to collide with incoming ICBMs.)

Kennedy devoted extensive time throughout early 1962 to conferring on the proper response to the Soviets' atmospheric nuclear testing of the year before. Glenn Seaborg, the chairman of the Atomic Energy Commission, became the president's key advisor on the subject of renewed American testing in the upper atmosphere and beyond. Considered one of the most esteemed chemists in American history, Seaborg was a Nobel Prize winner for his early-1950s work on ten transuranium elements. JFK trusted Seaborg because he was passionate about the peaceful uses of atomic energy. "[The president] made it clear," Seaborg recalled, "at every meeting during January, February, and March, that he had not yet made the definitive decision to resume tests."

According to Seaborg, Kennedy truly wanted to give Khrushchev every chance to agree to a nuclear test ban in 1962. There was no bluffing going on. Finally, in early April, the Kremlin announced its plan to renew testing at a later date. The president cringed with disapproval. Having failed to move U.S.-Soviet negotiations closer to a ban, and feeling backed into a corner, he reluctantly authorized a series of American atmospheric nuclear tests, the last of which would explode well into space. "On April 23, just two days before the opening of the test series, Secretary [of State Dean] Rusk called me to say that he had talked to the President who was at that time at Palm Beach," recalled Seaborg, "and the President . . . should get in touch with Chairman Khrushchev to suggest that after the completion of the U.S. atmospheric test series and the presumed second Russian test series that we all expected would be undertaken, the two countries should sign a treaty banning atmospheric tests. This was one of the earliest indications of his thinking in that regard." Disappointing peace activists, who had been counting on Kennedy to ban these high-radiation explosions, the administration was straddling the tricky line between the peaceful use of space and the air force's desire to make it into a Cold War battlefield.

When Kennedy learned that around two hundred executives from the defense industry, government, and local communities were meeting in Wilton, Connecticut, to discuss aerospace technology, he asked his advisor Roswell Gilpatric to deliver a major policy speech there. Gilpatric, who had served as undersecretary of the air force in the Truman administration, worked as a corporate lawyer in New York during the Eisenhower years. Now, as number two at the Pentagon, he had become indispensable to McNamara while earning the full confidence of the president. In fact, fearing that McNamara was inexperienced in foreign policy, JFK had handpicked Gilpatric to be his own private eyes and ears in the Pentagon.

The first fifteen minutes of Gilpatric's Wilton speech before these aerospace aficionados was boilerplate. But then, quite unexpectedly, he shifted gears and emphasized that the "long-standing proposal for cooperation with Russia was to ensure that space was used purely for peaceful purposes." This was still, Gilpatric said, "the national goal." Nevertheless, the United States would be "very ill-advised, if we did not hedge our bets. . . . We ought to be ready . . . [and] anticipate the ability of the Soviets at some time to use space offensively." Gilpatric added that while the Pentagon hadn't completed its strategic defense planning, he could foresee the development of "space systems which could be used to protect the peaceful or other defensive satellites now in operation."

Kennedy had adroitly used Gilpatric as a stalking horse to deliver a message to Khrushchev: the U.S. military-technology-industrial order was ready to meet any and all Soviet boasts and threats—even, if necessary, in outer space. In fact, the secretary of defense had already given the army top-secret authorization to proceed with the development of an antisatellite system under the name Mudflap. (This Defense Department directive was so secret that space historians question whether even the White House or air force was fully aware of it.) Gilpatric's comments, however, were widely read. "The furor that greeted these reports was immense," historian Paul Stares wrote in *Space Weapons and US Strategy*. "Not only did it appear to signal a reversal of the administration's position on the peaceful exploitation of space, but it also seemed to members of Congress that the Defense Department was competing with NASA."

On May 13, three days after Gilpatric's speech, Kennedy was at the first conference of the United Nations' new eighteen-nation Disarmament Commission, in Geneva, Switzerland, preparing to propose a four-point plan to ensure peace in space. The initial business of the conference was a discussion of Soviet and U.S. draft treaties to ban all nuclear testing, but the president's negotiations were soon bringing space into the discussion, and the domestic dustup over Gilpatric's Wilton speech may have been timed to enhance the American negotiators' proposals on space in general. For the time being, though, the USSR was unwilling to budge on nearly anything. "The reports from [our] representatives in Geneva during the spring of 1962," said Seaborg, "reflected a strong sense of frustration and discouragement."

While the Soviets blocked progress on Kennedy's efforts regarding the peaceful use of space, they accepted his open-handed proposal to cooperate in the management of weather satellites. In another friendly development, the Soviet cosmonaut Gherman Titov visited America with his wife, Tamara, in early May, and was greeted by a festive atmosphere of genuine admiration. The couple was hosted by John and Annie Glenn, who took them to see the memorials and museums on the capital's National Mall. The Titovs were mobbed by adoring fans and space enthusiasts everywhere they went around the Tidal Basin. On May 3, the two astronauts had a private meeting at the White House with the president, who welcomed them in a statesmanlike way, heavier on diplomatic protocol than raw enthusiasm. But as positive as the satellite cooperation and Titov's goodwill visit might have been, they proved only a temporary reprieve from the two superpowers' head-on competition for space dominance.

By the end of May, NASA was preparing for its next launch, which appeared at first glance to be a repeat of the Glenn mission: three orbits, with much the same equipment. However, the Mercury astronaut chosen, Scott Carpenter, would be expected to perform additional tasks while in space. No "man in a can" or a glorified chimpanzee, Carpenter would conduct a greater number of scientific experiments than had been attempted on any previous Mercury mission.

A native of Boulder, Colorado, Carpenter had won the coveted fourth astronaut slot after Deke Slayton was benched over medical concerns. A longtime navy test pilot known as the free spirit of the Mercury Seven, he was a natural flier, blessed with the ability to stay tranquil in high-tension situations. He had learned to fly planes over the Front Range of the Rockies, where weather patterns can shift dramatically. His Colorado upbringing may also have contributed to his ability to acclimate quickly to high-altitude situations without feeling lightheaded or oxygen hungry.

During the Korean War, Carpenter had flown numerous reconnaissance and antisubmarine missions along both the Siberian and Chinese coasts. After three deployments, he returned stateside to serve at the Naval Test Pilot School in Patuxent River and, later, at the Navy Line School in Monterey, California. If Shepard, Grissom, and Glenn were renowned as fighter pilots, all steady hands with good judgment, Carpenter stood out for his phenomenal visual capabilities. As legend had it, from a cramped plane cockpit, he could see the minutest details at the most extreme ranges, a real asset in the high-altitude surveillance business.

At 7:45 a.m. on May 24, President Kennedy watched from his White House bedroom as Carpenter's *Aurora 7* lifted off from Cape Canaveral; for the rest of the morning, the president darted out of meetings to catch further televised coverage. Television was indeed the magic machine of the era, bringing live events into the public's living rooms. Carpenter's three-orbit mission lasted almost five hours, his Mercury-Atlas 7 achieving a maximum altitude of 164 miles and an orbital velocity of 17,532 miles per hour. The number of flight "firsts" Carpenter accomplished included eating solid food in space and conducting successful experiments regarding liquid behavior in a weightless state. He took nineteen beautiful photographs of the flattened sun at orbit sunset. With his keen eyes, he identified particles of frozen liquid from outside the *Aurora* 7 and reported back to Mercury Control about the phenomenon. Chris Kraft, the Cape Canaveral flight director, thought that Carpenter conversed with himself too much, peering out the porthole, soaking up the sublime majesty of space. "He was completely ignoring our request to check his instruments," Kraft later wrote. "I swore an oath that Scott Carpenter would never fly again. He didn't."

In preparation for reentry, Carpenter made several mistakes, including one that wasted fuel. Finally, when he fired his retrorockets on reentry three seconds late, his capsule overshot the planned splashdown point by 250 miles. Carpenter had to escape the *Aurora* 7 before it sank, scrambling into an inflatable raft on a rough sea northeast of Puerto Rico. He was out of radio contact for thirty-nine minutes, with the ocean heaving mightily. Nobody knew whether he'd survived. After three hours lost at sea, he was saved by two frogmen who jumped from an SC-54 transport plane, finding him "smiling, happy, and not at all tired." He was retrieved by an H5S-2 helicopter and whisked to the deck of the USS *Intrepid*.

The breakdown of procedure on reentry embarrassed NASA brass. Fingers were pointed in various directions, and Carpenter took the brunt for being too fascinated with the beauty of space and negligent of technical procedures—which was somewhat unfair, since mechanical issues also played a part. As the second U.S. astronaut to orbit Earth, Carpenter had performed well, racking up NASA's fourth successful Mercury mission and moving the agency one more stride closer to the moon. Nevertheless, Chris Kraft, in his memoir, claims he drummed Carpenter out of the astronaut corps because of his poor performance on the Mercury mission.

An ebullient and grateful Kennedy made his now-traditional call, reaching Carpenter aboard the USS *Intrepid*. With heartfelt congratulations, he invited the astronaut to visit the White House down the line. However, no special ceremony was planned for Carpenter, indicating that the president and NASA were putting the brakes on the celebratory machine that had greeted the three previous Mercury astronauts. On June 5, Carpenter, his wife, Rene, and their four children spent about twenty minutes with Kennedy in the Oval Office. During the middle of a serious space-related conversation, one of Carpenter's young daughters belted out, "Where is Macaroni?" referring to the Kennedy family pony. Then her sister shouted accusingly, "And Caroline?" While the adults laughed, the two Carpenter kids truly wanted answers. "President Kennedy bent down to their faces," Carpenter recalled in his memoir *For Spacious Skies*. "Macaroni, America's most famous Pony, needed pasture and lived in the country, [the president] explained. The girls stood their ground unsmiling."

Even after the Carpenters' embrace by Kennedy, rumors continued to circulate that the astronaut had panicked during his mission. A decade later, Tom Wolfe came to Carpenter's defense in *The Right Stuff*. "One might argue that Carpenter had mishandled reentry," Wolfe wrote, "but to accuse him of panic made no sense in the light of the telemetered data concerning his heart rate and respiratory rate." A year and a half after his Mercury flight, Carpenter took a leave of absence from NASA to participate in a navy project called Sealab. He later returned to the space program, then retired early.

What Carpenter didn't know during his White House meeting was that Kennedy was grappling with a frightening nuclear issue. Twenty-four hours before he arrived, the United States had sent up an army Thor rocket equipped with a nuclear bomb set to explode at an altitude of thirty miles. It was part of the series of high-altitude tests grouped together as Operation Fishbowl, which would culminate with nuclear weapons detonated in outer space. The operation was in direct response to the August 30, 1961, announcement by the Soviets that they were ending their three-year moratorium on testing. During the first attempted launch, on June 2, radar lost track of the missile, and concerns over the safety of ships and aircraft on its trajectory led to the mission's being aborted, the rocket falling into the North Pacific near Johnston Atoll. Though the test itself hadn't occurred, the incident pointedly told the Soviet intelligence officers in the KGB that the United States was on the verge of matching and exceeding their own high-altitude nuclear testing from the previous year. Predictably, Moscow's diplomats at the Geneva disarmament talks were incensed, feeling double-crossed by the Kennedy administration. With bombastic overreaction, Khrushchev accused Kennedy of planning imminent attacks on Soviet cities from space.

For the time being, Kennedy's plan to coax the Soviets into a test ban agreement was shipwrecked. The American tests continued, and the next rocket, code-named Starfish Prime and manufactured by Los Alamos Scientific Laboratory, was launched on

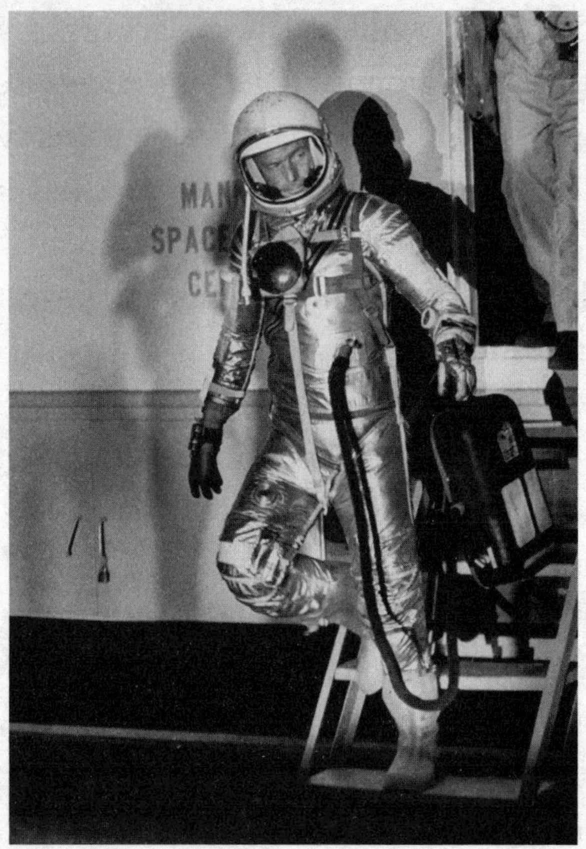

Mercury astronaut Scott Carpenter of Colorado dressed for spaceflight. He was a U.S. Navy officer and aviator. On May 24, 1962, Carpenter flew into space atop the Mercury-Atlas 7 rocket, becoming the second American (after John Glenn) to orbit the Earth.

July 9, 1962. Its W49 warhead detonated at a record altitude of nearly 250 miles, well into outer space, but its power was such that the enormous fireball could be seen flashing like heat lightning, even through heavy cloud cover, as far away as Honolulu. It was by far the largest nuclear weapon detonated to date, one hundred times more powerful than the bomb dropped on Hiroshima.

By design, Starfish Prime augmented the radiation found in the Van Allen belts, collections of charged particles circling Earth and held in place by the planet's gravity, which had been discovered in 1958 by James Van Allen of the University of Iowa. The U.S. military believed that the megabomb could be used to make the Van Allen belts destructive to dangerous Soviet satellites. Only after the massive Starfish Prime explosion did NASA study whether the fortified Van Allen belts could also prevent a successful lunar mission. Kennedy, asked beforehand whether the test could adversely affect the belts, replied jauntily that "I know there has been disturbance about the Van Allen belt, but Van Allen says it is not going to affect the belt." In truth, years passed before Earth and its atmosphere, and its belts, recovered from the seismic effects of Starfish Prime. There are numerous astrophysicists who believe they will never again be the same.

With the successful Glenn and Carpenter orbits balanced against the spectacularly ill-conceived experiment of Starfish Prime, the topic of space was a complicated one for the president. The flagging effort to bring nuclear disarmament to outer space weighed on JFK's mind, and his first priority was to make sure space remained safe enough for an astronaut to fly through, and for a world population to live beneath. On the other hand, Kennedy did have some notable successes to boast of. Since his May 1961 moonshot challenge, the United States had launched at least fifteen orbital satellites. And on August 27, NASA's *Mariner 2* probe took off from Cape Canaveral, bound for Venus on a three-and-a-half-month scientific mission to measure planetary temperatures and the interplanetary magnetic field, and to take other readings en route. On December 14, 1962, *Mariner 2* became the first spacecraft to fly past another planet, its perfect trajectory confirming Kennedy's earlier description of it as "the most intricate instrument in the history of space science."

What Kennedy kept in mind at all times was how NASA could help fuel technological innovation in the private sector. He was locked into Section 203(a)(3) of the Space Act of 1958, which charged NASA with the mandate to "provide for the widest and most practical appropriate dissemination of information concerning its activities and results thereof." To Kennedy this meant shortening the time gap between NASA's discovering new knowledge and its effective adaptation in the consumer marketplace. In June 1962 Kennedy approved the NASA Technology Utilization Program (originally called the Industrial Application Program) to quickly transfer new knowledge derived from places such as the Jet Propulsion Laboratory and Marshall Space Flight Center into the American manufacturing-innovation order. On July 10, in the spirit of the Technology Utilization Program, the United States made the first direct TV connection between continents after the launch of *Telstar 1*, an American Telephone and Telegraph Company (AT&T) active repeater satellite orbited by a NASA vehicle. Although the relays didn't function properly at first, the coming months saw two successful test transmissions between receiving ground stations at Andover, Maine, and Goonhilly Downs, England. Then, on July 23, *Telstar* began transmitting regular civilian TV broadcasts between the United States and Europe. "Telstar was the first true communications satellite," historian Michael J. Neufeld has explained, "one that could receive a signal from the ground, amplify it, and then immediately retransmit it," enabling live overseas broadcasts of events such as the Olympic Games and European elections.

What differentiated *Telstar* from other Kennedy-era satellites was that its funding emanated principally from the private sector; AT&T boasted that the innovation was a tribute to the American free-enterprise system. And while NASA wasn't directly responsible for the global commercial satellite system, it partnered with other groups, private and public, to make it a reality.

In an effort to prove that he was both pro–air force and pro-NASA Kennedy presented four X-15 pilots the Robert J. Collier Trophy in a Rose Garden ceremony. All the leaders in the aerospace industry attended, as did top Pentagon officials. Kennedy purposely had Webb and air force secretary Eugene Zuckert pose together in a photo op. The four honored pilots were Major Robert M. White, Joseph A. Walker, A. Scott Crossfield, and Commander Forrest Petersen.

Around the time of the Collier Trophy ceremony and *Telstar*, NASA publicly announced the specific framework for the future Apollo lunar mission, formally adopting John Houbolt's vision of a lunar orbit rendezvous, with one manned spacecraft circling the moon while a smaller craft detached for descent to the surface. With that framework in place, planning for future missions could be organized on a more detailed time line, bringing Apollo-Saturn into better focus. Feeling great about how his administration was shaping up, Kennedy installed a secret taping system in the White House, ostensibly to assist his future memoir. Space history would greatly benefit from the taped conversations Kennedy had in 1962 and 1963, particularly with James Webb.

NASA's long, drawn-out Apollo plan didn't thrill everyone. That summer of 1962, Senator Barry Goldwater gave a major speech on space policy, his first since Kennedy's moonshot oration before Congress. His preference was to build up the air force, not NASA moon ships. Goldwater, who had inherited a thriving Phoenix, Arizona, department store in 1930, had developed from an anti-New Dealer into a conservative Republican by the time of World War II. Joining the U.S. Army Air Corps, he flew cargo into war zones worldwide, making frequent runs between the United States and India. This assignment expanded to Goldwater's piloting over "the Hump" of the Himalayas to deliver much-needed supplies to Chiang Kai-shek's army in China. There were also flights to Nigeria, the Azores, Tunisia, and South America. Goldwater remained in the Arizona Air National Guard and Air Force Reserve after V-J Day, where he became a major general. All told, he flew more than 165 different planes, including the B-52 Stratofortress. Nobody could accuse him of not understanding military aviation.

Goldwater won a Senate seat in 1952, upsetting the incumbent Democrat, Ernest McFarland, then the Senate majority leader. Goldwater's political influence was essential to the creation of the U.S. Air Force Academy in Colorado Springs (the visitor center of which is named after him), and by mid-1962 he was being mentioned as a possible 1964 presidential candidate, in part because he was a quintessential hawk. Given his bedrock devotion to the air force, Goldwater thought that civilian space should indeed be militarized. It bothered him that NASA was always trying to placate the United Nations instead of militarily winning the Cold War.

New York liberal governor Nelson Rockefeller was also considered a serious GOP presidential contender for 1964, standing at the opposite end of the Republican continuum from Goldwater. Richard Nixon, the center-right pragmatist, was now running for governor of California, and few thought he would be game for another presidential run after his 1960 loss. At the time of the Starfish Prime detonation, Rockefeller was leading Goldwater by a wide margin in opinion polls. As the Gallup poll showed and as Goldwater understood, Kennedy at that point remained unburdened by any meaningful negative issues—and so, on July 17, the Arizona senator set out to give the president one. "The clock has already run too long a course without our pursuing more vigorously a military space program," Goldwater told the National Rocket Club. "How can we guarantee that space will be used for peaceful purposes without having the means to defend such a doctrine? It is our view that international law or agreement cannot exist without the physical means to enforce it."

Hoping to score midterm election points, the articulate Goldwater also accused the Kennedy administration of "gambling with national survival" by making military objectives in space secondary to peaceful scientific accomplishments. The Arizona conservative-libertarian had an alternative: "The armed forces should already be planning the development as soon as possible of a completely integrated space warfare system. Perhaps I should say, a super-system, since it will be far more comprehensive than other so-called systems."

Goldwater's criticism of Kennedy appeared in newspapers across the country. His "gambling with national survival" theme gathered a measure of traction, and in August he and fellow Republicans in the Senate opened a debate on the subject of space militarization. It wasn't exactly a prime spot on the calendar, tucked between summer recess and Labor Day, but they nevertheless made their prepared remarks in the full chamber. The administration was fully aware that Goldwater was looking for an equivalent to JFK's own charge of a "missile gap," which had so sorely dogged Eisenhower and Nixon. When the Republicans criticized the president for having actually "deterred" the armed services from preparing for space warfare, the White House offered figures from the new budget, showing that spending on military space programs had doubled in just one year, coming in at $1.5 billion, an enormous sum for 1962.

Dean Acheson once quipped that if you hurled a brick down a blind alley at night and heard a loud squawk, you knew you'd hit a cat. In this spirit, Goldwater's brick had clearly made an impact on Kennedy. On September 5, the White House announced that the president would visit space facilities in Alabama, Florida, Texas, and Missouri. In addition, Kennedy authorized Roswell Gilpatric to rewrite his upcoming speech in South Bend, Indiana, the new remarks directly rebuking Goldwater's hawkish appraisal. "The United States believes that it is highly desirable for its own security and for the security of the world," Gilpatric said in a kind of Kennedy Doctrine, "that the arms race should not be extended into outer space, and we are seeking in every feasible way to achieve that purpose." After that, it was left for the media-savvy Kennedy himself to bring the force of his optimistic personality to bear on the NASA–versus–air force view of space, which he could do as no other. And, perhaps more importantly, to prepare for budget discussions.

On September 11, President Kennedy began his space tour in Huntsville, Alabama. It was the day after Supreme Court justice Hugo Black ordered that the University of Mississippi allow James Meredith, a twenty-nine-year-old African American and air force veteran, to attend fall classes at its segregated Oxford campus. That decision would roil the school and the nation at large, as Mississippi governor Ross Barnett ordered state troopers to bar Meredith from enrolling. The White House was working hard to change the Old South, partly by using NASA to bring high-tech jobs and a futuristic thinking to backward regions slow to abandon violent and self-defeating prejudice. Determined to win the Oxford showdown, the Kennedy administration hoped that the recent selection of Hancock County, Mississippi, on the banks of the Pearl River near Louisiana, as the site of an Apollo engine-testing facility—these days, it's known as the John C. Stennis Space Center—might ameliorate the state's anti-federal stance.

Despite the nettlesome civil rights crisis in Mississippi, Kennedy didn't allow himself to be distracted from his central purpose in Huntsville: inspecting the development of the Saturn rocket, the moonshot vehicle. Rocket City, U.S.A., pulled out all the stops for Kennedy's visit, honoring him with Dixieland jazz and a twenty-one-gun salute. Visiting both the Marshall Space Flight Center and the Ballistic Missile Agency base, Kennedy was hosted by von Braun, the city's most famous citizen. When posing for photographs together, they looked like twin glamour doppelgängers cut from the same cloth. Pointing to a drawing of a Saturn rocket, the proud rocketeer enthused to his boss, "This is the vehicle designed to fulfill your promises to put a man on the moon in this decade." Then, looking at the Saturn model first and then flashing his eyes at Kennedy, he added, "And, by God, we'll do it!"

Von Braun gave the president a guided tour of a well-protected bunker from which they witnessed the static firing of a Saturn C-1 booster. Karl Heimburg, the Test Lab director, told JFK that this booster produced 1.3 million pounds of thrust, dwarfing anything the Soviets could muster. When the rocket's engines fired up, on time to the millisecond, Kennedy's jaw dropped. "Just as the last echoes reverberated among the huge test stands and blockhouses of the Marshall Center Test Lab, Kennedy grasped von Braun's hand impulsively and congratulated him warmly," von Braun's friend and biographer Erik Bergaust recalled. "In a rare gesture of credit-sharing, von Braun waved his hand toward the team members nearby, in a display of appreciation."

Even though NASA had already determined that lunar orbit rendezvous (LOR) was the approach it would use for landing on the moon, there remained dissension in the ranks. As von Braun explained the LOR concept, Kennedy slyly retorted, "I understand that Dr. Wiesner doesn't agree with this," then called his science advisor over for an impromptu debate, with Webb and Seamans joining in. White House aides, Vice President Johnson, and the press looked on in consternation. After five minutes, the debate between Wiesner and Webb got heated and Kennedy stepped in to play referee. "Well," he said, "maybe we'll have one more hearing and then we'll close the books on this issue."

At Marshall, Kennedy also visited the principal hangar where a Saturn 1 booster was housed. As von Braun used rocket models to give Kennedy a short, impromptu lecture, the camera-courting president positioned himself at the best angle to allow photographers to capture both him and the giant rocket behind him. Von Braun, however, kept moving closer to the president, inadvertently botching the photo op. One journalist blurted out, "Look at von Braun trying to upstage the President!" The assembled officials broke out in laughter. Miscues aside, it was evident at Huntsville that Kennedy had grown personally fond of von Braun. In the way he exuded controlled optimism while maintaining a subtly detached air, von Braun was perhaps closer to Kennedy in persona than any Democratic senator or congressman of the era.

The next leg of Kennedy's space tour was to the NASA Launch Operations Center on Cape Canaveral. Quite spontaneously, Kennedy asked von Braun to travel there with him on Air Force One, so they could talk shop. He boarded without even a travel kit. On the plane, a reporter asked Kennedy who was going to win the LOR debate. "Jerry's going to lose it, it's obvious," JFK joked, referring to Jerome Wiesner. "Webb's got all the money, and Jerry's only got me." Kennedy was being funny; weeks earlier, LOR had been chosen, and there was no turning back.

At Cape Canaveral, the president inspected perhaps the most sophisticated high-technology government facility in the world in a two-and-a-half-hour walk-around. To his pleasure, the massive NASA funding he'd pushed for was being put to full use, with lunar expedition planning clearly in high gear. Watching the president move quickly around the facility, inspecting the assembly shop and educating himself about pre-determined launch loads, spacecraft component separation, heat shield verification, and other minutiae of space preparation, one missile technician cried out, "Who said John Glenn is the fastest American alive? Jack Kennedy has him beat a mile!"

Even in Florida's stultifying coastal humidity and mosquito hordes, Kennedy didn't wilt. Beaming confidence and joviality, he gave a brief pep talk, proclaiming, "We shall be first!" On a helicopter tour with Gordon Cooper and Gus Grissom as seatmates, as ABC News space correspondent Jay Barbree recalled, the two Mercury astronauts pointed out the key infrastructure highlights of the growing moonport. "They showed him where one day a monster called Saturn V would stand on its launch pad," Barbree wrote. "Here the name Apollo was gaining substance with every passing day."

President John F. Kennedy (*center*) with Vice President Lyndon Johnson (*right*), listening intently to Dr. Wernher von Braun (*left*) explain the models of the Saturn C-1 booster rocket during their inspection tour of the Space Flight Center.

PART IV

Projects Gemini and Apollo

President John F. Kennedy spoke before a huge crowd at Rice University's football stadium on September 12, 1962, in Houston, Texas, on the nation's space effort.

17

"We Choose to Go to the Moon"

Rice University, September 12, 1962

This generation does not intend to founder in the backwash of the coming age of space. We mean to be a part of it—we mean to lead it. For the eyes of the world now look to space, to the moon and the planets beyond, and we have vowed that we shall not see it governed by a hostile flag or conquered, but by a banner of freedom and peace.

—JOHN F. KENNEDY, 1962

Staying on schedule, President Kennedy left Cape Canaveral late on the afternoon of September 11, bound for Houston. The holiday atmosphere of the "space tour" was such that in a city of just over nine hundred thousand people, some three hundred thousand turned out to greet JFK upon his arrival. Addressing the adoring crowd, he said, "I do not know whether the people of the Southwest realize the profound effect the whole space program will have on the economy of this section of the country. The scientists, engineers, and technical people who will be attracted here will really make the Southwest a great center of scientific and industrial research as this nation reaches out to the moon. In this place in America are going to be laid the plans and designs by which we will reach out in this decade to explore space." Among other things, Kennedy's positive words were intended to repair damage from a gaffe Robert F. Kennedy had made while visiting the University of Indonesia, in Jakarta, the previous February. Asked by a student about the Mexican War of 1846–48, RFK replied, "Some from Texas might disagree, but I think we were unjustified. I do not think we can be proud of that episode." Many Texans were furious at this sentiment, forcing the president to control the damage rather than risk losing votes in 1964.

Accompanying President Kennedy in Houston were Vice President Lyndon Johnson, Defense Secretary Robert McNamara, Navy Secretary Fred Korth, Congressman Albert Thomas, and NASA administrator James Webb. Von Braun was also still in the entourage, constantly reassuring reporters that the moon flight would happen within the decade. "The people here realize the effect the space age will have on their city," Kennedy told a crowd of well-wishers. "It is most appropriate that the manned spacecraft center should be located here in Houston, identified as the most progressive city in the area. From this place will be made the plans to take Americans to the moon—and bring them back."

Kennedy spent that evening on the sixth floor of the Rice Hotel in downtown Houston. This was where he'd stayed exactly two years before, while preparing his speech to the Greater Houston Ministerial Association, which successfully inoculated him from concerns over his Catholicism. This time, as he was ushered in, he was handed a welcome-back telegram from the Ministerial Association that read, "May God continue to guard and guide you in the leadership of our nation." Anybody who telephoned the Rice Hotel after the president checked in was met with a switchboard operator saying, "White House."

When Kennedy awoke on September 12, he read the very positive lead editorial in the *Houston Press* about the economic benefits of Project Apollo. "Sixty firms have moved into Houston–Harris County or expanded their activities here as a result of the opening of the National Aeronautics and Space Administration's (NASA)

Moonshot Command Post and Research Center," the editorial boasted. "Twenty-five to thirty more have plans for Houston offices as yet officially undisclosed." There was nothing Kennedy liked more than to read stories about how NASA was spurring business enterprise in the southern states.

After breakfast, Kennedy headed to Rice University, where a sun-drenched crowd of forty thousand eagerly waited at the football stadium to hear him speak about going to the moon. Even at ten in the morning, the weather was blazingly hot and humid, and the throngs of spectators fanned themselves madly as JFK arrived at the podium in a dark blue suit, with a white handkerchief in his breast pocket, white shirt, and blue tie, posing for photographs with NASA administrators before speaking. His hair shone reddish in the sunlike glare of camera lights. The *Houston Press* reported that "everyone perspired" in the "roaster" of a stadium, built in 1950, but the upbeat feeling was like that of a high-octane campaign rally. Only a single "I like Ike" sign, raised above the crowd's heads, was a manifestation of protest.

The president's speech had been drafted by Ted Sorensen, with important contributions from various NASA advisors and wordsmiths. Space-related articles in *National Geographic* were also consulted. But if the key phrasing of the speech belonged to Sorensen and NASA scribes, the spirit of raw aspiration was pure JFK. Knowing that his moonshot speech to Congress the previous year had been very buttoned down, Kennedy had decided that the Rice address would be a stem-winder, filled with the kind of soaring rhetoric that had thrilled the world in his "Ask Not" inaugural address. At Rice he would tie his patriotic belief in American exceptionalism directly to his prioritization of the manned space effort. A copy of the Rice speech, now kept at the John F. Kennedy Presidential Library and Museum in Boston, shows all the president's last-minute handwritten tweaks. Because the Rice Owls were usually slaughtered every fall by the Texas Longhorns in a lopsided college football rivalry, Kennedy personally added a comical line in promoting his Apollo moonshot: "And they may as well ask: Why climb the highest mountain? Why 35 years ago fly the Atlantic? Why does Rice play Texas?"

Dr. Kenneth Pitzer, president of Rice University, introduced Kennedy to the roaring crowd, naming the president a visiting professor and declaring the afternoon as the opening salvo of Rice's semicentennial year. Pledging to expand Rice's science programs to meet America's Space Age needs, Pitzer committed the university to providing graduate-level instruction in geomagnetism, Van Allen radiation, auroras, and atmospheric structures. (The next year, Rice would become the first university to announce the opening of a graduate school in space science.) The *Houston Press* went so far as to deem Rice the "educational pilot plant" of NASA.

When Kennedy took the rostrum, smiling boyishly, the crowd went wild. He looked suntanned and relaxed, undisturbed by the heat. Sitting directly behind him onstage were Lyndon Johnson, Albert Thomas, and other government officials. While those who sat in the stands—including ten thousand Boy Scouts and Cub Scouts—heard the president's message perfectly, the dignitaries ensconced on the speaker's platform behind him could barely understand a word. For the TV cameras filming the speech in color, Kennedy was downright effervescent. "I can remember it clearly today," Bob Gomel, then a *Life* magazine photographer, recalled fifty years later. "He has his fist clinched on the podium, and his delivery was so dynamic."

Speaking with poetic grace, perfect timing, and flashes of Harvard wit, Kennedy delivered an oratorical masterpiece. There was purposeful masculinity to his well-crafted words. Positioning science and technological research at the forefront of American life, he began his speech by praising Houston as the $123 million home of NASA's Manned Spacecraft Center. "We meet at a college noted for knowledge, in a city noted for progress, in a state noted for strength, and we stand in need of all three," he said. "For we meet in an hour of change and challenge, in a decade of hope and fear, in an age of both knowledge and unforgettable ignorance. The greater our knowledge increases, the greater our ignorance unfolds."

What made the speech so exquisite was Kennedy's reflecting on fifty thousand years of recorded history, from cavemen to jet pilots to astronauts. The president mocked those timid citizens who wanted to stay still on Earth a little longer, who didn't aim for the moon, joining ranks with "those who resisted the horseless carriage and Christopher Columbus." Discoveries such as Newton's law of universal gravitation were evoked, as were such inventions as the steam engine, electric lights, and the telephone. "This is a breathtaking pace, and such a pace cannot help but create new ills as it dispels old, new ignorance, new problems, new dangers," Kennedy said. "Surely the opening vistas of space promise high costs and hardships, as well as high reward. So it's not surprising that some would have us stay where we are a little longer to rest, to wait. But this City of Houston, this state of Texas, this country of the United States, were not built by those who waited and rested and wished to look behind them. This country was conquered by those who moved forward—and so will space."

The words were vintage Kennedy, a distillation of his evolved thinking on space since *Sputnik*. At their core was an insistence that the United States, no matter the financial cost, had to dominate space. "The exploration of space," he said, "will go ahead whether we join in it or not, and it is one of the great adventures of all time. . . . Those who came before us made certain that this country rode the first waves of the industrial revolution, the first waves of modern invention, and the first waves of nuclear power. And this generation does not intend to founder in the backwash of the coming age of space. We mean to be a part of it. We mean to lead it."

Displaying what astrophysicist Neil deGrasse Tyson later called "politically uncommon fiscal candor," Kennedy laid out the costs of the moon program. "This year's space budget is three times what it was in January 1961," he told the crowd, "and it is greater than the space budget of the previous eight years combined. That budget now stands at $5.4 billion a year—a staggering sum, though somewhat less than we pay for cigarettes and cigars every year." Bringing the abstract down to earth, the president said that space expenditures would soon rise further, "from 40 cents per person per week to more than 50 cents a week for every man, woman, and child in the United States." Although he acknowledged the risk of the undertaking, he believed America's hopes for peace and security rested on its seizing world leadership in space.

The heart and soul of the Rice speech connected NASA to both America's frontier tradition and the concept of American exceptionalism. Pride, prestige, and national defense were major factors, and beating the Soviets was a geopolitical imperative, but the United States couldn't be defined by its Communist adversary. Instead, Kennedy explained at Rice, going to the moon presented the grand historic challenge of an unexplored frontier, and was the noblest illustration of the American pioneer spirit in the twentieth century. In winged words, the president delivered one of his most timeless sentiments, placing Apollo among mankind's noblest aspirations: "We choose to go to the moon—we choose to go to the moon in this decade and do the other things, not because they are easy, but because they are hard, because that goal will serve to organize and measure the best of our energies and skills, because that challenge is one that we are willing to accept, one we are unwilling to postpone, and one which we intend to win."

Coming toward his conclusion, Kennedy posed an exciting challenge to the nation. "Many years ago, the great British explorer George Mallory, who was to die on Mount Everest, was asked why did he want to climb it. He said, 'Because it is there.' Well, space is there, and we're going to climb it, and the moon and planets are there, and new hopes for knowledge and peace are there. And, therefore, as we set sail, we ask God's blessing on the most hazardous and dangerous and greatest adventure on which man has ever embarked."

At those words, the stadium erupted in applause. The president had once again thrown down the gauntlet, tying the nation's very heart to the goal of reaching the moon. With its grand gestures and motivational words, his speech ranks among the most inspiring ever delivered by an American president, and those in the audience felt a part of history. Terry O'Rourke, a camera-carrying high school sophomore, had ridden his bike to Rice to hear his hero speak, and experienced a transformative moment. "I remember the times, it was before Mustangs and miniskirts," O'Rourke recalled in 2002. "The Cold War was real. It was scary. John F. Kennedy did something. He took the horror of the Cold War and made something beautiful, *a dream for all of us.* He was like a coach giving calls to the team. He was young, but he knew what the hell he was doing."

As Kennedy exited the stage, he nodded at Webb and the other effusive NASA administrators with a wide grin. "All right," he said. "Now you guys do the details!"

It was instantly clear that Kennedy had scored a winner. Broadcast on the radio and nightly news broadcasts, the speech had an immediate effect beyond Rice Stadium, and his declaration that "We intend to be first" was discussed in TV and print media as a significant doubling-down on his original moonshot appeal to Congress. The echoes of the oration continued reverberating in the culture, until by the twenty-first century it ranked as one of the high points of JFK's presidency. Film clips from the address have been played so many times on TV over the decades that people are often tricked into thinking they remember the speech's galvanizing importance on the day it was delivered. "I certainly remember it," Neil Armstrong said of Kennedy's Rice speech in 2001, "but it's a bit hazy because I've heard recordings of it so many times since, that you're not certain whether you're remembering or you're remembering what you're remembering. . . . And, of course, it's been colored by the fact I read so many stories of how that process actually occurred and what led to his conclusion to do that."

The event remains a high-water mark in the history of Rice University and of Houston. Rice class of '63 graduate Paul Burka, an attendee who went on to become executive editor of *Texas Monthly* magazine, believed the "We choose to go to the moon" speech was eternal because Kennedy "encapsulates all of recorded history and seeks to set it in the history of our own time." Rice student Jacob Scher, who also witnessed the presidential appeal firsthand, said Kennedy "blew me away." Others, like Rice professor of chemistry Robert Curl, who won the Nobel Prize later in his career, recall being startled by the largesse of Project Apollo. "I came away in wonder that he was seriously proposing this," Curl recalled. "It seemed like an enormous amount of money to spend on an exploration program."

As Kennedy traveled from Rice down the Gulf Freeway to the temporary Manned Spacecraft Center's Research Division complex on Telephone Road, more than forty thousand people stood along the route to wave and get a glimpse of him. Arriving at the center's Rich Building (one of twelve Houston-area NASA sites), he was greeted by five of the Mercury Seven: John Glenn, Alan Shepard, Gus Grissom, Gordon Cooper, and Deke Slayton. The astronauts escorted the president around the scientific displays and exhibits. Dr. Robert Gilruth, first director of the new Manned

Spacecraft Center, along with associate director Walter Williams and the heads of Projects Apollo, Gemini, and Mercury, gave JFK a forty-minute briefing on the progress of research for the moonshot.

President John F. Kennedy emerging from inside a model of the futuristic NASA spacecraft during his tour of the new Manned Spacecraft Center in Houston, Texas.

Enthralled by the space jewels he saw on his tour of the facility, Kennedy stayed twice as long as scheduled. At one point, he climbed into the cockpit of an Apollo Command Module mock-up, which had been designed to test the crew seats and instrument panels. Staring at a seventeen-foot-high lunar lander prototype, he seemed almost paralyzed with wonder. Shepard described to Kennedy the work of the "lunar excursion vehicle" (LEV; also called the "Lunar Bus"), praising the extraordinary effort of the Manned Spacecraft Center team. Surrounded by engineers, technicians, and computer specialists, Kennedy reiterated his pledge that the Apollo moon landing would happen within the decade. "To talk of placing a tremendous rocket outside the orbit of the earth, to send it to the moon to rendezvous, to go to the moon's surface, to put men on the moon, to take them off, to rendezvous again, and bring them back to earth safely, and to talk about doing that in the next 5 or 6 years indicates how far and how fast we have come and how far and how fast we must go," he said. "And back of all the extraordinary scientific and technical accomplishments which must be made to make this possible, of course, are the men who are involved, and particularly those who are at the point of the spear, those who must fly this mission into the most unknown sea."

At some level, both Kennedy's Rice speech and the spontaneous remarks he delivered in front of the Mercury astronauts were responses to his Republican critics. At Rice, his lofty rhetoric, elevating the moonshot into the pantheon of human exploration and scientific advancement, was a slap at Barry Goldwater's contention that NASA was a distraction from the more important military uses of space. On Telephone Road, his remarks were a counterweight to Dwight Eisenhower, who'd mocked Project Apollo's breakneck pace. With the midterm elections looming, the former president was, that same day, out politicking for the first time since leaving the White House, stumping for Republican senatorial candidates.

Eisenhower and Kennedy were at a unique juncture in their strained relationship. A few weeks before, in *The Saturday Evening Post*, Ike had levied what amounted to his own personal platform for the nation's future. Eisenhower didn't neglect space in his state-of-the-nation manifesto, but while proclaiming his pride in John Glenn, Scott Carpenter, and the other Mercury astronauts, he came down hard on Kennedy's multibillion-dollar budgetary expansion of NASA. In a repeat of his commentary at the Naval War College the previous year, Eisenhower lamented the amount of money that was, in his view, being wasted on Project Apollo. "By all means, we must carry on our explorations in space, but I frankly do not see the need for continuing this effort as such a fantastically expensive crash program," he wrote. "From here in, I think we should proceed in an orderly, scientific way, building one accomplishment on another, rather than engaging in a mad effort to win a stunt race."

Being criticized by the still popular Eisenhower put

Kennedy in a political bind: realistically, he was unable to ignore his predecessor's putdown, yet it would be foolhardy to confront it head-on. The Cold War had obviously put Americans in various camps, reflecting their attitudes toward Communist and especially Soviet influence in America. But even though that struggle overlapped with the space race in crucial ways, NASA's space effort had yet to become a liberal or conservative issue as of 1962. In his speeches and press conferences, Kennedy strove to keep the Mercury, Gemini, and Apollo programs above that political fray, and when he described the effort to put a man on the moon, it was invariably presented as a uniting, bipartisan endeavor. Now Eisenhower's critical commentary, purposely given with the midterm elections in mind, gave millions of Republican-leaning Americans a fiscally pragmatic rallying cry: "Why the great hurry to get to the moon and planets?"

Eisenhower was undeniably right that removing the hurry from Project Apollo would lower NASA's budgetary expenditures, and that was a fiscal argument with enough conservative voter appeal to divide popular support for space exploration. As newspapers began quoting Ike's *Saturday Evening Post* opinions in editorials, Kennedy held his response for a moment when it would have the desired impact but also land softly enough that it wouldn't alienate his White House predecessor. With tensions rising over the increasing Soviet military presence in Cuba, Kennedy needed Eisenhower's (and by extension, the GOP's) support when it came to defending the Monroe Doctrine and countering the threats arising in Castro's Cuba.

After his Houston trip triumph, Kennedy made the last stop on his space inspection tour, visiting the McDonnell Aircraft factory in St. Louis, where the Gemini two-man space capsule was being built with Gus Grissom's supervisory help. The visit was planned so carefully that it felt scripted. Air Force One landed at the aircraft plant, touching down just in time for a shift change, meaning that nearly ten thousand McDonnell workers were on hand to represent the blue-collar face of the aerospace industry. Escorted by MIT-trained physicist and company founder James Smith "Mac" McDonnell, with Webb and von Braun also in the entourage, the president inspected a Gemini capsule mock-up, watched production in a dust-free "white room," and met with top company executives. The only impromptu segment of the two-and-a-half-hour visit was, oddly enough, the part that would normally have been written down in advance: Kennedy's two-minute speech at the factory. Reiterating the most salient points from his Rice speech, Kennedy brought the enormous budget for Mercury, Gemini, and Apollo down to kitchen-table level, reminding his audience that each American was contributing forty cents per week—still possibly a strain for middle- or lower-income families, but that was the price of a truly national project. Through their contributions, he reminded them, "Every citizen of this country has a stake and is participating in this effort."

In part, Kennedy's space tour was designed to highlight how NASA was supporting the New Frontier economy, spurring growth not only in the public sector but also in private companies across America. Visiting McDonnell Aircraft was a great way to get that message across. Although less than two dozen years old, McDonnell was already America's largest manufacturer of fighter jets—the two-seat F-4 Phantom II (a twin-engine, supersonic, long-range, all-weather fighter-bomber) was its newest wonder. The company, which had snared prime NASA contracts to develop and provide space capsules for both the Mercury and Gemini programs, was the biggest employer in greater St. Louis, and all McDonnell shareholders, even if they were Republican, could clearly see the benefits of Kennedy's moonshot gambit staring at them from the bottom line. Furthermore, Secretary of Defense Robert

McNamara had contracted McDonnell (which would merge with the Douglas Aircraft Company to form McDonnell Douglas in 1967) to produce various classes of missiles.

Another smart reason for Kennedy to have toured McDonnell Aircraft was that his administration was about to choose Grumman (the largest employer in Long Island, New York, whose specialty was jet aircraft such as the A-6 Intruder and the E-2 Hawkeye) to be the chief contractor on the Apollo lunar module that would bring the Apollo 11 astronauts to the moon. Kennedy and NASA would wait until November 7, after the midterm elections, to announce that Grumman had won the lucrative bid.

United Press International correspondent Alvin Spivak interpreted the president's words during his stops in Houston and St. Louis in purely political terms: "By forceful implication, he endeavored to meet the challenge of former President Dwight D. Eisenhower's question 'why the great hurry' in NASA's 'fantastically expensive space program aimed at the moon.'" Kennedy was aware that Congress was showing signs of balking at the ever-ballooning NASA budget, but his whirlwind tour through Alabama, Florida, Texas, and Missouri that September had given him (and a rapt TV,

McDonnell Aircraft contributed mightily to NASA's manned space efforts. On September 12, 1962, after his Rice University speech, President Kennedy flew to St. Louis, Missouri, to speak with company employees.

radio, and newspaper audience) firsthand knowledge of the way American taxpayers' "40 cents per week" was being spent. To most of the voting public, funding Kennedy's moonshot was a down payment on the future greatness of the United States.

Just two days after the Houston and St. Louis events, Kennedy attended the America's Cup race in Newport, Rhode Island. Fresh from calling space the "new ocean" at Rice, he was happy to return to the real thing, the Atlantic. "I really don't know why it is that all of us are so committed to the sea, except I think it is because . . . we all came from the sea," he said at a dinner for the America's Cup crews. "And it is an interesting biological fact that all of us have, in our veins[,] the exact same percentage of salt in our blood that exists in the ocean, and, therefore, we have salt in our blood, in our sweat, in our tears. We are tied to the ocean. And when we go back to the sea, whether it is to sail or to watch it[,] we are going back from whence we came."

Kennedy didn't have the luxury of meditating on either the old ocean or the new ocean for long, because two of the most consequential long-term crises of his White House tenure had deteriorated during his robust space tour. In Mississippi, state officials were still defying the Supreme Court's order to desegregate the University of Mississippi, while in Georgia, arsonists had burned two African American churches that had been active in the voter registration movement. Civil rights leaders such as Martin Luther King Jr. and Ralph Abernathy were expressing their outrage. With great force, Kennedy condemned Jim Crow laws. In late September, Attorney General Robert Kennedy ordered U.S. Marshals to help facilitate James Meredith's registration at Ole Miss. When segregationists triggered a riot on the campus, JFK ordered in federal troops to restore law and order.

Even as opposition to the civil rights movement was threatening to blow open American society from within, another crisis was doing the same from without. In the aftermath of the Bay of Pigs invasion the previous year, the United States had adopted various measures to destabilize Castro's regime, including assassination plots, sabotage of the Cuban economy, and contingency plans to blockade the island. Convinced that a full-scale invasion from America was inevitable, Castro agreed to a Soviet offer to deploy intermediate-range nuclear missiles in Cuba. During the late summer of 1962, the USSR increased the delivery of military systems to the Caribbean island. In response, Kennedy asked Congress for standby authority to order 150,000 reservists to active duty for a year. Three days later, in the midst of JFK's space tour, the Soviet Union lambasted the request as "an act of aggression," warning that it would thwart any incursion into Cuba with war—even, it was implied, nuclear war.

Ensconced in the White House, Kennedy remained firmly committed to preventing acts of aggression against the United States by or through Cuba. At a press conference on September 13, he addressed the situation directly. "Let me make this clear," he said. "If at any time the Communist build-up in Cuba were to endanger or interfere with our security in any way, including our base at Guantanamo, our passage to the Panama Canal, our missile and space activities at Cape Canaveral, or the lives of American citizens in this country . . . then this country will do whatever must be done."

Group portrait of the second group of men selected to be astronauts in NASA's Project Gemini space program, 1963. *Bottom row, left to right:* James Lovell (1928–), Jim McDivitt (1929–), and Pete Conrad (1930–99); *second row:* Elliot See (1927–66) and Thomas Stafford (1930–); *third row:* Edward White (1930–67) and John Young (1930–2018); *fourth row:* Neil Armstrong (1930–2012) and Frank Borman (1928–)

18
Gemini Nine and Wally Schirra

The Gemini astronauts would in effect open the doorway to a moon landing—an event not measured then in decades but a mere handful of years.

—FRANCIS FRENCH AND COLIN BURGESS,
IN THE SHADOW OF THE MOON (2007)

The week President Kennedy was touring America's space facilities and giving his soaring speech at Rice University, thirty-two Project Gemini finalists were at their homes scattered around the country, nervously waiting to see who would make the final cut. Those thirty-two candidates had been selected after rigorous examination that past July, from a larger pool of more than two hundred fifty applicants, all of them white men. NASA's evaluation of the semifinalists was grueling, with the prospects required to show extraordinary psychological and physical fortitude as well as quick problem-solving skills. Nine slots were available for NASA's second astronaut group—two more than for Mercury because Gemini's launch schedule would be much more compressed: twelve missions within 603 days (one every 60 days), versus Mercury's six missions over 451 days (one every 112 days). Colin Burgess, a space historian who befriended the men who were selected to the "New Nine," as well as others who came close, concluded that "NASA was not looking for another bunch of accomplished stick-and-rudder test pilots with limited academic credentials. They wanted intellectual giants to help to solve the complex problems of space exploration. All nine of those new astronauts had excelled academically at whatever institution they attended."

James Webb was indeed adamant about the need for academic excellence, with his ideal candidates having both high-performance jet experience and first-rate records from great university engineering programs like Purdue, Caltech, Princeton, UCLA, Georgia Tech, and Michigan. Kennedy, known for staffing his entire administration with so-called whiz kids, heartily approved of Webb's recruitment priorities. The final list of would-be Gemini astronauts included six civilians, a nod to the president's emphasis on the peaceful nature of America's space program. On average, the Gemini "New Nine" astronaut was thirty-two and a half years old, about two years younger than the average Mercury astronaut was when selected. At five feet nine inches, the composite selectee was an inch shorter and also four pounds lighter than his Mercury predecessors.

On September 14, 1962, while Kennedy was at the America's Cup race in Newport, the nine men chosen to be Gemini astronauts received phone calls from NASA—not from Webb, but from Deke Slayton, one of the current Mercury Seven astronauts. Neil Armstrong, who already worked for NASA evaluating high-altitude aircraft at Edwards Air Force Base, got a call from Slayton. With little preamble, he asked Armstrong if he still wanted to be an astronaut.

Armstrong, the man who had once described his engineer personality as "born under the second law of thermodynamics, steeped in the steam tables, in love with free-body diagrams, transformed by Laplace, and propelled by compressible flow," was ready for the moment—or for rejection. Ever since Glenn's unforgettable *Friendship 7* mission, Armstrong had grown excited about traveling to space. Not long before the call from Slayton, Armstrong had analyzed his chances and concluded that he could be passed over for any number of reasons, all of them beyond his control. Now, though pleased by his selection, he reacted with little apparent emotion. "Yes, sir," he said matter-of-factly. With that, Slayton instructed Armstrong to report to Houston, where he was to check into the Rice Hotel under the code name "Max Peck," rendezvous with the other eight selectees, and get ready for the big announcement three days later.

On September 16, the Gemini Nine, having successfully evaded reporters, met in secret at Ellington Air Force Base, southeast of Houston. Having seen the two-person Gemini prototype in St. Louis, Kennedy had lit a fire to fast-track Project Gemini and announce the chosen astronauts. Bob Gilruth, director of the Marshall Space Flight Center, told the astronauts that there would be eleven or twelve manned Gemini flights, and that one of the nine of them would inevitably be the first man on the moon. "There'll be plenty of missions," Gilruth promised, "for *all* of you."

With great fanfare, the Gemini Nine astronauts were presented to the public on September 17 at the University of Houston's eighteen-hundred-seat Cullen Auditorium. It was the Mercury Seven rollout of 1958 all over again. Cheers erupted when the roster was read: civilian pilots Neil Armstrong and Elliot M. See Jr.; air force officers Frank Borman, James McDivitt, Thomas Stafford, and Edward White; and naval aviators Charles Conrad, James A. Lovell, and John W. Young. Some in the press dubbed the New Nine the "Kennedy moon corps." Armstrong biographer James Hansen later wrote, "In the opinion of individuals responsible for the early manned space program, it was unquestionably the best all-around group of astronauts ever assembled."

Gemini was a workhorse project tasked with solving in space the challenges of actualizing Kennedy's moonshot. The Gemini Nine were the lucky test pilots given the first opportunity to experiment with the cutting-edge aerospace hardware, to perfect the complicated techniques of rendezvous and conducting docking maneuvers in Earth's orbit. They would learn how to spacewalk, to master guided reentry, and to maneuver spacecraft into higher and lower orbits.

Congress began debating an increase in the NASA budget just a few days after the Gemini Nine's introduction. With the 1962 midterm elections just weeks away, Democrats wanted to show progress on Kennedy's moon pledge, but costs were rising, as was congressional concern. By September, Gemini's projected price tag had risen from $530 million to $745 million, due to the need to design and develop new systems rather than reuse those from Project Mercury. The New Nine rollout was one administration gambit to avoid a cost-analysis debate in Congress. Another was also on deck: in just three weeks, Mercury astronaut Wally Schirra would be lofted into space for the fifth manned Mercury mission, scheduled to make six orbits of Earth. If he was successful, steadfast manned space Democrats expected to reap rewards on Election Day.

For a year and a half, President Kennedy had been inspiring Americans to slip the bonds of Earth and literally reach for the moon, pushing past limits that had constrained mankind through its entire history. For some, however, there was still a glass ceiling beyond which they weren't permitted to go.

Inspired by Kennedy's vision, women were lobbying hard to join America's astronaut corps, applying to NASA, working behind the scenes, and inundating the White House with enlistment pleas. For example, Kennedy received a letter from Susan Marie Scott of Kentucky asking to audition for Gemini. His secretary, Evelyn Lincoln, forwarded the letter to O. B. Lloyd Jr., director of NASA's Office of Public Services and Information. In his response to Scott, Lloyd wrote,

"Many women are employed in the program—some of them are in extremely important scientific posts. But we have no present plans to employ women in spaceflights. There are no women pilots to our knowledge, who have the degrees of scientific and flight training required for the success of those missions. Since there is no shortage of qualified male candidates, there is no need to train women for space flight at this stage in the program."

As Scott found out, it was impossible for women to be genuinely considered for Project Gemini. An applicant needed a superior recommendation from a military or scientific employer and a laudable record as a test pilot. Service in the Korean War was also looked upon favorably. All the criteria were tilted toward male applicants only. "NASA did not state gender in its selection requirements, but more than two decades of discrimination by the military didn't give the agency any qualified choice other than men," Francis French and Colin Burgess wrote in *Into the Silent Sea.* "Not only did the military still bar women from flying high-performance aircraft except in extremely rare circumstances, but civilian companies rarely hired women pilots either, let alone trained them as test pilots."

While African Americans' struggle for civil rights

was beginning to shake the pillars of Jim Crow, female pilots were struggling for equality in their fight against NASA's patriarchy. They did, however, have one offbeat ally in Dr. William "Randy" Lovelace II. An aeromedicine pioneer, Lovelace was chairman of the Special Advisory Committee on Life Science and from his office in New Mexico had conducted the intensive medical examinations that had helped winnow down the first class of likely astronauts to the final Mercury Seven. Although holding what were then traditional views on women's roles in American society, Lovelace believed that women were better equipped physiologically for NASA space travel because they were, on average, shorter and smaller than men, needed less food and oxygen, and had better blood circulation and fewer cardiac problems.

After meeting American aviator Geraldyn "Jerrie" Cobb in 1959, Lovelace invited her to take the same tests as the Mercury astronauts—and was amazed at her aptitude. Beginning in 1960, Lovelace had begun testing this hypothesis at his privately financed Lovelace Foundation for Medical Education and Research, in Albuquerque. Intrigued as to whether Cobb was an anomaly, Lovelace accepted a financial gift from the fabled pilot Jackie Cochran, the leader of the Women

Airforce Service Pilots (WASP) program in World War II, to examine eighteen other seasoned female pilots at his New Mexico clinic for secret testing. These women, all dexterous airplane pilots with commercial ratings, were put through a series of rigorous tests on centrifuges to simulate the pressure of launch and reentry. They graduated with flying colors. When word of the tests leaked to the media, the top twelve, along with Jerrie Cobb, were christened the "Mercury 13." These women pilots ranged in age from twenty-three to forty-one, and ran the gamut from flight instructors to homemakers and from scientists to bush pilots.

Just before these women finalists were to gather at the Naval School of Aviation Medicine, in Pensacola, Florida, for advanced aeromedical examinations, they received telegrams informing them that their program was being effectively shut down. NASA hadn't certified gender in Lovelace's work. In fact, when NASA leadership learned of the experiment, they made it abundantly clear that the agency wasn't going to employ women astronauts. The impetus for NASA's decision was Kennedy's May 25, 1961, moonshot announcement, and Webb's belief that all the agency's astronaut-training energy had to be targeted toward that lunar objective. In other words, it wasn't the time for a shift on gender. That didn't stop the American press from lionizing these women's test results. Some of the choice headlines read: "Astrogals Can't Wait for Space," "Spunky Mom Eyes Heavens," and "Why Not 'Astronauttes' Also?"

Fueled by the media attention and desire to shatter the glass ceiling, Janey Briggs Hart refused to give up her space dream easily. The wife of Michigan senator Phil Hart and the mother of eight children, Hart orchestrated a letter-writing campaign to the White House, arguing that women deserved to be included in the NASA space program. The pressure became so intense that Kennedy booted the issue to Congress, where a special subcommittee of the House Committee on Science and Astronautics convened. On the first day of testimony, Hart and the other women pilots spoke righteously on behalf of women's equality, and they were gaining momentum. Then, a most unlikely spoiler appeared before the subcommittee: John Glenn, who echoed Webb's contention that funding women in space drained money needed for the moonshot, and was generally a waste of tax dollars. "I think this gets back to the way our social order is organized," Glenn testified. "It is just fact. The men go off and fight the wars and fly the airplanes and come back and help design and build and test them. The fact that women are not in the field is a fact of our social order."

The Mercury 13 pilots were devastated that Glenn, whom they all admired, was opposed to female astronauts, putting the U.S. space program on the wrong side of history. The following year, the Soviets did what the Americans wouldn't, making Valentina Tereshkova the first woman in space. At heart, this was a stunt to one-up the Americans. Lifting off aboard *Vostok 6* on June 16, 1963, Tereshkova became a global hero after making forty-eight Earth orbits over the course of seventy hours, at one point coming within three kilometers of Cosmonaut Valery Bykovsky, who'd launched aboard *Vostok 5* just two days earlier. Frustrated that NASA had flummoxed his Mercury 13 project, Lovelace kept fighting a rearguard action from his home in New Mexico, hoping for the inclusion of a woman astronaut on the Gemini roster. But in December 1965, he and his wife were killed in a plane crash, depriving the Mercury 13 of their most devoted advocate. It would be nearly twenty years before Sally Ride became the first American woman in space, lifting off aboard the Space Shuttle flight STS-7—the *Challenger*—on June 18, 1983.

From a wide-lens historical perspective, the Mercury 13 were fighting an uphill battle for job equality that was central to the women's movement. During the Kennedy years, women earned only 60 percent of the average wage for men. For the exact same job, a working man earned $5,147, to $3,283 for women; it was imperative that this gap be closed. While Kennedy was progressive on such women's issues as day care centers, fair employment, and college admissions, he behaved as if the U.S. military workforce, with space exploration folded in, were a male prerogative. Nevertheless, Kennedy's Equal Pay Act of 1963 (which prohibited arbitrary discrimination against women in the workforce) was a step in the right direction. Frances Perkins, America's first woman cabinet secretary (under FDR), believed the law, along with the 1963 publication of Betty Friedan's *The Feminine Mystique*, was the opening salvo of the modern women's movement.

The idea of an African American, Hispanic, Native American, or Jewish American astronaut quite simply wasn't in NASA's organizational plan for the early 1960s. WASP supremacy still held sway. Many of NASA's facilities were in the Deep South and Southwest, where racial segregation roadblocks were only starting to be dismantled. For example, when the black mathematician Julius Montgomery was hired at Marshall Space Flight Center in Huntsville, he faced harassment from NASA employees who were members of the Ku Klux Klan. White employees wouldn't say hello or even look at Montgomery. But in the end he prevailed by performing flawlessly. After the Civil Rights Act of 1964, NASA finally made long-overdue strides toward having a diversified, multicultural workforce.

Compared to the other Gemini recruits and the stymied women of Mercury 13, Wally Schirra seemed, at age thirty-nine, a grand old man of space exploration. But after a series of delays, he was finally poised to become America's newest space hero, scheduled to launch on October 3, 1962, aboard a Mercury-Atlas 8 (MA-8).

Schirra was born into an aviation family in New Jersey in 1923. His father, Walter M. Schirra Sr., flew bombing sorties over Germany during World War I for the Royal Canadian Air Force. After the war, the elder Schirra made a living barnstorming at county fairs in New Jersey and Pennsylvania. Florence Schirra, the astronaut's mother, joined her husband doing wing-walking stunts to awestruck crowds. By the time Wally was fifteen, he was already flying airplanes by himself.

After excelling in high school, Schirra attended the U.S. Naval Academy in Annapolis from 1942 to 1945 and served during the final months of the war aboard a navy cruiser. After the war, he married Josephine "Jo"

Cook of Seattle (stepdaughter of Admiral James L. Holloway) and trained as an aviator at the Naval Air Station in Pensacola, earning his wings and joining Fighter Squadron 71 in 1948. Other pilots were in awe of Schirra's natural ability and technical skills. Determined to shatter aviation records, he became only the second navy pilot to log one thousand hours in jet aircraft. When the Korean War erupted, Schirra was seconded by the navy to the air force, where he became operations officer with the 154th Fighter-Bomber Squadron. Between 1951 and 1952, Schirra flew ninety combat missions, usually in an F-84 Thunderjet, downing one MiG-15 and inflicting serious damage on two others. After receiving the Distinguished Flying Cross and the Air Medal with oak leaf cluster for distinguished wartime service, he moved on to a career as a test pilot and aeronautical engineer.

In the late 1950s, Schirra bounced around navy test schools, participating in the unveiling of the Sidewinder missile and the F7U-3 Cutlass jet fighter. At one test, Schirra fired a Sidewinder missile, and the projectile doubled back and started trailing his jet, forcing him into evasive maneuvers. Nobody ever doubted that Wally had the right stuff. Once chosen as a Mercury Seven astronaut, he earned the reputation of being the comedian of the bunch, loving to muck it up with reporters. "Levity is the lubricant of a crisis," he explained of his prankish nature. In truth, his carefree personality belied an extremely careful and diligent work ethic. "My rambunctious approach to the off-duty aspect of life may have fooled some people, but this was not a game, I often said to myself. This is for real. I was not interested in the glamour of being a space hero. Instead, I was interested in getting up and getting back."

Set to become the fifth American in space and to make the country's third orbital spaceflight, Schirra named his spacecraft *Sigma 7*—"7" for the Mercury Seven and "Sigma" after the Greek symbol for the sum of the elements of an equation, a mark long adopted for engineering excellence. "Not a fancy name like Freedom or Faith," he recalled. "Not that I didn't appreciate those names, but I wanted to prove that it was a team of people working together to make this vehicle go. . . . I thought that it was a very well-made machine, and very, very carefully designed."

During his nine-hour-plus mission on October 3, Schirra reported back to Mercury Control everything he did or encountered, his voice relayed via the *Telstar* satellite to TV and radio audiences around the world. As Schirra checked off his list, it became clear to Flight Director Chris Kraft at NASA that MA-8 was going to be the smoothest mission yet. After orbiting six times, *Sigma 7* splashed down northeast of Midway Island in the Pacific, where Schirra was retrieved by Navy SEALs from the recovery ship USS *Kearsarge.* "In mission control, I winked at Deke Slayton and lit up my now traditional cigar," Kraft wrote in *Flight.* "Schirra was the perfect astronaut and he'd just carried out a perfect mission."

At the White House, a busy Kennedy couldn't watch the whole nine-hour MA-8 mission, but he received a constant flow of updates. "The President was always extremely interested in these flights," Evelyn Lincoln recalled, "and there was a great deal of excitement around the White House. Commander Walter Schirra was in the space ship early in the morning, waiting. I had my television set tuned in for the event. What a relief when he got into orbit." A delighted and breathless Lincoln kept her boss apprised of the flight every half hour.

Rather than scientific experimentation, the mission of *Sigma 7* focused mostly on engineering: the performance and operation of the spacecraft, the capabilities of global spacecraft tracking and communication systems, and the effects of prolonged microgravity on Schirra himself. However, Schirra had also carried a special two-and-a-half-pound handheld camera while aboard, recording the marvelous imagery of the star-filled adventure. With preparation considered meticulous even for a well-trained NASA astronaut, Schirra piloted his mission so free of flaws that reporters considered the feat almost mundane. "I ate and I wasn't hungry," he said. "The tubes of peaches, meat, and vegetables were tasty." He even refined a method of cruising that saved enough fuel to repeat the six orbits, had he been allowed and had there been enough oxygen on board. Slayton and Shepard later wrote that Schirra's efficiency "would have turned a robot green with envy."

NASA officials congratulated themselves that it had been the right mission at exactly the right time. There were plenty of reasons for American pride in *Sigma 7.* Nevertheless, its nine-hour flight still paled in comparison to the latest Soviet space accomplishments. Two months before Schirra's flight, the USSR had smashed the record for human endurance in space when cosmonaut Andriyan Nikolayev piloted *Vostok 3* through sixty-four Earth orbits over the course of a three-day, twenty-two-hour, and twenty-eight-minute flight. Nikolayev also made a creditable attempt at staging a space rendezvous with *Vostok 4*, which had been launched one day after his own flight. In the United States, space enthusiasts were acutely aware that America's deficit in the space race was not closing.

Eleven days passed before arrangements were made for Schirra to visit the White House. In the interim, he was honored fifty different ways in Houston. At a press conference at Rice University, Schirra boasted that the flight had been free of problems, that he'd experienced no difficulties with weightlessness, and that Mercury was now ready for a full one-day mission. Finally, on October 16, Schirra, Jo, and their two young children arrived at the White House at meet the president.

At 9:25 a.m., the Schirras were led into the Oval Office. Five-year-old Suzanne Schirra lit up the room as she gazed wide-eyed and bashful at the handsome president and said quietly, "I know who you are!" Kennedy responded with the delight he naturally found in children, taking special care to entertain the girl and her twelve-year-old brother, Walter III. Sitting in his favorite seat, a rocking chair, he chatted with each member of the family. JFK's composed ability to contain his darkest concerns that morning was so effective that Schirra detected nothing out of the ordinary. At ten o'clock, the astronaut and his family were politely escorted out. The Associated Press reporter covering the half-hour visit described the president's demeanor as "homey" and "relaxed." But this time, his elusive charm masked a darker reality.

Twenty-five minutes before the Schirras had arrived, McGeorge Bundy had asked to see the president in the White House family quarters. The national security advisor disclosed that the CIA had analyzed surveillance photos of Cuba taken on October 14, confirming the construction of launch bases for Soviet nuclear missiles just a hundred miles from American soil. It was the news that Kennedy had dreaded.

Starting at 11:30 a.m., Kennedy spent hours in the Cabinet Room with the senior advisors who comprised the Executive Committee of the National Security Council (ExComm). Deputy CIA director General Marshall Carter showed Kennedy the top-secret U-2 photos taken over Cuba, pinpointing fourteen canvas-covered missile trailers. As various possibilities for a U.S. response were proposed and discussed (including air strikes, invasion, and naval blockade), it became clear that any military response could easily result in all-out war. The president calmly wondered aloud about Khrushchev's ballsy gamble. "Why would the Soviets permit nuclear war to begin under that sort of half-assed way?" he mused.

As ExComm members and their staffs worked to

game out America's options, Kennedy attempted to carry on with his other presidential duties, including a campaign swing through the Midwest on behalf of Democratic candidates. Seemingly undistracted, he pulled off speeches in Cincinnati, Ohio, and Muskegon, Michigan, with aplomb. At a hundred-dollar-a-seat fund-raiser in Chicago, however, speaking before five thousand people and a large broadcast audience, a distracted JFK skipped over large sections of his prepared text, ending well short of his allotted time. With dead air looming on televisions across metropolitan Chicago, Mayor Richard Daley rushed to the podium and hastily called for a benediction. Claiming a head cold, Kennedy canceled the rest of his Midwest tour and returned to Washington.

On October 18, the National Photograph Interpretation Center advised the administration that two medium-range ballistic missile sites in Cuba could be operational within weeks. Two days of ExComm meetings in the White House ensued, and on October 22, an unruffled Kennedy made the missile crisis public in a televised address. Stating firmly that the United States would never tolerate Soviet offensive weapons in the Caribbean, the president announced the implementation of a naval quarantine designed to prevent any

Soviet ships carrying offensive weapons from reaching Cuba.

Like the rest of the country, officials and staff at NASA wrestled with anxiety as the long-burning fuse of the Cold War seemed close to reaching its charge. "For almost two weeks," recalled Assistant Flight Director Gene Krantz, "the space program was understandably preoccupied with the blockade and possible invasion of Cuba, which could presage an all-out nuclear conflict with Russia." Krantz, an air force veteran like many of his NASA colleagues, had already been notified that his reserve unit was on standby, and he could be called up at any time.

On October 28, after thirteen nerve-racking days and a complex series of both public and back-channel communications, Kennedy and Khrushchev agreed on measures to bring a peaceful resolution to what became known as the Cuban Missile Crisis. The Soviet premier agreed to remove the missile bases from Cuba and allow on-site verification by UN inspectors. Kennedy promised that America wouldn't invade Cuba, and secretly committed to removing U.S. Jupiter missiles from Turkey within a year. Essesntially, a trade of the two sets of weapons—U.S. Jupiters and Soviet IRBMs—ultimately ended the crisis.

A gaggle of customers in a California appliance store gather in the electronics department on October 22, 1962, to watch President John F. Kennedy deliver a televised address to the nation on the subject of the Cuban Missile Crisis.

The searing experience of contemplating nuclear war in real time had tempered both Khrushchev and Kennedy, who'd been stunned by how ready his army and air force officers were to go to war. "You will never know," Kennedy confided to Ambassador John Kenneth Galbraith, "how much bad advice I had." The Cold War still held both leaders in its iron grip, but the experience of living with the specter of annihilation for thirteen terrifying days motivated the two men to open up a telephone hotline between Moscow and Washington, as well as to begin a thoughtful correspondence aimed at lessening global tensions. Chastened, they embarked on negotiations for what became the Limited Nuclear Test Ban Treaty, signed the following August to restrict atmospheric and underwater nuclear tests.

From the beginning of his administration, Kennedy's attitude toward space exploration had evolved, although not in a straight line. Never blinded by the lure of the stars, he instead balanced his idealism with pragmatism, seeking a mission that would renew America. World War II and the Cold War, he knew, had aged the country. With instincts reinforced by his own life experiences, he realized that the United States needed youth and new frontiers. It needed energy, originality, optimism, and a sense of both individual achievement and teamwork. The Mercury, Gemini, and Apollo programs gave all that in spades. JFK always knew that his greatest potential legacy could accrue from ending the Cold War with the Soviets, but the moon mission was his backup plan, giving the possibility of a huge geopolitical win—an appealing prospect to a man who'd been raised to believe that second place was for losers.

But in the aftermath of the Cuban Missile Crisis, the president began questioning both his own moon mandate and the reasoning underlying the whole U.S.-Soviet space race. Growing tension existed in American space circles between those with scientific goals and those who simply wanted to beat the Soviets to the moon. Both sides had points to make, but with an annual government allocation that had grown from $500 million to $3.7 billion in just two years, they were encountering resistance in Congress. On November 5, the *New York Times* put the problem succinctly into a headline: "Space Goals Put Strain on Budget; NASA, for First Time, Must Tailor Projects to Funds." The post-Gagarin days of blank checks were coming to an end, and NASA faced the prospect of its manned space program starving its work on communications satellites, meteorite studies, Mars probes, and other unmanned space science missions—the one area where the United States was verifiably ahead of the USSR.

The space agency was suddenly scrutinized for expenditures. Worsening the situation were rumors that Brainerd Holmes, NASA's director of the Office of Manned Space Flight, was campaigning for a steep hike in the budget in order to put a man on the moon by 1966, four years ahead of the Kennedy-imposed deadline. If the expanded funding wasn't forthcoming, then Holmes advocated concentrating all NASA's assets only on the moonshot.

Some members of Congress felt they were being blackmailed by the forty-year-old Holmes, a formidable figure who had proved himself as something of a marvel in the complex field of missile development at RCA in the 1950s. Because his launches, however complex, always worked the first time, he was known as "One-Shot" Holmes. The year before, he'd quit the corporate world for the high-pressure job in Washington, DC, responsible for America's efforts to put a man on the moon. On his desk, Holmes kept a rocket-shaped toy bank given to him by a friend, which he claimed would "keep me thinking of the taxpayers' money."

Now, in late 1962, Holmes suggested accelerating the Apollo program despite the added cost. He had heard President Kennedy, during his Rice University speech, call the challenge of reaching the moon "one that we are willing to accept, one we are unwilling to postpone, and one which we intend to win." To One-Shot Holmes, that meant giving Projects Gemini and Apollo everything NASA had, and then some.

Interestingly, Holmes's immediate superior, James Webb, heard the Rice speech differently. After listing some of the scientific and unmanned space shots on which NASA was working, Kennedy had actually said that the manned moon mission was "one which we intend to win, along with the others." Refusing to forfeit the "others" even for the sake of Apollo, Webb was irritated by Holmes's confusing press comments. Because of Holmes's record of success, however, and the respect Webb accorded him, his stance had to be taken seriously. Webb wrote to the president describing the cost of moving up the date of the moonshot. Kennedy, wanting all the facts before weighing in on the increasingly contentious issue, asked David Bell at the Bureau of the Budget to give a candid assessment of the situation at NASA.

Meanwhile, it was still election season. On November 6, the Democrats lost seats in the House of Representatives, but maintained a comfortable majority (258 to 176), while in the Senate, their majority grew. Kennedy had avoided the midterm election jinx that has historically plagued presidents. In this election, the ranks of Kennedy Democrats in the Senate had grown. This was stupendously good news for the administration and for NASA. Most gratifying of all to the president was that his youngest brother, Edward "Ted" Kennedy, won a special election in Massachusetts to represent the state as its junior senator. It was the seat JFK had held prior to his election as president.

On November 13, David Bell presented the results

of his assessment on NASA spending to the president, reporting that the agency was managed quite well by Webb and Dryden, with Projects Gemini and Apollo generally on track and waste being kept to a minimum. But one conclusion stood out: according to Bell, there was no moon race with the Soviets. For all their *Sputnik*s and *Vostok*s, Bell saw zero evidence that the Soviets were constructing facilities for boosters and capsules capable of taking cosmonauts to the moon. Khrushchev's space efforts were geared more toward an eventual USSR space station, not a moon walk.

While Bell's report was largely supportive of the planned Apollo spending, its contention that the United States was the sole competitor in the moon race effectively bolstered Webb's position. Poking a hole in Holmes's moon-only argument, the report stressed the equal importance of "programs for scientific investigations in space, in which the United States from the start has been recognized as the world leader."

"NASA takes the view," Bell noted, that if reductions were to be made to the agency's budget, they should be applied "at least in part to the manned lunar landing program."

On November 21, 1962, Kennedy summoned his space advisors to the White House for a frank discussion of the merits of space exploration and the moonshot. The transcript of the resulting arguments that day is invaluable to historians. While the president mostly listened, his comments mirrored his record over the course of his administration, seeing the adventure of space exploration from several strategic viewpoints. All the while, JFK looked for ways to tie NASA's massive effort more closely with the economic health of America. In the end, he made his convictions and priorities clear.

The participants in the Cabinet Room included Webb, Wiesner, Bell, Dryden, Seamans, and Holmes. After a general discussion of a proposed $440 million supplement to the NASA budget, Webb allowed that in terms of accelerating the moon program, "we're prepared to move if you really want to put it on a crash basis." Kennedy asked Webb pointedly whether he thought of the moon mission as the "top-priority program of the Agency." Sensing that Kennedy was in no mood for vague indirection or double-talk, Webb responded bluntly:

James Webb: No, sir, I do not. I think it is one of the top-priority programs, but . . . [s]everal scientific disciplines that are very powerful begin to converge on this area.

President Kennedy: Jim, I think it is the top

> priority. I think we ought to have that very clear. Some of these other programs can slip six months, or nine months, and nothing strategic is going to happen. . . . But this is important for political reasons, international political reasons. This is, whether we like it or not, in a sense a race. If we get second to the Moon, it's nice, but it's like being second any time. So that if we're second by six months, because we didn't give it the kind of priority, then of course that would be very serious. So I think we have to take the view that this is the top priority with us.

Although Bell had presented the president with CIA U-2 reconnaissance discounting the possibility of a Soviet moonshot, that hadn't been enough to shake JFK's focus and commitment. Having never lost an election himself, and having just retained his party's congressional majority and elected his brother to the Senate, Kennedy had no intention of being second. And he never forgot that the Apollo moonshot wouldn't happen without continuing public support demonstrated by robust budgets and tireless insistence on beating the Soviets. Webb continued to argue for prioritizing other, unmanned scientific ventures by appealing to Kennedy's old-style, Ivy League faith in the academic elite:

Webb: The people that are going to furnish the brainwork, the real brainwork, on which the future space power of this nation for twenty-five or a hundred years are going be to made, have got some doubts about it and . . .

Kennedy: Doubts about what, with this program?

Webb: As to whether the actual landing on the moon is what you call the highest priority.

Kennedy: What do they think is the highest priority?

Webb: They think the highest priority is to understand the environment and . . . the areas of the laws of nature that operate out there as they apply backwards into space. You can say it this way. I think Jerry [Wiesner] ought to talk on this rather than me, but the scientists in the nuclear field have penetrated right into the most minute areas of the nucleus and the sub-particles of the nucleus. Now here, out in the universe, you've got the same general kind of a structure, but you can do it on a massive universal scale.

Kennedy: I agree that we're interested in this, but we can wait six months on all of it.

The six-month time frame was Kennedy's way of expressing priority: which NASA projects could wait six months and which might fail due to that much delay. Webb next tried to convince the president that the drawn-out time frame of the moonshot would eventually cost it public support, while a schedule of exciting if unmanned experiments would generate broader enthusiasm in Congress and beyond. Also, if evidence emerged that the Soviets weren't actually racing America to the moon, the funding for Project Apollo would be put in a stranglehold. An animated disagreement ensued, ending only when JFK pulled rank. It remains unclear, however, whether Webb was speaking from the heart or playing devil's advocate to draw Kennedy out.

Kennedy: I would certainly not favor spending six or seven billion dollars to find out about space no matter how on the schedule we're doing. I would spread it out over a five- or ten-year period. But we can spend it on . . . Why aren't we spending seven million dollars on getting fresh water from saltwater, when we're spending seven billion dollars to find out about space? Obviously, you wouldn't put it on that priority except for the defense implications. And the second point is the fact that the Soviet Union has made this a test of the system. So that's why we're doing it. So I think we've got to take the view that this is the key program. The rest of this . . . we can find out all about it, but there's a lot of things we can find out about; we need to find out about cancer and everything else.

Webb: But you see, when you talk about this, it's very hard to draw a line between what . . .

Kennedy: Everything that we do ought to really be tied into getting onto the Moon ahead of the Russians.

Webb: Why can't it be tied to preeminence in space, which are your own . . .

Kennedy: Because, by God, we keep, we've been telling everybody we're preeminent in space for five years and nobody believes it because they have the booster and the satellite. We know all about the number of satellites we put up, two or three times the number of the Soviet Union . . . we're ahead scientifically. It's like that instrument you got up at Stanford which is costing us a hundred and twenty-five million dollars and everybody tells me that we're the number one in the world. And what is it? I can't think what it is. [Interruption from multiple speakers: "The linear accelerator."] I'm sorry, that's wonderful, but nobody knows anything about it!

Webb pointed out that only a full range of progress could usher in major advancements in space, but Kennedy bluntly wrested the argument back to his way of thinking:

Kennedy: We ought to get it, you know, really clear that the policy ought to be that this is the top-priority program of the Agency, and one of the two things, except for defense, the top priority of the United States government. I think that that is the position we ought to take. Now, this may not change anything about that schedule, but at least we ought to be clear, otherwise we shouldn't be spending this kind of money because I'm not that interested in space. I think it's good; I think we ought to know about it; we're ready to spend reasonable amounts of money. But we're talking about these fantastic expenditures which wreck our budget and all these other domestic programs and the only justification for it, in my opinion, to do it in this time or fashion, is because we hope to beat them and demonstrate that starting behind, as we did by a couple years, by God, we passed them.

In their final exchange of the day on space, Webb unwittingly made Kennedy's point for him.

Webb: In Berlin you spent six billion a year adding to your military budget because the Russians acted the way they did. And I have some feeling that you might not have been as successful on Cuba if we hadn't flown John Glenn and demonstrated we had a real overall technical capability here.

Kennedy: We agree. That's why we want to put this program . . . That's the dramatic evidence that we're preeminent in space.

When listening to this White House conversation, it is important to remember that Webb spoke nonstop and was difficult to turn off. When Kennedy cut Webb off by saying, "I'm not that interested in space," he was cutting to the chase. Unlike with his public speeches, Kennedy wasn't in a "New Ocean" feel-good science state of mind. He wanted to drill home to Webb that the moonshot should be sold as a serious Cold War national security priority.

In other words, Kennedy wasn't merely arguing for the Apollo moon program. With a sweeping sense of history, he was arguing for a new era in which technological superiority *was* power. For the same reason that emperors of old paraded their armies in the streets, the president's moon program was a showcase for America's technological might, and its contracts with corporations such as McDonnell Aircraft, North American Aviation, Boeing, Chrysler, and others were a collaborative government–private sector project for the new technologies that would guarantee the American century. To voters, he presented NASA's moonshot as a proof of national greatness. But at the same time, in internal discussions, he described it as a negotiating weapon in the Cold War struggle with Khrushchev: Mercury, Gemini, and Apollo as Olympian deterrents to contain Soviet expansionism.

On September 12, 1962, President John F. Kennedy visited the Manned Spacecraft Center in Houston, one of twelve NASA sites in east Texas. He was presented with a model for the Apollo command space capsule.

19
State of Space Exploration

Kennedy provided the inspiration and financial support for the space program—and spurred rapid innovation. While trying to solve the problems of manned space flight, scientists laid the foundations for satellite television, global positioning systems, microchips, solar panels, carbon monoxide detectors and even the Dustbuster.

—WALTER ISAACSON, 2018

President Kennedy began 1963 in a boastful mood. Two weeks before the New Year, NASA's *Mariner 2* had succeeded spectacularly, becoming the first space vehicle to make meaningful contact with one of Earth's neighboring planets.

Both the United States and the Soviet Union had wanted to be the first to achieve this landmark, and Venus, being on average the closest planet to Earth, was the logical target. In February 1961, the Soviets had launched the first Venus probe, *Venera 1*, but a communications failure sent it off course, and it missed by 62,000 miles. A subsequent attempt at a Venus landing was made in late August 1962, with *Sputnik 19*, but the craft failed to escape Earth's orbit. *Mariner 2*, a 447-pound probe packed with measuring instruments sitting atop a two-stage Atlas-Agena rocket, was America's response. Designed by NASA's Jet Propulsion Laboratory, it was completed on an astonishingly breakneck schedule and launched from Cape Canaveral just two days after *Sputnik 19* and five days after an American predecessor, *Mariner 1*, failed in takeoff.

On December 14, after 110 days in space, *Mariner 2* arrived at its destination, coming within 21,607 miles of Venus, scanning the planet with its pair of radiometers, and sending back valuable new information. Scientists had long presumed that Venus had a relatively benign environment, one that might even support life of the type found on Earth. *Mariner 2* erased any such delusions. Its readings indicated blistering temperatures and extremely high atmospheric pressure. Continuing toward the sun, the probe also sent back new knowledge on solar wind and interplanetary dust.

As the first successful interplanetary probe, *Mariner 2* marked a turning point, giving astrophysicists and astronomers a firsthand look at space and opening a new era of exploration that would see probes traveling as far as Pluto. Applauded by space scientists around the world, it was a win for the purely scientific side of space exploration, but it also put the United States firmly ahead of the USSR in the race to explore the solar system.

Kennedy was well aware of an interplanetary mission's public relations value. On January 17, 1963, as a parallel to the visits of Mercury astronauts after their successful spaceflights, Kennedy fêted the lead scientists behind *Mariner 2* at the White House. Standing next to a model of the probe, the president called the voyage "an extraordinary technical accomplishment by the United States," one that indicated that there was a "broad spectrum of mastery in the field of space, other than the effort of the human probe." The scientists surrounding JFK were still, in fact, in the thick of analyzing the data from *Mariner 2*, and the full results would not be announced until late February. Had science been the only consideration for the meeting, Kennedy might have waited until then, but he'd brought the scientists to the White House that day for a specific reason: he was submitting his next federal budget to Congress that very day, and asking Congress to appropriate yet more money for the NASA manned space program.

The total budget was larger than any proposed by any president in history (even by FDR at the height of World War II) and included a conspicuous increase for Apollo and Gemini. The overall earmark for space was $5.7 billion, of which $4.2 billion applied to 1964 and $1.5 billion to future years. The total for space in 1964 would rise 75 percent from 1963 and more than 300 percent over what had been budgeted for 1962. No mere blip after a decimal point, Kennedy's space program would account for more than 3.5 percent of the nation's total spending. By 1966, a whopping 5.5 percent of the federal budget would go to the moon program.

James Webb, a masterly Capitol Hill appropriations fund-raiser, was quick to promote the New Frontier budget as "austere," framing the increase as a reasonable continuation of Congress's nearly unanimous commitment in 1961 to put an American on the moon within a decade. Being number one in space was expensive,

Kennedy said, and NASA would only continue to grow as the United States jockeyed to "maintain a position of world leadership in the exploration and utilization of space." Showcasing the recent success of *Mariner 2*, Kennedy's budget message read like a "State of Space" white paper:

> Efforts are being concentrated in the continued development of the complex Apollo spacecraft and the large Advanced Saturn launch vehicle needed to boost the Apollo to the moon. A lunar orbit rendezvous approach will be used to accomplish during this decade the first manned lunar landing. Under this technique the Apollo spacecraft will be boosted directly into orbit around the moon, where a small manned lunar excursion module will be detached and descend to the surface of the moon. It will later return to the orbiting Apollo which will return to the earth.
>
> The recent Mariner flight past Venus attests to the progress we are making in unmanned space investigations. Development of geophysical, astronomical, meteorological, and communications satellites will also continue. This budget provides for strong research efforts aimed at developing the

technology needed for advanced space missions, including future manned space flight and unmanned explorations of Venus and Mars.

Kennedy's hopes for heading off a budget battle were short-lived. In Congress, a strong cadre of fiscally conservative Republicans set out to block NASA's astronomical growth, supported by a cabal of concerned Democrats. Former president Eisenhower continued to snipe that "anybody who would spend $40 billion in a race to the moon for national prestige is nuts." In *Fortune*, journalist Arthur Krock charged the Kennedy administration with manipulating the press to cheerlead for NASA and other New Frontier initiatives. Furthermore, Krock charged Kennedy with exhibiting a "bristling sensitiveness to critical analysis" unmatched in U.S. presidential history. He even went so far as to suggest that JFK's public relations blitz, anchored in flattering newspaper editors and TV moguls, was counter to Thomas Jefferson's understanding of freedom of the press. "We have had limited success in managing news," Kennedy sardonically countered Krock, "if that is what we have been trying to do."

Krock's criticism about White House's press cooption was particularly biting because he'd been a reliably pro-Kennedy reporter throughout the president's political career. Regarded as the dean of Washington journalists after winning the Pulitzer Prize for Correspondence in 1935 and 1938, Krock had promoted Kennedy's first book, *Why England Slept* (1940), far and wide. Then, in 1946, at the request of his old sponsor, confidant, and friend Joseph Kennedy Sr., he'd helped promote the young JFK in his first run for Congress. Somewhere along the line, Krock's relationship with the elder Kennedy soured, and now newspapers all over America were quoting his criticisms of the president. "The official [White House] release of information in the areas of nuclear and space exploration are not determined on whether the American public that pays is entitled to the facts," Krock wrote. "Nor is safeguarding the national security the determining factor, though this is always the explanation for concealment. The controlling policy factor is whether the release will or will not improve our 'world image,' and give this government a lead in the psychological sector of the cold war."

While JFK might not have disagreed with that last premise, he would certainly have protested Krock's contention that he was disregarding national security, a preoccupation that was in fact central to his policy directives. To Kennedy, who was presiding over the most perilous period of the Cold War, "peacetime" didn't exist in the usual sense, and every measure of national accomplishment (technological, military, economic, social, and moral) needed to be weaponized in the competition for geopolitical influence. Under the president's direction, certain activities at NASA may have been concealed or, at other times, sugarcoated, but Krock was exaggerating the White House's media manipulation. From Kennedy's perspective, NASA *had* to generate a plethora of positive publicity in order to keep congressional appropriations rolling. The world could see the results of what had already been accomplished: in only two years as president, having given NASA the funding it needed, Kennedy had taken the United States from launching its first Mercury astronauts to exploring Venus, revolutionizing satellite communications and meteorological technology, and even creating new scientific disciplines such as bioastronautics and space medicine. Now Americans had to understand the importance of NASA's next space steps, and that meant widespread publicity. It meant a new monthly column by von Braun in *Popular Science*, to help explain Apollo's objectives. It meant photo spreads in *Life* magazine to highlight the family lives of the Gemini astronauts. It meant ticker-tape parades and Oval Office receptions for returning space heroes. As Arthur C. Clarke wrote and as Kennedy surely understood, "The eyes of all ages are upon us now, as we create the myths of the future at Cape Canaveral in Florida and Baikonur in Kazakhstan. No generation has been given such powers, and such responsibilities. . . . If our wisdom fails to match our science, we will have no second chance."

The reality that eluded critics such as Eisenhower and Krock was that Kennedy's New Frontier of technology had seized young Americans' imaginations, and the further that NASA progressed toward the moon, the more their imaginations would soar with it. When Eisenhower gave his farewell address, for example, there were virtually no computer science programs at American universities. But by 1963, such departments had been established at Stanford, MIT, Carnegie Mellon, the University of Utah, the University of Illinois at Urbana, and the University of California, Berkeley. To be sure, some impetus for the surge came from federal funding provided by the National Defense Education Act of 1958, which itself was a response to *Sputnik* and the perceived education gap between U.S. and Soviet universities. As we have seen, Senator Kennedy had been a supporter of that legislation, calling the race for Cold War advantage "a race of education and research," and now he was confident that advanced computer technology would extend the American edge over the USSR in communications satellites and space probes such as *Mariner 2.* Under Kennedy, it became tantamount to a national duty for students to study physics, mechanical engineering, and computer programming, and NASA's high-profile advances were a constant reminder of what was possible. "Remember when NASA was advertising Tang as its big contribution to the civilized world?" recalled Bob Taylor of the Pentagon's information processing agency, referencing the orange powdered drink that became associated with NASA astronauts. "Well there was a better example" in the computer science advances the agency made possible.

Early in April 1963, under intense congressional pressure to slash $700 million from the Apollo program, Kennedy asked Lyndon Johnson, in his role as head of the National Aeronautics and Space Council, to conduct a thorough review of NASA expenditures, objectives, and programs. Even though JFK and LBJ weren't personally close, they were in complete alignment on the moonshot goal, and the vice president went right to work. What Kennedy treasured about Johnson was that he was in cahoots with the Southern bloc of Democratic senators such as Walter George, Richard Russell, and Herman Talmadge—powerful allies of Project Apollo. Before the review was completed, Johnson advised his boss to constantly argue that "our space program has an overriding urgency that cannot be calculated solely in terms of industrial, scientific, or military development." If lawmakers criticized NASA, LBJ suggested they be called out as soft on communism. Timidity wouldn't be tolerated. Johnson wanted Kennedy to go directly to the American people, alerting them that the "future of society is at stake" if the NASA budget was reduced.

On May 13, Johnson wrote Kennedy a personal note and attached a thoughtful memo that would serve as New Frontier artillery against lawmakers, journalists, and pundits who questioned the prioritization of Gemini and Apollo as national security imperatives. Johnson's report brilliantly laid out how the moon landing would reap massive long-term benefits in international prestige, scientific breakthroughs, and economic benefits nationwide:

I. BENEFITS TO NATIONAL ECONOMY FROM NASA SPACE PROGRAMS

1. It cannot be questioned that billions of dollars directed into research and development in an orderly and thoughtful manner will have significant effect upon our national economy. No formula has been found which

attributes specific dollar values to each of these areas of anticipated developments, however, the "multiplier" of space research and development will augment our economic strength, our peaceful posture, and our standard of living.

2. Even though specific dollar values cannot be set for these benefits, a mere listing of the fields which will be affected is convincing evidence that the benefits will be substantial. The benefits include:

a) Additional knowledge about the Earth and the Sun's influence on the Earth, the nature of interplanetary space environment, and the origin of the solar system as well as of life itself.

b) Increased ability and experience in managing major research and development efforts, expansion of capital facilities, encouragement of higher standards of quality production.

c) Accelerated use of liquid oxygen in steelmaking, coatings for temperature control of housing, efficient transfer of chemical energy into electrical energy, and wide-range advances in electronics.

d) Development of effective filters against detergents; increased accuracy (and therefore reduced costs) in measuring hot steel rods; improved medical equipment in human care; stimulation of the use of fiberglass refractory welding tape, high energy metal forming processes; development of new coatings for plywood and furniture; use of frangible tube energy absorption systems that can be adapted to absorbing shocks of failing elevators and emergency aircraft landings.

e) Improved communications, improved weather forecasting, improved forest fire detection, and improved navigations.

f) Development of high temperature gas-cooled graphite moderated reactors and liquid metal cooled reactors; development of radioisotope power sources for both military and civilian uses; development of instruments for monitoring degrees of radiation; and application of thermoelectric and thermionic conversion of heat to electric energy.

g) Improvements in metals, alloys, and ceramics.

h) An augmentation of the supply of highly trained technical manpower.

i) Greater strength for the educational system both through direct grants, facilities and scholarships and through setting goals that will encourage young people.

j) An expansion of the base for peaceful cooperation among nations.

k) Military competence. (It is estimated that between $600 and $675 million of NASA's FY [fiscal year] 1964 budget would be needed for military space projects and would be budgeted by the Defense Department, if they were not already provided for in the NASA budget.)

While Kennedy was very grateful to Johnson for his reassuring recommendations, he also turned to von Braun for a booster shot of fortitude. In early May, the president made a return visit to Huntsville for an earth-shaking static firing of a Saturn booster stage. From a safe bunker, he watched the locked-down rocket roar, his greenish-gray eyes dancing like those of a delighted boy. "That's just wonderful!" he shouted. "If I could only show all this to the people in Congress!"

If Kennedy couldn't bring a Saturn rocket to Capitol Hill, he could at least project the excitement of another space shot, one that occurred the following week when Gordon Cooper took off aboard *Faith 7* for the sixth and final launch of Project Mercury. Cooper, a native Californian and veteran air force pilot, was already a familiar face to the TV-watching public, having served as capsule communicator (CAPCOM) for Alan Shepard's first suborbital spaceflight and for Scott Carpenter's *Aurora 7* mission. And Cooper could certainly barnstorm, a fact he showed off two days before liftoff by buzzing the NASA offices at Cape Canaveral in an F-102 jet. That stunt didn't gain him any brownie points with Webb, but his gallant performance in space on May 15 more than made up for this irrepressible horseplay.

If Americans' interest in Mercury had waned as launches, orbits, and splashdowns became semiregular events, Cooper's spaceflight reignited the fascination by adding both duration and suspense. As planned, the journey would begin on May 15 and last almost a day and a half—more than all previous Mercury missions combined. As had become his habit, Kennedy watched TV coverage of the liftoff from his White House bedroom and Oval Office side room. Seven hundred fifty miles to the south, Cooper pretended he was on his own, actually falling asleep in the cockpit as he awaited countdown. At 8:06 a.m., *Faith 7* executed a flawless takeoff, starting Cooper on a flight that would last more than thirty-four hours.

Throughout his mission, Cooper was connected by radio with two NASA flight directors (working in shifts) and with his fellow astronauts, and if all had gone according to plan, he may have been best remembered for an in-flight calmness that bordered on ennui. Instead, on the afternoon of May 16, during Cooper's third-to-last orbit, Americans got a dose of tense drama when a short circuit deactivated the spacecraft's automatic altitude and flight-control system. Instead of relying on autopilot for reentry, Cooper would have to take her down the old-fashioned way. "I had to initiate retrofire," Cooper later explained, "use the window view for attitude reference, and control the spacecraft with the manual proportional system." Despite carbon-dioxide levels rising, the unflappable Cooper's reentry and splashdown were even more accurate than those of previous Mercury missions. With precision akin to hitting a floating bull's-eye, Cooper brought *Faith 7* down just 4.5 miles from the designated prime recovery ship, 81 miles south of Midway Island.

Six days later, Major Cooper was standing in the White House Rose Garden to receive his NASA Distinguished Service Medal from the president. The public ceremony was far more lavish than some of the recent homecomings from space, and Kennedy's remarks on the occasion were clear, historically based, and partial to lunar expedition:

> I know that a good many people say, "Why go to the moon," just as many people said to Lindbergh, "Why go to Paris." Lindbergh said, "It is not so much a matter of logic as it is a feeling."

> I think the United States has committed itself to this great adventure in the sixties. I think before the end of the sixties we will send a man to the moon, an American, and I think in so doing[,] it is not merely that we are interested in making this particular journey but we are interested in demonstrating a dominance of this new sea, and making sure that in this new, great, adventurous period the Americans are playing their great role, as they have in the past.

Flashing his trademark humor, along with his personal identification with the astronauts, JFK also made reference to Cooper's piloting skills and meditative calm at the end of his mission. "One of the things which warmed us the most during this flight," Kennedy said, "was the realization that however extraordinary computers may be, that we are still ahead of them and that man is still the most extraordinary computer of all. His judgment, his nerve, and the lessons he can learn from experience still make him unique and, therefore, make manned flight necessary and not merely that of satellites. I hope that we will be encouraged to continue with this program." That hope would be tested throughout late spring, as Kennedy faced the first serious opposition to the expansion of America's space program.

After Cooper's mission, the Kennedy administration revved up the publicity machine with creative verve. The quarter of a million people who lined the streets of Washington watching Cooper travel from the White House to the U.S. Capitol constituted one of four grand public receptions staged for the astronaut that week, along with others in Honolulu, Cocoa Beach, and New York, where people turned out in droves for a ticker-tape parade. At the Capitol, Cooper spoke to a joint session of Congress, receiving a standing ovation and delivering an eloquent speech that included a prayer he said he'd composed while in orbit. Although addressed to God, his prayer could just as well have been a pitch to the congressmen and senators who held NASA's purse strings. "Help us in our future space endeavors," he began, "that we may show the world that a democracy really can compete, and still are able to do things in a big way, and are able to do research, development and can conduct many scientific and very technical programs." One senator called Cooper's prayer "one of the most impressive things" he'd heard in seventeen years in Congress. Nonetheless, it did little to sway certain fiscally conservative members of Congress intent on slashing NASA's budget.

That same month, Jack and Jackie Kennedy hosted

the Mercury astronauts at the White House for drinks, hors d'oeuvres, and storytelling. And astronauts had an ulterior motive: a futile effort lobbying for more Mercury missions. "Is it true you're all Republicans?" the president asked the gathered spacemen. "I don't know *what* the hell we are," Gus Grissom replied, to laughs. Pointing to his Oval Office rocking chair, the president asked Gordon Cooper to "Take a swing in *this* capsule"; Cooper happily agreed. Photographers clicked away. By the time the astronauts left the White House, their admiration for Kennedy had grown by leaps and bounds. However, once back in Houston for a meeting, the Mercury Seven found that they were in trouble for not having cleared their informal Oval Office chat with NASA officialdom.

A few weeks later, after the hoopla diminished, the Senate Republican Policy Committee circulated a scathing attack on the Gemini and Apollo programs. Questioning the costs of expediting them, the statement questioned whether "other aspects of human needs should be bypassed or overlooked in the one spasmodic effort to achieve a lunar landing at once." The senators suggested that the excess funds could be better used for education, health care, and other challenges closer to home. Other criticisms from Republicans rose on military grounds, charging that while the USSR was almost certainly bolstering its ability to weaponize space, the United States was not. "To allow the Soviet Union to dominate the atmosphere 100 miles above the earth's surface while we seek to put a man on the moon could be, in the opinion of many, a fatal error. . . . Intrinsic prudence, according to some, demands that we concentrate on the development of families of missiles operating in the suborbital and orbital areas rather than to devote such a large proportion of our efforts to lunar shots." The point that drew the greatest attention, though, was that NASA was robbing the best and brightest Harvard, MIT, and Caltech computer specialists for its moon challenge, depriving other fields of needed expertise.

Kennedy and Johnson were right to be concerned about NASA budget slashes. Even though Cooper had kept a finger in the dike for a few weeks, by summer NASA had clearly lost some of its razzle-dazzle glamour on Capitol Hill. On July 1, the *Washington Post* reported that "the United States space program is receiving the first searching review of its aims and activities since its inception five years ago."

The House Committee on Science and Astronautics combed through NASA's projects and identified $490 million in cuts. Although the Senate Republican Policy Committee had suggested a slowdown of the moon program to save money, the House committee had carefully left intact most projects that were directly moon related, even as it failed to take into account certain technical aspects of mission planning. The space agency resisted with special vigor when the committee concluded that funding for unmanned lunar probes could be withheld given that manned flights would bring back the same data. A NASA official countered that it was impossible to build a landing craft without first knowing the characteristics of the moon's surface. Pushing back from the bottom line, Webb warned that cuts of more than $400 million would impede the effort to land an American on the moon within Kennedy's schedule.

Not to be outdone in the prolonged funding dispute, the pro-space lobby regularly trotted out senators and representatives to praise NASA in unqualified terms. Why punish the one federal agency, they would ask, that was overperforming? On the House floor, Congressman James Fulton of Pennsylvania elevated NASA administrators and astronauts into the pantheon of explorers alongside Columbus, Hudson, de Soto, Crockett, Boone, and Lewis and Clark. All these legends were considered "nuts" in their day, Fulton said, and he argued that the new breed of "nuts" in Kennedy's New Frontier "will lead a great America in the conquest of outer space." Fulton ended his appeal by saying he favored increased funding for Project Apollo "because it is in keeping with the pioneer spirit of this great nation."

Out in the media landscape, where he'd become a familiar figure, von Braun also came to Kennedy's aid in the budget debate. Building off his friendships with Walt Disney, Walter Cronkite, and other opinion makers, von Braun routinely leaked the false premise that the Soviets were winning the space race—a contention Khrushchev was also spinning weekly, for his own purposes. But in point of fact, the only thing NASA officials feared more than a competitive Soviet moon program was no lunar effort from them at all. If the USSR were to have admitted that they were actually behind the United States in space and missile technology, NASA's exorbitant funding might have dried up in Congress.

Due to the Soviets' secrecy, their actual intentions concerning the moon remained murky. Although Bureau of the Budget chief David Bell had concluded the previous November that there was no solid evidence for a USSR moon program, America's intelligence services were unwilling to express such certainty. "We cannot say definitely at this time that the Soviets aim to achieve a manned lunar landing ahead of or in close competition with the United States," read a CIA National Intelligence Estimate dated December 5, 1962, "but we believe the chances are better than even that this is a Soviet objective." The report went on to reiterate that there was no *firm* evidence of any planning for a Soviet manned moon mission, whether in competition with the United States or not, but the agency couldn't rule out that such a program existed.

When asked about racing to the moon against America, Kremlin officials routinely denied any such objectives. Nevertheless, Sergei Korolev and others in the Soviet space world intensely desired to defeat the United States in a head-to-head moon race. Korolev persisted in proposing various schemes to accomplish a lunar landing. With great conviction, Korolev wanted to develop a huge booster designed to hoist seventy-five tons into orbit. This ultimately became the N1 moon rocket. Korolev, a fine salesman, persuaded Khrushchev in 1962 to approve its development.

Bell's report, if made public, would have blown a giant hole through the New Frontier narrative that the moon was the ultimate trophy in the superpower rivalry between free-world democracy and expansionist communism. As history has shown, although the Soviets were struggling in the rocketry realm, going to the moon remained an imperative. In early 1963, however, the Soviets were experiencing serious setbacks just trying to send an unmanned Luna vehicle to the moon, despite the fact that they'd landed a less sophisticated probe on the surface three and a half years before. In the first four months of 1963, three Luna craft were launched unsuccessfully. The first failed to escape Earth's orbit, the second couldn't find its orbit of the moon, and the last missed the moon altogether. Similarly, *Mars 1*, an unmanned flyby probe launched toward the red planet in November 1962, disappeared into space due to antenna malfunction, after flying more than sixty-six million miles. While proposals for manned missions to both the moon and Mars continued to be bandied about at the Soviet advanced research institution known as OKB-1, the dismal results of these unmanned missions put a damper on enthusiasm, although Korolev continued to push for landing cosmonauts on the moon before the Americans.

While American agencies remained in the dark about true Soviet lunar intentions, the State Department was becoming increasingly anxious over the lack of progress toward nuclear treaties and agreements. One observer said that Soviet foreign policy was "in a state of perfect inertia. It isn't moving for good or ill." Americans had reason to hope that was the case. Concerning a proposal to ban nuclear weapons in space, the United States feared the Soviets were stalling until they had a nuclear bomb in orbit. On April 29, Secretary of State Dean Rusk abruptly canceled any further approaches to the Soviets about joint space collaborations, and the next week, on May 8, he gave the president a proposal specifically addressing the rampant fears, titled "U.S. Reaction to Soviet Placing of a Nuclear Weapon in Space." Kennedy was more than receptive to the proposal's core recommendation for implementation of "an active anti-satellite capability at the earliest possible time, nuclear and non-nuclear." As military historian Paul Stares pointed out, the new policy was predominantly a preemptive move against any Soviet decision to arm space, with or without a treaty. "Moreover," he writes, "it would provide insurance for possible domestic criticism that the administration had not taken necessary precautions." Defensive or not, the plan still constituted a significant step in the militarization of space.

Greater progress was being made toward terrestrial nuclear arms control. On June 10, 1963, Kennedy delivered a remarkable commencement address at American University in which he announced that later that summer the United States, Great Britain, and the Soviet Union would begin three-way talks aimed at reducing the proliferation of nuclear weapons and limiting atmospheric and underwater nuclear tests. It was a big step forward, but in fact only a piece of the larger picture JFK wanted to present with his speech, which encompassed "a topic on which ignorance too often abounds and the truth is too rarely perceived—yet it is the most important topic on earth: world peace."

Partially inspired by an open letter published by a consortium of college professors in the *New York Times* pleading for a nuclear test ban treaty with Russia, Kennedy's speech was a clarion call, decrying the stockpiling of nuclear weapons, the potential for devastation and radiation poisoning, and the massive, wasteful expense. Calling on all nations and all peoples to "examine our attitude toward peace itself" and to cast off assumptions that peace is impossible, the speech advocated for a reexamination of American attitudes toward the USSR and the Cold War, and for trading existential struggle for peaceful competition. Eschewing pie-in-the-sky concepts of "universal peace and good will of which some fantasies and fanatics dream," it called instead for "a more practical, more attainable peace . . . a series of concrete actions and effective agreements which are in the interest of all concerned." While acknowledging that "no treaty . . . can provide absolute security against the risks of deception and evasion," it can "offer far more security and far fewer risks than an unabated, uncontrolled, unpredictable arms race."

"The United States, as the world knows, will never start a war," Kennedy concluded. "We do not want a war. We do not now expect a war. This generation of Americans has already had enough—more than enough—of war and hate and oppression. We shall be prepared if others wish it. We shall be alert to try to stop it. But we shall also do our part to build a world of peace where the weak are safe and the strong are just. We are not helpless before that task or hopeless of its success. Confident and unafraid, we labor on—not toward a strategy of annihilation but toward a strategy of peace."

On the other side of the world, Khrushchev told aides that Kennedy's speech was "the best speech by any president since Roosevelt." But just as the Soviet leader was feeling good about Kennedy's peace overtures, the U.S. president departed on the eighth international trip of his presidency. Visiting Cologne, Frankfurt, and Wiesbaden in West Germany, JFK enthused about NATO, and his meetings with the West German chancellor were meant to display U.S. resolve to keep NATO as its top foreign policy priority. On June 26, Kennedy capped his West Germany trip by delivering a rousing address to 450,000 people crowded into an enormous plaza in West Berlin, delivering his now-famous "Ich bin ein Berliner" speech, advocating "the right to be free" over "the failures of the Communist system."

The Berlin speech echoed through the Cold War landscape, heartening anticommunists everywhere. The West Berlin audience went wild with admiration for Kennedy's bold loyalty to them; in a speech of fewer than seven hundred words, he'd assuaged their worst fears of abandonment while also shining a beacon to those on the other side of the wall, declaring that "freedom is indivisible, and when one man is enslaved, all are not free." The speech immediately lifted the morale of those in the beleaguered city and informed the Soviet Union that despite the president's recent call for a more peaceful competition, the battle for hearts and minds around the world would remain fierce.

In Berlin, on June 26, 1963, President Kennedy speaks before the Brandenberg Gate, telling the crowd in this divided German city, "Ich bin ein Berliner."

On November 18, 1963, President John F. Kennedy watches the launch of a Polaris missile from the USS "Observational Island" military vessel off the coast of Cape Canaveral.

20
"The Space Effort Must Go On"

The uncertainty in the United States at this time about the exact character of a Soviet program to send men to the moon is in retrospect understandable, since the situation in the Soviet Union was both complex and confusing.

—JOHN M. LOGSDON, *JOHN F. KENNEDY AND THE RACE TO THE MOON* (2010)

Five years after *Sputnik* and the creation of NASA and two years after Kennedy's moonshot plans sailed through Congress on a May breeze, Project Apollo should have been an unassailable colossus, a point of national pride that was too big to fail. Project

Mercury had been a stellar success, pushing America's space capabilities forward in a remarkably short time span. In St. Louis, the larger, two-seat Gemini capsule was being built on Mercury's model, and it would soon be launched into orbit atop a Titan II with 430,000 pounds of thrust. North American Aviation was designing and building major elements of the three-man Apollo spacecraft in Downey, California, that would soon inaugurate trips around Earth and then around the moon. Grumman, of Bethpage, New York, was developing the LEM. And MIT in Cambridge, Massachusetts, was designing a vast array of new computer technology for NASA. At the Marshall Space Flight Center, the Huntsville engineers were constructing preliminary stages of the massive Saturn V rocket to power those Apollo missions, while NASA engineers and scientists elsewhere were perfecting the lunar orbit rendezvous approach that would eventually bring American astronauts to the moon's surface. Apollo was clearly gathering momentum. If the program was successful, the United States would leapfrog over the Soviets in overall space achievement. America's space probes had a far greater record of success, its satellite programs were more advanced, and, crucially, it had the money to keep moving forward. No one was certain of the record on the direction of the Soviet space program, but Khrushchev's boasts of its superiority continued to intimidate most people in the West.

In the United States, pro-space politicians and administrators were facing widening pushback over NASA's astounding costs, as other economic and military priorities were crowded out of budgets. The fact that Senator Robert Kerr of Oklahoma, NASA's biggest budget booster, died of a heart attack on January 1, 1963, suddenly made Webb's lobbying on Capitol Hill much harder. The new head of the Senate Space Committee, Senator Clinton Anderson (D-NM), was more interested in environmentalism and nuclear power than an Apollo moonshot. Meanwhile, conservative congressmen continued efforts to chip away at the agency's funding, supported by former president Eisenhower, who complained to House Minority Leader Charles Halleck that "a spectacular dash to the moon" would "vastly deepen" America's debt and be a "tax burden," and represented "the antithesis of fiscal soundness." Civil rights leaders such as Ralph Abernathy pointed to all the societal good that the same budget could do in fighting urban poverty in cities such as Detroit, New York, and Chicago. Criticism was mounting, too, from parts of the scientific community, which recognized that unmanned, research-oriented probes offered superior return on the dollar compared with manned missions. What perturbed Webb the most was that Secretary of Defense Robert McNamara was telling Washington lawmakers that Project Apollo had no "substantial military value," thus making NASA a bull's-eye for conservative-minded budget hurdles.

Throughout the summer of 1963, infighting over budgets roiled NASA, contributing to the resignation of D. Brainerd Holmes, director of the Office of Manned Space Flight and a reliable moonshot advocate. "The man-in-space program is in the throes of a management crisis," Richard Witkin wrote in the *New York Times*, "that is deeply worrying a growing number of highly placed officials." In NASA administrator Webb's view, the greater worry was that if Congress cut NASA's funding by any more than 5 percent, the moonshot could not occur within a decade—and Republicans in Congress were seeking cuts twice that size. When a top-tier NASA engineer was asked by a Republican skeptic what would be discovered on the moon, he sarcastically snapped, "Russians."

There was no question that Kennedy, even more than Webb, insisted that U.S. dominance in all aspects of space be the primary objective of NASA, regardless of whether Soviet cosmonauts were moon bound or not. His philosophy was put on public display at a July 17, 1963, press conference. Apollo was a marvelous alternative to all-out war with the USSR or future proxy wars such as Korea. "The point of the matter always has been not only of our excitement or interest in being on the moon," he said, trying to mute critics, "but the capacity to dominate space, which would be demonstrated by moon flight, I believe, is essential to the United States as a leading free world power. That is why I am interested in it and that is why I think we should continue."

In Moscow, Khrushchev had his own headaches and systemic problems. Recent high-profile successes had been offset by equally dismal failures. Equally dispiriting was that Sergei Korolev, director of the Soviet rocket program, was suffering from several serious health problems and required frequent hospitalization. (He would die of botched surgery in January 1966.) Khrushchev's budget difficulties made NASA's woes seem pedestrian. The fact was, competing with the United States in space was a backbreaking endeavor for the Soviets, with financial limitations stymieing any thought of competing on a moonshot in real life rather than on the battlefield of propaganda. "He told me many times we cannot compete with America when their economy is several times bigger than ours, so we have to choose some directions," recalled Sergei Khrushchev, the leader's son. "And his direction was housing and agriculture, not the moon. If he could be the first on the Moon for free he would be happy to do this, but he was not ready to pay for it."

Hunting for a possible way out of the space-race paradigm, the Kremlin began sending out feelers that summer. During a visit to the Soviets' top-secret Star City space-training facilities, British physicist and radio commentator Sir Bernard Lovell was asked to meet with members of the Soviet Academy of Sciences about a sensitive topic. Was the United States serious, they asked, about a joint moon mission with the USSR?

Lovell, creator of the Jodrell Bank Observatory near Goostrey, Cheshire, had been commentating on astrophysics in Great Britain since the 1920s. Often critical of American claims of space superiority, he was known to be deeply impressed with the Soviets' Kosmos series of satellites, whose purpose—military or scientific?—had perplexed U.S. intelligence and piqued its suspicions. Flexing his international clout, Lovell wrote to Webb on July 23, 1963, specifying that he had been requested to do so by those in the Kremlin who were "interested in an international program to get men to the moon."

Although Kennedy and Khrushchev had uncomfortably batted around the idea of a joint American-

Soviet moon mission at their June 1961 Vienna summit, this back-channel suggestion came as a shock to Webb, who was immediately apprehensive. He wasn't alone. Despite the urge toward peaceful coexistence in JFK's recent American University speech, many in the administration, including Secretary of State Dean Rusk, had no interest in initiating joint space talks with the Soviets, both for strategic and political reasons. State Department diplomats insisted it was too dangerous to act on Lovell's feeler until (and unless) it was more formally communicated. Were the proposal to leak beyond the small coterie who knew about it (which included Johnson, National Security Advisor McGeorge Bundy, assistant NASA deputy administrator Hugh Dryden, and select State Department officials), they worried it could damage the president's 1964 reelection effort.

Kennedy wasn't so sure. Over the next months, as the space program remained in flux over congressional funding and U.S.-Soviet diplomacy wrestled with decisions of global consequence, he began to balance Apollo's potential value as a Cold War checkmate against the potentially greater value of using it as a bargaining chip toward peace. If conducting a joint mission meant helping to end the Cold War and ushering in an era of peace, the president was ready to put aside his pride and perhaps clasp hands with his Russian counterpart.

Although Kennedy was pragmatic to the core, his peace bell had been rung back in June, when he attended an eighteen-minute attack simulation at the North American Air Defense (NORAD) operational center and United States Space Command, an airspace monitoring and early-warning facility located deep inside Cheyenne Mountain, near Colorado Springs. As JFK watched a simulated Russian strike on the United States that afternoon, he turned solemn and was visibly shaken. "The bombers were stopped, but the intercontinental missiles came on and erupted in white ovals as they struck American cities," wrote *Time* White House correspondent Hugh Sidey, who had accompanied the president to Colorado. "Muttered one Air Force officer: 'We have no way to stop them.'"

Kennedy was haunted by the insider knowledge that if such a devastating thermonuclear attack were actually to occur, America would have no choice but to launch its own nuclear missiles, setting off a worldwide armageddon. During Kennedy's first year in the White House, the Soviets tested the "Tsar Bomba," the largest nuclear weapon ever detonated, at Novaya Zemlya. If such a bomb were to hit New York or Chicago, the death toll would be in the millions. Faced with that doomsday scenario, Kennedy viewed the prospect of giving up a chunk of national prestige by pursuing a joint U.S.-Soviet moon mission as eminently sensible. Saving the world may have been the paramount consideration, but it wasn't the only one. Kennedy knew that a joint venture would save massive amounts of money for both superpowers. It would give NASA access to the Soviet space program, pulling back an Oz-like curtain that the CIA had been trying to pull back for years, with very limited success. If the collaboration came to pass, it would also bring both Kennedy and Khrushchev eternal acclaim, and almost certainly a shared Nobel Peace Prize.

But the president knew that the risks were also enormous. Any suggestion of collaboration would inevitably invite attacks suggesting that JFK was opening himself and the country to a KGB trick meant only to trip up ongoing U.S. moonshot plans. Allowing the Soviets access to the NASA space program would also risk exposing hard-earned satellite secrets, private-sector research, classified intelligence, and computer technology, and it could put the Pentagon's military programs at risk. At the nuts-and-bolts level, the challenges of merging the very different engineering approaches and technologies of the United States and the USSR would likely prove difficult and would amplify mission risk.

While the Kennedy administration recognized the technological virtues of the *Sputnik*s between 1957 and 1961 and the *Vostok*s between 1961 and 1963, the idea that NASA could easily integrate the varied Soviet components into Apollo was unlikely. Furthermore, the payload capacity of von Braun's Saturn V dwarfed any rockets the Soviets were developing.

The timing, however, made the bold idea somewhat appealing. On August 5, after years of hard work and immense frustration, the Soviet Union, United States, and Great Britain had finally agreed on language for the Limited Test Ban Treaty, signed in Moscow, prohibiting nearly all nuclear weapons testing, with only underground tests excepted. Having so recently faced the real possibility of nuclear conflict over the Cuban Missile Crisis, Kennedy was relieved and even ecstatic to be able to offer a measure of relief to a worried world. "It is rarely possible," he said in urging congressional ratification of the treaty, "to recapture missed opportunities to achieve a more secure and peaceful world." History, Kennedy knew, had seen many episodes where events outran diplomacy, igniting war between nations despite their leaders' desire to avoid it. Kennedy was determined not to let that happen in the nuclear age.

Change was in the air that summer. On August 28, a coalition of civil rights groups staged the March on Washington for Jobs and Freedom, drawing a crowd of some 250,000 marchers to the nation's capital, an event that culminated in Martin Luther King Jr.'s historic "I Have a Dream" speech from the steps of the Lincoln Memorial, one of the signature moments of the civil rights movement. At a White House meeting after the march, Kennedy spoke to King about the power of nonviolent protest. Although still wary of the potential for civil rights demonstrations to spark violence, the president had been impressed with King's speech, which aligned with his own mounting desire to put the power of his office behind the causes of peace and human rights. In an official statement, Kennedy praised "the deep fervor and quiet dignity" of the thousands of activists "both Negro and white," adding that the desire for equality was "neither novel nor difficult to understand."

In late August and September 1963, Kennedy was in the rare position of presiding over a time when the arc of the moral universe seemed to be bending toward justice, and maybe even global peace. By averting military showdowns in Cuba and Berlin, he'd received global kudos as a peacemaker, and his leadership on the Limited Test Ban Treaty received similar acclaim. Dr. Albert Schweitzer, who'd won the Nobel Peace Prize in 1952 for his "Reverence for Life" philosophy, wrote Kennedy from Gabon, Africa, describing the treaty as "one of the greatest events, perhaps the greatest" in history. "Finally," he wrote, "a ray of light appears in the darkness in which humanity was seeking its way." If Kennedy could now persuade his nation to truly join hands with its communist adversary and begin exploring not only outer space but their shared humanity, he would be a statesman for the ages.

Congress still had to ratify the nuclear test ban treaty, which wasn't a given that fall of 1963. Senator Barry Goldwater, always the hawk, tried to argue that the test ban was weak on Soviet verification, but his skepticism was a minority opinion. After a month of hearings, the Senate Foreign Relations Committee approved the treaty, followed three weeks later by the full Senate. On October 7, a proud JFK added his signature.

The Cold War, the president believed, was at a potential thawing point. Some national security analysts were telling him that the Soviets were on the verge of abandoning plans to send a human to the moon, so Khrushchev didn't have very much to lose. Below the radar, back-channel communications had continued since Lovell's July outreach to Webb. On September 11, Dryden attended a prearranged meeting in New York City with the Soviets' authority on space negotiation, Anatoli Blagonravov, who with the Kremlin's approval once again mentioned joining forces for a moonshot. In late August and early September, eager to judge whether this was just talk among lower-level officials, Kennedy asked an array of Russian experts if they thought the Kremlin proposal was serious; many did.

With great force and conviction, Webb took the more tradional Cold Warrior position. He was convinced that Project Apollo was on track, aimed squarely at the moon, and needed no Soviet help whatsoever. To Webb, the most disturbing possibility was that the Soviet overture was a sly red herring, a KGB disinformation campaign aimed at tricking Congress into cutting NASA's budget. "No bucks, no Buck Rogers," as the saying went—but also "no race, no rush." Even though the overall federal budget had been passed on August 11, Congress had the prerogative to make adjustments, meaning that the question of NASA's budget dragged on, becoming a cauldron of hot debate and crowded hearings as summer turned to fall. In that contentious atmosphere, Webb and other top NASA officials considered the timing of the Soviet proposal to be inconvenient at best and destructive at worst. By early September, Johnson and McNamara had advised Kennedy to steer well clear of the Lovell-Blagonravov proposal, and warned against any public discussion of a joint mission.

On September 18, 1963, Kennedy had a private conversation with James Webb in the Oval Office about the Apollo program. Thinking of the future, the president wondered aloud whether the U.S. moon landing could happen during his tenure in the White House, were he to be reelected in 1964. "No, no," Webb said. "We'll have worked to fly by though while you're president[,] but it's going to take longer than that." Webb then described in no uncertain terms what NASA would be able to deliver to the country by 1968: "A basic ability in this nation to use science and very advanced technologies to increase national power." That, according to Webb, would be the single most important achievement of the New Frontier space program.

The president, as was his style, probed Webb further about how NASA was using science and advanced technology to increase national power vis-à-vis the Soviet Union. At one juncture, Kennedy point-blank asked, "Do you think the lunar, the manned landing on the moon is a good idea?" Webb, for the umpteenth time in the past nineteen months, reassured him with a "Yes, sir, I do."

Listening to the tape of this Kennedy-Webb conversation provides a marvelous window into the president's ability to simultaneously look at space from political, scientific, economic, and national security perspectives. He's concerned that the moonshot has lost its glamour. At one juncture, the president asks, "If I get reelected, I'm not—we're not—go[ing] to the moon in my—in our period are we?" Webb again answered, "No," and Kennedy replied, his voice slightly sulky, "We're not going . . . yeah." Behind closed doors, talking only to Webb, Kennedy reveals his understandable concern that he gambled by putting the lunar voyage at the heart and soul of the New Frontier. But the conversation also divulges his resolve not to retreat from his brazen pledge of May 25, 1961. Webb does his part to buck up JFK's once-soaring faith in all things NASA. "Why should one spend that kind of dough to put a man on the moon?" Kennedy asks, and answers himself. "But it seems to me . . . we've got to wrap around in this country, a military use for what we're doing and spending in space. If we don't, it does look like a stunt." Webb tells Kennedy, "I predict you are not going to be sorry, no sir, that you did this." Kennedy agrees, recognizing that his American moonshot would be a huge part of his legacy. "I think," Kennedy said, ending the conversation about his reelection fortunes in 1964, "this can be an asset, this program."

At the time of the Webb meeting Kennedy had already decided to raise the idea of a joint U.S.-Soviet moonshot in his upcoming UN speech. While others within the administration deemed that idea ludicrous at best and anti-American at its core, Kennedy had moved beyond the simple Cold War polarity of his congressional years. As he'd made clear at American University, his thoughts were on a far larger geopolitical realignment, and his hopes were pinned on Soviet-American disarmament talks. To get there, he needed to be flexible. Even though he himself had framed winning the race to the moon as essential for America's national pride, he was willing to possibly sacrifice the win if doing so built a bridge for peace.

Quite simply, Kennedy, unlike many of his top foreign policy advisors, had stopped viewing the U.S.-Soviet rivalry as a terminal condition. National Security Advisor McGeorge Bundy thought this was the best state of mind for the president to have. He wrote Kennedy a memo arguing that the moonshot could be used for two purposes: to press for joint space initiatives with the USSR or be a "spur" to unleash American technological dynamism. Bundy didn't want Kennedy to feel boxed in by his "end of the decade" challenge. In this memorandum, Bundy mused that "if we cooperate, the pressure comes off" regarding a moon landing by 1970. "We can easily argue that it was our crash effort in '61 and '62," he wrote, "which made the Soviets ready to cooperate."

On September 19, the day before Kennedy was to speak at United Nations headquarters in New York, Soviet foreign minister Andrei Gromyko made the first move toward rapprochement in an uncharacteristically forthright address to the General Assembly. Startlingly, he was poised to move on the issue of banning nuclear arms in space, disagreement over which had vexed U.S.-Soviet negotiators since 1957. Stating in the clearest terms that "the Soviet Union is ready" to "ban the placing into orbit of objects with nuclear weapons on board," Gromyko dropped his nation's longtime insistence that the United States remove short- and medium-range missiles from foreign bases as part of any agreement.

As Kennedy prepared to take his turn addressing the UN delegates, Bundy pressed him to remain mum on the subject of a joint moon initiative. The moonshot was America's destiny alone, he argued. Also, in the wake of his "Ich bin ein Berliner" speech, NATO allies would think the president a turncoat if he suddenly considered such a bizarre collaboration with the Kremlin. However, like Albert Einstein, JFK believed that imagination can be more important than knowledge. Why not float a trial balloon at the United Nations and inventory the response?

When Kennedy arrived at UN headquarters, he was suffering such acute back pain that his physician, Admiral George Burkley, was accompanying him. He limped noticeably as he approached the podium at the General Assembly. Even with regular cortisone shots, his Addison's disease was getting the best of him.

At the outset of his September 20 address, Kennedy quoted from the letter Albert Schweitzer had written him heralding the Limited Nuclear Test Ban Treaty, which had by that time already been signed by more than one hundred countries. "Today," he said, "the clouds have lifted a little so that new rays of hope can break through." Celebrating those rays without ignoring the remaining thunderclouds would be the topic of his address. When Kennedy listed the recent positive diplomatic developments between the United States and Soviet Union, he was met with rounds of thunderous applause from the delegates. "The world has not escaped from the darkness," he continued. "The long shadows of conflict and crisis envelop us still. But we meet today in an atmosphere of rising hope, and at a moment of comparative calm. My presence here today is not a sign of crisis, but of confidence."

About halfway through his address, Kennedy pivoted to space, disregarding the advice of Bundy, and of all the other skeptics in his administration. Ted Sorensen, who'd helped write the address, had been privy to its controversial language in advance:

> In a field where the United States and the Soviet Union have a special capacity—in the field of space—there is room for new cooperation, for further joint efforts in the regulation and exploration of space. I include among these possibilities a joint expedition to the moon. Space offers no problems of sovereignty; by resolution of this Assembly, the members of the United Nations have foresworn any claim to territorial rights in outer space or on celestial bodies, and declared that international law and the United Nations Charter will apply. Why, therefore, should man's first flight to the moon be a matter of national competition? Why should the United States and the Soviet Union, in preparing for such expeditions, become involved in immense duplications of research, construction, and expenditure?

> Surely we should explore whether the scientists and astronauts of our two countries—indeed of all the world—cannot work together in the conquest of space, sending some day in this decade to the moon not the representatives of a single nation, but the representatives of all of our countries.

Nobody had seen Kennedy's unprecedented proposal coming. Representatives from around the world cheered enthusiastically, many seeing the idea as a salve for strained East-West relations and a step toward de-escalating the Cold War. In the wake of the pacifistic nuclear test ban and hawkish "Ich bin ein Berliner," JFK was believable in both roles: the peacemaker and the strong voice for Western democratic values. The historian Walter A. McDougall suggested that Kennedy's UN speech (and other statements about U.S.-Soviet cooperation in space) "were just exercises in image-building."

Beyond the UN General Assembly Hall, reactions were less enthusiastic. In Washington, the speech landed with a jolt, stunning senators and congressmen of both parties. White House advisors, the Joint Chiefs, and the CIA were flabbergasted. This wasn't a trial balloon in a freewheeling press conference, but an outright proposal to Khrushchev, with the world as a witness. In response, however, Khrushchev was silent. When asked about Kennedy's overture, the strong-minded Gromyko demurred with a polite comment about the president's fine overall dramatic delivery. The Soviets' official TASS news agency was less diplomatic, lampooning aspects of Kennedy's speech.

With no response from Moscow, Kennedy quickly backed away from his daring overture. At his next press conference, on October 9, he was asked whether he would continue to champion a joint U.S.-Soviet mission to the moon. Characteristically, the president explained his current attitude in some detail:

> We have received no response to our—to that proposal, which followed other proposals made on other occasions. As I said, our space program from the beginning has been oriented towards the peaceful use of space. That is the way the National Space Agency was set up. That is the position we have taken since my predecessor administration. I said this summer that we were anxious to cooperate in the peaceful exploration of space, but to do so, of course, requires the breakdown of a good many barriers which still exist. It is our hope those

barriers, which represent barriers of some hostility, some suspicion, secrecy and the rest, will come down. If they came down, of course, it would be possible for us to cooperate. So far, as you know, the cooperation has been limited to some exchange of information on weather and other rather technical areas.

We have had no indication, in short, that the Soviet Union is disposed to enter into the kind of relationship which would make a joint exploration of space or to the moon possible. But I think it is important that the United States continue to emphasize its peaceful interest and its preparation to go quite far in attempting to end the barrier which has existed between the Communist world and the West and to attempt to bring as much as we can the Communist world into the free world diversity which we seek. So the matter may come up, but I must say we have had no response which would indicate that they are going to take us up on it.

Throughout the NASA bureaucracy, morale was shaken for weeks over the thought that Kennedy was using Project Apollo as a bargaining chip with the Soviets. A group of GOP hawks in Congress moved to forbid the use of NASA funds for any cooperative effort toward a moon mission with the USSR. CIA chiefs lamented Kennedy's pell-mell, flip-flopping attitude. Columnists in major newspapers attempted to explain the president's turnaround, offering every possible point of view. None of them need have bothered. The proposal seemed dead on arrival.

Throughout 1963, Kennedy had been frustrated by the lack of confirmable intelligence on whether the Soviet Union was still planning a moonshot. CIA director John McCone could never provide Kennedy with a confirmation one way or another. It was a Cold War mystery. Not until October 1 did JFK get proper intelligence community feedback. In a classified CIA document titled "A Brief Look at the Soviet Space Program," Kennedy was informed that Khrushchev was "unquestionably planning manned lunar landings . . . but there is no evidence that the program is proceeding on a crash basis." Such vagueness frustrated Kennedy. The implication by default was that there was indeed a race to the moon. Unbeknownst to the CIA was that the Soviets were already designing a huge booster to carry out a lunar landing. While there were timetable and budget debates going on among the Kremlin leadership in 1963, by 1964 the Soviet Union officially approved a moon voyage. Contemporaries of Kennedy who argued that the United States wasn't truly in a moon race of some sort were proved wrong in the annals of history.

If Kennedy was frustrated by the CIA's lack of sureness, the agency was starting to worry about Kennedy's untrustworthiness. Too often for its liking, JFK had sent up peace flares trying to end the Cold War, in disregard of the intelligence agency's recommendations. His comments about a joint U.S.-Soviet moonshot were seen as reckless. The Pentagon wasn't militarizing space just so that Kennedy, playing God, could hand Khrushchev an early Christmas gift. A queasy feeling was circling around the corridors of the CIA that Kennedy could not be trusted with national security secrets. On October 3, 1963, just after Kennedy received "A Brief Look at the Soviet Space Program," and Arthur Krock wrote a column in the *New York Times* quoting sources at the CIA as saying, "If the United States ever experiences [an attempted coup to overthrow Kennedy], it will come from the CIA and not the Pentagon." Krock led readers to believe that Kennedy's life might be in peril, and that the CIA represented a "tremendous power and total unaccountability to anyone."

When media interest over a joint moonshot quickly receded, Congress renewed the battle over the NASA budget. Those who wanted to remove as much as a billion dollars from NASA and use it for air force space projects continued their campaigns. Budget hawks who wanted to strip NASA simply to save money continued their policy march. From his Pentagon perch, Secretary of Defense McNamara continued to carp that the moonshot didn't help national security an iota. Others were agitated by reports that aerospace industry suppliers were producing substandard prime components for some NASA vehicles. NASA's doubters might have succeeded in slashing the budget on one pretext or another, but for the quiet power of Texas congressman Albert Thomas, who blocked the way, protecting space investments especially in his Houston district. In the end, the initial request of $5.7 billion was knocked down to $5.3 billion, a number that Webb considered rock bottom for a successful Apollo mission by 1970. And the Space and Information Division of North American, Inc., went on a public relations blitz to help NASA, proudly announcing that Apollo capsules were already being "hand-tooled" in sync with Kennedy's moonshot deadline. "I don't know of any technical problem that will stop us from reaching the moon in this decade," Harrison Storms, president of the division, told the *New York Times*. "We have answers to all the main [technical] questions."

On October 25, Kennedy finally received an official

answer from the Kremlin about his UN proposal of a joint moonshot, and it was as surprising in its own way as Kennedy's overture had been. With a tone of mockery, Khrushchev not only rejected the idea of cooperating on a lunar voyage but announced that his country wasn't interested in a moonshot at all. If Khrushchev was telling the truth, this apparent collapse of the competition gave NASA's congressional opponents new ammunition. Why continue to fund a hell-for-leather moon race if no one else was running? The calculated answer that quickly emerged was that Apollo had to continue on schedule to show that America wasn't influenced in the least by Moscow's ploys. But when a supplementary appropriations bill was passed in December, the NASA budget would be reduced by another $300 million. Still, Khrushchev helped NASA out more than he ever realized.

Instead of sticking to his story, Khrushchev became his own worst enemy. On November 1, at a Moscow reception for Prince Souvanna Phouma of Laos, he boasted that his country had just launched the earth-controlled Flight 1 unmanned satellite, whose aim was to perfect rendezvous protocol in space. This was the same type of technology Project Gemini was trying to master. When asked about Kennedy's UN offer for a joint U.S.-Soviet lunar voyage, Khrushchev unleashed a barb at Kennedy that backfired: "What could be better than to send a Russian and an American to the moon together, or better yet, a Russian man and an American woman?" he said, to general laughter. This dig at the fact that NASA had yet to put a woman in space, as the Kremlin had with Valentina Tereshkova, came only one week after the Soviets declared they were bowing out of the Cold War competition to the moon. It seemed that Khrushchev was yanking America's chain on all accounts. The Soviet leader, trying to be cagey, neither accepted nor rejected Kennedy's joint space proposal. Furthermore, if the Russians were engaged in perfecting a Gemini-like rendezvous, they were likely still secretly planning a Soviet moonshot. Inadvertently, Khrushchev's joke gone awry also helped NASA push through its recommended budget on Capitol Hill.

Despite vocal domestic opposition, Kennedy's UN proposal proved a propaganda windfall for democracy versus communism, winning hearts and minds much as his Peace Corps idea had a few years before, and leaving a humanitarian glow in its wake. Dean Acheson had famously deemed the president's successful navigation of the Cuban Missile Crisis "an homage to plain dumb luck." Whether by luck or design, the net effect of Kennedy's UN speech was to make America look like a peaceful giant on the world stage, and apparently cause the USSR to quit the moon race—all while NASA protected most of the funding it needed for Project Apollo.

Although talk of a joint U.S.-Soviet moon mission had gone nowhere, Kennedy didn't quite let the dream go—and neither did the fickle Khrushchev, who made vague comments on October 31 indicating some residual interest. The Soviet premier's remarks were far from firm, but JFK reacted to the slight sense of encouragement by issuing a National Security Action Memorandum directing Webb to create proposals on accommodating cooperation with the Soviet Union in outer space, including on lunar landings.

That early November, Kennedy had his last private meeting with Webb. "I have to tell you, I think that the Secretary of Defense will not want to support the [space] program as having substantial military value," Webb grumbled to the president about McNamara's downplaying of Project Apollo's value. "So you're going into a campaign [1964] with me saying it has very important technological benefits for the military, and the Secretary of Defense being unwilling to say it." Kennedy knew that Webb and McNamara were at odds. "Well, is there anything personal between you?" Kennedy began. "Don't let it get personal." This was wise advice. If the moonshot were to be accomplished, NASA and the Department of Defense would have to get on the same page, promoting the effort as the "American way" in action. To Webb's delight, Kennedy firmly sided with him over McNamara in the feud between the two loyal New Frontiersmen.

21
Cape Kennedy

The great achievement of the men on the moon is not only that they made history, but that they expanded man's vision of what history might be. One moon landing doesn't make a new heaven and a new earth, but it has dramatized the possibilities of doing so.

—JAMES RESTON, *NEW YORK TIMES*, 1969

On November 16, 1963, John F. Kennedy traveled to Cape Canaveral to be briefed on Project Apollo and evaluate the progress being made on the Saturn C-I booster rocket. The first two-stage Saturn was slated to test-launch before Christmas—the date was eventually moved to January 29, 1964—marking the inaugural time the upper stage would be tested. Walking around under the behemoth, Ray-Ban sunglasses covering his eyes, head tilted upward in astonishment at the sheer height and girth of the engineering marvel, JFK beamed with pride. The Saturn C-I would soon carry the biggest payload that any nation had ever launched into space, breaking the record held by *Sputnik 7* and *8*. Robert Seamans, the deputy administrator of NASA, who was with the president on the inspection tour, felt that "for the first time [Kennedy] began to realize the dimensions of these projects."

While Kennedy chatted with the ebullient von Braun, who was also at Florida's Space Coast that day, trading jokes and gossip, the president felt his spirits soar. They had come a long way since 1953, when they first met in New York City for *Time* magazine's "Man of the Year" television show, to making space travel the talk of America.

Furthermore, Kennedy was pleased that von Braun wasn't tolerating Jim Crow segregation or white supremacist banter at the Marshall Flight Space Center in Alabama; if only his friend Senator George Smathers of Florida would be similarly enlightened on civil rights.

The highlight of Kennedy's Cape Canaveral visit was taking a helicopter ride with Gus Grissom and Gordon Cooper around the Atlantic beachfront and mangrove thickets along the Banana River and the Intracoastal Waterway to see the new launch complex. Goofing around with the astronauts, clowning as if they were school chums, the president de-stressed from the pressures of official Washington. Jokes were made about the nightlife of Cocoa Beach and the number of contractors getting rich from the NASA space program. The final frontier, Kennedy enthused, truly was outer space, for, as Arthur C. Clarke wrote, NASA was building "the myths of the future at Cape Canaveral." The easy camaraderie JFK had with the Mercury Seven, whom he honored with the Collier Trophy that October for aeronautical and aerospace excellence at a White House Rose Garden ceremony, felt uncontrived. He and his "knights of space" were tied together by the going-to-the-moon sweepstakes, which was anchored around American innovation audacity and a fierce determination to win the Cold War rivalry against the Soviet Union. From Kennedy's aerial perspective that day, Cape Canaveral had grown into a teeming technological campus invested in the future. "We gave him a first-class bird's-eye view of the new Moonport," Cooper recalled, "where one day in the not-so-distant future a Saturn would sit with a manned Apollo space craft atop it."

As Kennedy climbed the stairs of Air Force One, ready to fly south to Palm Beach, he suddenly reversed course, trotting back down to the tarmac to have a private word with Seamans. "Now, be sure that the press really understands [Saturn]," the president told the NASA leader. "I wish you'd get on the press plane that we have down here and tell the reporters there about payload."

When JFK arrived at Palm Beach that evening for dinner with his ailing father, who had suffered a stroke two years before, he was filled with excitement about all things Apollo. As if brainwashed by NASA's public affairs office and von Braun, the moonstruck president wanted to talk about little else besides space. No longer was he a bit worried about the United States' being behind the Soviets in payload-lifting capacity. The president's visit to Cape Canaveral had reinforced his faith in exactly what a gargantuan, well-funded, centralized federal government project could achieve in breakneck time. "Learning as he stood before the Saturn 1 booster on November 16 that the United States was about to take the lead in lift capability," the historian John Logsdon explained in a 2011 article in the *Space Review*, "seems to have convinced the president that the space program was on a positive path."

After his Florida visit, Kennedy headed to Texas for a two-day, five-city, pro-space-image-building swing. Many pundits believed that the post–Civil War Democratic "solid South" might turn against the president in 1964 because of his Justice Department's fulsome embrace of the nonviolent civil rights movement led by Martin Luther King Jr. At scheduled appearances in San Antonio, Houston, Fort Worth, Dallas, and Austin, JFK planned to preach the gospel of Apollo, which was popular with conservative and moderate Democrats, while avoiding the minefield of civil rights liberalism. His Texas stops would not only mark the first "pre-campaign" trip for the Kennedy '64 reelection effort, but also offer a welcome change of scenery for both Kennedys. Jackie Kennedy had given birth prematurely that August. The baby, named Patrick Bouvier Kennedy, died of respiratory problems less than two days later. The First Lady, devastated by the death, had spent most of the ensuing months in seclusion, mothering Caroline and John-John, nearly six and three, respectively. The Texas trip was her grand reentry into public life, her time to reconnect romantically with her husband, to recapture the glow of their genuine love story.

On November 21, Kennedy spoke at the Brooks Air Force Base for the dedication of the Aerospace Medical Health Center, in San Antonio. The complex of buildings was erected not far from where celebrated aviators Charles Lindbergh and Claire Chennault learned to master the skies while training at Kelly Field and Randolph Field. With Vice President Lyndon B. Johnson,

Governor John Connally, and Senator Ralph Yarborough in the welcoming party, the president praised the cutting-edge medical innovations being pioneered at the well-funded air force facility. "Many Americans make the mistake of assuming that space research has no value here on earth," he said. "Nothing could be further from the truth. Just as the wartime development of radar gave us the transistor, and all that it made possible, so research in space medicine holds the promise of substantial benefit for those of us who are earthbound." The term *space medicine* had been coined in 1948 by Dr. Hubertus Strughold, a former Nazi physician who immigrated to the United States after World War II as part of Operation Paperclip. Mainly working out of Randolph Air Force Base, Texas, Strughold believed that space exploration would revolutionize health care in America.

What Kennedy understood was that astronauts weren't the only ones helped by space medicine research—*everybody* was. From Mercury through Gemini to Apollo, space medicine contributed to radiation therapy for the treatment of cancer; foldable walkers (constructed from lightweight metal developed by NASA); personal alert systems (devices worn by individuals who might require immediate emergency medicine or safety assistance); CAT and MRI scans (devices used by hospitals to look inside the human body, first developed by NASA to take better pictures of the moon); muscle-stimulant devices (to prevent muscle atrophy in paralyzed patients); advanced types of kidney dialysis machines; and dozens of other protocols and innovations.

Full of marvel, Kennedy launched into a succinct dissertation on how medical space research would lead to better overall health care in America. He drove home the salient point that the auxiliary technological benefits of the $25 billion NASA moonshot were profound and diverse. “Examinations of the astronauts’ physical, and mental, and emotional reactions can teach us more about the differences between normal and abnormal, about the causes and effects of disorientation, about changes in metabolism which could result in extending the life span,” the president boasted. “When you study the effects on our astronauts of exhaust gases which can contaminate their environment, and you seek ways to alter these gases so as to reduce their toxicity, you are working on problems similar to those in our great urban centers[,] which themselves are being corrupted by gases and which must be clear.”

Whatever new medical devices were commissioned by NASA and the air force to monitor an astronaut’s heart and brain wave activity, Kennedy said, would soon find application in general hospitals; they absolutely did. The implantable heart defibrillator, a tool to constantly monitor heartbeats, was developed by NASA and could deliver a shock to restore heartbeat regularity. Likewise, ophthalmologists would be able to help patients with eye defects using new laser-light protocols developed for space travel. Even simple *everyday* health-related objects (such as the special foam used for cushioning astronauts during liftoff, which was starting to be used in pillows and mattresses at hospitals to help prevent ulcers, relieve pressure, and ward off insomnia) were already benefiting patients. "This space effort must go on," JFK exhorted. "The conquest of space must and will go ahead. That much we know. That much we can say with confidence and conviction."

Speaking about the "new frontier of outer space," Kennedy ended his San Antonio oration by linking space, medicine, and the moonshot with an anecdote by Frank O'Connor, the Irish author of the recent memoir *An Only Child.* With great relish, the president recounted that O'Connor wrote about "how, as a boy, he and his friends would wander the countryside, and when they came to an orchard wall that seemed too high and too doubtful to try and too difficult to permit their voyage to continue, they took off their hats and tossed them over the wall—and then they had no choice but to follow them. This Nation has tossed its cap over the wall of space, and we have no choice but to follow it," Kennedy said. "Whatever the difficulties, they will be overcome."

At JFK's side in San Antonio was Gordon Cooper, perhaps the most popular of the Mercury Seven, after John Glenn. Children pined for the astronaut's autograph. Local medical professionals clambered for a photo. In front of reporters, Kennedy and Cooper walked together, shaking hands hard and fast. At one point they joked about who had a darker Florida tan. "We were at Brooks Air Force Base in San Antonio and the president came over and asked me if I could go to Dallas with him the next day," Cooper recalled. "He said he could use a 'space hero' with him on the trip. I couldn't make the trip because some important systems tests were scheduled at the Cape for the next day: November 22, 1963."

From San Antonio that afternoon, the Kennedys flew to Houston, where more than ten thousand people greeted them at the airport. Thirty thousand more lined the route down Main Street to Texas Avenue, just to glimpse the effervescent First Couple waving from an open car. Upon the president and Mrs. Kennedy's arrival at the downtown Rice Hotel, thousands of well-wishers mobbed them at the roped-off lobby entranceway. Bringing the Manned Spacecraft Center to Houston had made both Kennedys beloved in

Space City, U.S.A. The hotel suite reserved for them had been specially decorated with masterworks of art loaned by local collectors, and their refrigerator stocked with caviar, champagne, and Heineken (the president's favorite beer). This wasn't the president's first time in this hotel. Two of the great speeches of his career (his defense of Catholicism to the Greater Houston Ministerial Association in 1960 and the "We choose to go to the moon" oration at Rice University in 1962) were delivered after first checking in to the Rice Hotel.

Bill Kilgarlin, the chairman of the Democratic Party in Harris County, Texas, expressed his hope to the press that *all* Houstonians abandon partisan political differences to embrace the First Couple with the utmost respect. This sentiment was echoed by the county's Republican chairman, George H. W. Bush, himself a future U.S. president. Bush asked Houstonians, specifically Republicans, to extend a "warm and cordial welcome" to JFK for bringing the Manned Spacecraft Center to Houston, and he warned against protests. "I would strongly condemn this," Bush said. "It would be a disgrace to the President and the high office he holds. There may be some nuts around who might do something, but they won't be Republicans."

In the early evening, the Kennedys, followed by a pack of politicians, aides, and journalists, spoke at a

League of United Latin American Citizens banquet. No sooner had they entered the Rice Hotel's ballroom than a standing ovation erupted. The gala, carried live on local Houston television, culminated in the First Lady addressing the largely Hispanic crowd in Spanish to hearty applause and shouts of "Olé!" as a mariachi band played up-tempo Mexican folk songs such as "Volver, Volver" to close the event.

Later that same night, the Kennedys were the special guests of honor at a testimonial dinner honoring Congressman Albert Thomas, held at the Houston Coliseum. Thomas's support of the moonshot had been crucial to the president's selling NASA's lunar ambitions on Capitol Hill and jump-starting Houston's current economic boom. Insisting that Apollo was on track, that the moon was within reach, JFK praised Thomas for having lassoed NASA's Manned Spacecraft Center for Houston.

Kennedy and Thomas were right to celebrate. Starting in 1961, Houston had become synonymous with the very term *progress*: plans were now under way to build an Astrodome (the world's first air-conditioned stadium) and the Galleria (a modern indoor shopping mall with an ice rink in the center). Sleek new "space age" edifices such as the Humble Oil Building were erected. Almost overnight, the NASA connection had lured corporations such as IBM and Bellcom to Houston to participate in the Apollo challenge. These Fortune 500 giants opened up Houston facilities, which created more jobs and generated more revenue to employees. By the time the Manned Spacecraft Center opened in Houston in 1962, more than 125 space-oriented firms and corporations had created offices in the Clear Lake area, including Honeywell, North American Aviation, General Electric, Lockheed Electronics Company, Sperry Rand, and Texas Instruments. As for the law firm orchestrating the construction of the Manned Spacecraft Center, it was—no surprise—Brown and Root, the big project builders synonymous with Albert Thomas.

When Kennedy went off script to praise the city's iconic role in space exploration, he tripped over one of his words. Referring to Thomas's tireless efforts to bring NASA to Houston, he said, "He has helped steer this country to its present position of eminence in space. Next month when the United States of America fires . . . the largest payroll—*payload*—into space giving us the lead." As he recovered in the midst of this awkward sentence, the president instantly picked up on his own flub and gave the audience a dose of humor: "It will be the largest payroll, too!" he exclaimed. "And who should know that better than Houston. We put a little of it right here."

There was no question that Texas was one of the big beneficiaries of the administration's space program. The fingerprints of Lyndon Johnson were all over the gains for the state. Twenty-three sites had been considered by NASA, but Houston won out. Florida, to many objective space beat reporters, was the obvious place to have based the Manned Spacecraft Center. But Johnson—with help from Southern congressmen Overton Brooks, Olin Teague, and especially Albert Thomas still chairing the House Appropriations Committee—had executed an end run and scored big. "He [Kennedy] and I had a big argument about it, big fight," Senator George Smathers of Florida recalled. "Johnson tried to act like he didn't know. . . . It never made sense to have a big operation at Cape Canaveral and another big operation in Texas. But that's what we got, and we got that because Kennedy allowed Johnson to become the theoretical head of the space program."

After the Houston Coliseum event, a limousine took the Kennedys to the airport, for a 10:30 p.m. departure to Fort Worth. The long hours they had spent in San Antonio and Houston were memorable for the sincere enthusiasm they engendered from the public. The president's full-bore commitment to Projects Gemini and Apollo was evoked at every event, as though he were pointedly reclaiming the New Frontier goal of manned flight to the moon and proudly giving it his signature once again. Unbeknownst to reporters, in his alligator-hide briefcase Kennedy carried a speech draft that, after some tweaking, he planned to deliver to the Dallas Citizens Council the next afternoon, at the city's Trade Mart. It was anchored in large part around the New Frontier's uncompromising belief in American space supremacy:

> We have regained the initiative in the exploration of outer space, making an annual effort greater than the combined total of all space activities undertaken during the fifties, launching more than 130 vehicles into earth orbit, putting into actual operation valuable weather and communications satellites, and making it clear to all that the United States of America has no intention of finishing second in space.
>
> This effort is expensive—but it pays its own way, for freedom and for America. For there is no longer any fear in the free world that a Communist lead in space will become a permanent assertion of supremacy and the basis of military superiority. There is no longer any doubt about the strength and skill of American science, American industry, American education, and the American free enterprise system. In short, our national space effort

represents a great gain in, and a great resource of, our national strength.

The Kennedys spent a comfortable night at the Hotel Texas in Fort Worth. On the morning of November 22, over coffee, the president read a *Houston Chronicle* column joking that he might be met with violent protest in Dallas if he dared speak about anything except the joys of nautical sailing. Dallas, unlike Houston and San Antonio, didn't benefit from NASA contracts; JFK was rather unpopular with archconservatives and libertarians there. "There will sure as shootin' be some who leave and let go with a broadside of grapeshot in the presidential rigging," the column by Saul Friedman threatened. In disgust, Kennedy threw the newspaper aside. Unlike San Antonio and Houston, with their many throngs of enthusiastic Kennedy supporters, Dallas was deemed by Friedman "a mecca for those who see all sorts of evils, conspiracies, and acts of treason in the federal government—and especially the administration of President John Kennedy." Playing off this disturbing column, in defiance, the president added a line to his scheduled Trade Mart speech, shaming voices "in the land—voices preaching doctrines wholly unrelated to reality." He added his hope that "fewer people" would listen to the John

Birch Society–type paranoia and instead unify around manned spaceflight.

After breakfast with the local Chamber of Commerce, in a gesture toward what looked good on TV, Jack and Jacqueline Kennedy flew from Fort Worth's Carswell Air Force Base to Dallas's Love Field, even though it was only a thirteen-minute flight. The president, full of enthusiasm for NASA, sat on the plane with Representative Olin Teague. "He wanted to go to the Cape for the Saturn launch in [January]," Teague recalled of their in-flight conversation. "He thought the space program needed a boost and he wanted to help." On a more somber note, Kennedy told Teague he was planning to meet in Dallas, for the first time ever, the family of Wilford "Bud" Willy, the navy aviator who, with Joe Kennedy Jr., was killed in the air during World War II trying to destroy a Nazi missile compound in France. Sadly, Kennedy never had the chance to commiserate with the Willy family or evoke space exploration at the Trade Mart or watch the Saturn 1B launch from Cape Canaveral in January 1964.

On November 22 at 11:55 a.m. CST, Jack and Jackie's motorcade left Love Field in Dallas for a ten-mile trip through downtown in a convertible with the top down. The First Lady was ensconced at her husband's left in the third-row seats in the presidential limo with

Texas governor John Connally and his wife, Nellie, seated in front of them. Vice President Johnson and his wife, Lady Bird, followed in another vehicle. If Gordon Cooper had taken the president up on his offer in San Antonio, he would have been riding with them on that fateful day. Thirty-five minutes later, three shots were fired as JFK's convertible passed the Texas School Book Depository, across from Dealey Plaza in downtown Dallas. Bullets entered Kennedy's neck and head as he collapsed toward his wife. The governor was shot in the back but survived. At one in the afternoon, Kennedy was declared dead at Parkland Memorial Hospital. His lifeless body was rushed to Love Field and carried onto JFK's beloved Air Force One. Before the plane left for Washington, DC, Lyndon Johnson took the oath of the U.S. presidency, administered by U.S. district judge Sarah Hughes. "That was a bad day, I'll tell you," Bob Gilruth of NASA recalled about the somber mood in Houston. "We cried. A lot of us stood in front of the television there and cried."

John Glenn was driving home from an exercise at Texas's Ellington Air Force Base when he heard about Kennedy's shocking death on the radio. His wife, Annie, at a department store in Houston, soon rendezvoused with her husband in tears. The Kennedys had become like family to them. "I called Bobby and Ethel [Kennedy] that night," Glenn recalled, "and later represented the astronauts at the President's funeral. In the days that followed, as the initial shock and grief receded, Annie and I sat back as we had after Pearl Harbor and assessed our responsibilities to each other and to the country, and what we might do. It was a time for soul searching."

Just three months after Kennedy's death, John Glenn retired from NASA, announcing that he would run for a U.S. Senate seat in Ohio. "I always believed that serving in high public office and having the opportunity to help determine the future of the country was one of the greatest positions that anyone could aspire to," Glenn explained. "Now, with JFK's assassination, it was more important than ever before for good people to enter public life." While Glenn lost in 1964, he eventually ended up serving in the Senate from 1974 to 1999, and even ran for U.S. president himself, in 1984.

Robert Kennedy continued to push for Apollo following his brother's death. Instead of sending soldiers to die in Vietnam, he believed, America needed to launch astronauts to the moon. It would be hard to exaggerate how personally close Bobby became with John Glenn during the Johnson years. Together with their wives they rafted down the Colorado River and cruised Florida waters. "We loved John because he was just a natural," Ethel Kennedy recalled. "There was nothing phony baloney about him." When Bobby announced he was running for president in early 1968, Glenn became one of his top campaign surrogates. When RFK was assassinated on June 5, 1968, in Los Angeles, Ethel, who had been at her slain husband's side, telephoned Glenn back East, asking him to rush to McLean, Virginia, to help calm and comfort her children then living at Hickory Hill. "When we lost Bobby, John immediately went to the house to be a force of light for the children," she recalled. "They all loved him. He stayed with them for days."

Over the decades following RFK's death, Glenn kept in close contact with NASA, and in 1998, at age seventy-seven, he got his wish to go back into space. Though it had been thirty-five years since *Friendship 7*, Glenn was chosen to be part of the crew of the space shuttle *Discovery*. His participation allowed NASA scientists to study the impact of space travel on senior citizens. "Even on my *Discovery* voyage," Glenn confessed, "I thought of myself as an ancient member of Kennedy's space corps."

When Alan Shepard learned of Kennedy's death, his devil-may-care attitude vanished. His first dejected instinct was to cancel all appointments and watch the TV coverage at his Houston home with his wife, Louise. That evening, he was slated to attend dinner at the

Vanderhoefs' mansion in the upscale River Oaks part of town but canceled. The Houston socialite Peter Vanderhoef recalled that his friend Alan was "too shook up to eat." Shepard had advertised Kennedy as a fellow "space cadet"; now he was gone. "He was devastated by Kennedy's death," Shepard's biographer Neal Thompson wrote. "And more than a little worried. *What will this do to the space program*?" In the immediate years following Kennedy's death, Shepard tried to earn an assignment for an Apollo moon mission, even though his health (Ménière's disease) prohibited it. His persistence, in the end, paid off. In 1969 he underwent a new surgical technique for Ménière's, and it worked miracles; NASA put him back in the mix. From January 31 to February 9, 1971, he commanded *Apollo 14*, NASA's third successful lunar landing mission. His golfing on the moon was broadcast live on color television. At age forty-seven, he remains the oldest man to have walked on the moon.

Scott Carpenter moped around "in a pall of grief" after hearing about the Dallas killing. Like Glenn and Shepard, the fourth Mercury astronaut in space grieved with his wife, Patty. "Before the assassination and before the war escalated in Vietnam a two-term Kennedy administration was a safe political assumption," Carpenter recalled in his memoir. "It was logical to assume that NASA would get the country to the moon and back and even that John [Glenn] might have a shot at going. These hopes and assumptions presented among other things a comfortable family timetable for the Glenns. John could take part in the lunar expeditions and then turn to politics, a natural move for the Ohio-born Democrat—a move urged on him by two Kennedys, John and Robert. The man was, after all, the most popular and recognized man on the planet—a priceless asset for the Democratic administration. . . . The assassination changed everything."

In Moscow that evening, Nikita Khrushchev had just finished dinner when his telephone rang with the message that Kennedy had been shot. The Soviet premier turned ashen and crestfallen. When a short while later Foreign Minister Gromyko called back to offer details of the assassination in Dallas, he openly wept. He felt shortchanged without exactly knowing why. The KGB told Khrushchev it believed right-wing extremists, angry at the prospect of U.S.-Soviet détente, were responsible for the murder. Around that time, Robert Kennedy had been covertly working to set up a Kennedy-Khrushchev summit in coming months. The hope was that a NATO–Warsaw Pact nonaggression treaty could be brokered and, perhaps, joint efforts in outer space agreed upon. Khrushchev told his son that he was willing to take risks with President Kennedy, but with Lyndon Johnson as president, "everything will be different."

Neil Armstrong was in a sports car on Interstate 10, en route from Pensacola to Houston, when he heard, also on the radio, that Kennedy had died. Crushed by the news, he went blank with sorrow. Although he had never met the slain president, his life since Glenn's orbit was dedicated to fulfilling Kennedy's moonshot challenge of May 25, 1961. In the coming years, as Armstrong trained for his moon mission, first going into space on *Gemini VIII* in 1965 and then famously on his Apollo 11 mission with Buzz Aldrin and Michael Collins in 1969, he embraced Kennedy as a true visionary leader. Armstrong, who was reluctant to give interviews, revealed years later that Kennedy's moonshot, in his opinion, had only a "50-50 chance" of an astronaut landing safely on the lunar surface; nevertheless, he was 100 percent into the challenge whatever the odds. Even though Armstrong was nominally a Republican, he never forgot Kennedy's leadership in space, and praised the thirty-fifth president whenever asked, until his own death in 2012.

What Kennedy had miraculously done was bring together Americans on the political right and left in a collective we're-all-in-it-together endeavor of great scientific merit. Just days after her husband's assassination, Jackie Kennedy met with President Johnson in the Oval Office, along with the new First Lady, Claudia "Lady Bird" Johnson. As a widow, her face impassive, Jackie had one must-do request: renaming the site of NASA's space launches in Florida, from Cape Canaveral to Cape Kennedy. She reminded the new First Couple of her husband's May 25, 1961, pledge to Congress and his September 12, 1962, recommitment at Rice University. In a 1974 interview at the John F. Kennedy Presidential Library and Museum in Boston, she recalled telling the Johnsons that afternoon that her primary post-Dallas worry was that American citizens would forget about the moon pledge. "I kept thinking, that's going to be forgotten, and his dreams are going to be forgotten, and I thought, 'Well, maybe they'll remember someday that this man did dream that.'"

There was no need for her to worry. On November 28, President Johnson fulfilled Jacqueline Kennedy's wish at the end of his Thanksgiving message to the nation. Via Executive Order 11129, NASA's Civilian Launch Operation Center on the Florida coast, he said, would be renamed the John F. Kennedy Space Center. Taking the realization even a step further, Johnson ordered that the military-run Cape Canaveral station (station number one of the Atlantic Missile Range) also be part of the new Kennedy Space Center. To help differentiate the double honor, the Air Force renamed the military launch site Cape Kennedy Air Force Station. These were the first official actions taken by the U.S. government to permanently honor John F. Kennedy. All the Mercury and Gemini astronauts were thrilled by the designation, but privately, they worried that Johnson wouldn't be as fully supportive of American moonshots, that unbearably Congress would direct Apollo funding elsewhere. "The loss of John Kennedy had been an incalculable one to the space program," Gordon Cooper believed. "While Lyndon Johnson assured everyone that he was equally supportive, we knew he didn't have the total commitment that JFK had."

When Wernher von Braun first learned of Kennedy's assassination, he was devastated. Like the other NASA mainstays, he stayed glued to CBS News that grim weekend, listening to his friend Walter Cronkite extrapolate on the Dallas tragedy and basking in a slow-burning rage. As fate would have it, von Braun had been scheduled to dine with President Kennedy at the White House three days later, on November 25 with his wife, Maria. Instead, November 25 was the day of the president's funeral, and von Braun put an *X* on his calendar. In his daily journal of December 9, 1961 (kept by his executive assistant, Bonnie Holmes), he said that the "most important thing" for Kennedy's moonshot to be achieved was to "make sure Mr. Webb holds the line." With Kennedy now gone, Webb and von Braun, sometimes at odds with each other, forged an iron knot in the heart of NASA. "What a waste," von Braun told Holmes of JFK's death. "What a tragic loss of a friend and a great leader." According to Holmes, it was the "one time I ever saw him actually cry. He was very moved."

Refusing to be beaten down or derailed, von Braun continued to work on his Saturn V rocket, burning the midnight oil to keep up the pace, using Kennedy's death as a spur to push even harder at the Marshall Space Flight Center to fulfill the moonshot dream. The most meaningful way to honor the martyred president, he realized, was to keep Project Apollo on track at his "Huntsville School." The other honorifics were secondary to the moonshot itself.

Shortly after president Kennedy was buried at Arlington National Cemetery, on November 25, 1963, the space program achieved a major milestone: orbiting its first hydrogen-fueled rocket. The Atlas-Centaur rocket (both stages designed by General Dynamics Astronautics) brought America that much closer to the moon. This rocket had been assigned by Kennedy and Webb the important task of eventually "soft landing" an unmanned surveyor craft on the moon to record what the surface was like. While Kennedy was being eulogized for being the youngest individual and first Catholic elected to the American presidency, molding a sweeping civil rights bill, launching the Peace Corps, and negotiating the Nuclear Test Ban Treaty during the most tension-filled years of the Cold War, his championing of NASA and his dream of going to the moon were treated as the most magic moments.

At the time of Kennedy's death, the public most closely associated NASA with the manned spacecraft programs: Project Mercury (1958–63), which tested the ability of one man in up to several hours in Earth's orbit; Project Gemini (1962–66), in which two-man crews in one spacecraft were assigned a variety of tasks, including rendezvous and docking in Earth's orbit with a target vehicle and moving around outside the spacecraft itself; and Project Apollo (1960–72), in which three-man crews were set on progressively more ambitious missions, culminating in lunar landings. Although the road was bumpy, the American public had collectively decided to honor the martyred president by funding Apollo throughout the 1960s and early 1970s. Just as Kennedy had instructed, NASA astronauts were training to reach the moon by the end of the decade.

The Apollo 11 rocket on its mobile launch platform just after rollout from the Vehicle Assembly Building at the Kennedy Space Center, Cape Canaveral, Florida, on its way to Launch Complex 39A on May 17, 1969. Apollo 11, the first manned lunar landing mission, was launched on July 16, 1969, with astronauts Neil Armstrong, Edwin "Buzz" Aldrin, and Michael Collins on board.

CBS News anchor Walter Cronkite holds up the *New York Times* during the Apollo 11 telecast, July 20, 1969, at 10:11 p.m.

Epilogue: The Triumph of Apollo 11

"The *Eagle* has landed!"

NEIL ARMSTRONG, JULY 20, 1969

For the next four years after Dallas, President Johnson and NASA administrator James Webb defended Kennedy's Apollo moonshot pledge even as the Vietnam War and great social programs such as Medicare and Medicaid vied for funding and attention. There was no learning curve necessary for Lyndon Johnson when it came to prioritizing the American moonshot as part of the Great Society. NASA during the LBJ-Gemini years tested von Braun's first Saturn 1B launch vehicle and then his immense Saturn V, the launch vehicle for lunar missions. Johnson's attitude was perfectly summed up in his Atlantic University dedication speech in Boca Raton, Florida. "We cannot be the first on earth," he said on October 25, 1964, "and second in space." Kennedy's pledge of putting a NASA astronaut on the moon remained a national mandate even as a counterculture youth revolution rocked the nation. Just nine days after his Atlantic University speech Johnson defeated Barry Goldwater in the U.S. presidential election. In space-budget terms NASA (pro-Johnson) had defeated the air force (pro-Goldwater).

From 1964 to 1969, whenever Congress considered gutting the Apollo programs, Johnson evoked the martyred JFK with don't-you-dare political mastery. In an ironic way, Kennedy's death guaranteed Apollo lunar landing appropriations; if he had lived, Congress might have forfeited the budget, deeming the "end of the decade" schedule for the moonshot impossible. Instead, NASA continued to employ thirty-six thousand people, hired four hundred thousand contractors, and operated facilities worth $3.65 billion whose chief purpose was achieving Kennedy's moonshot.

President Johnson, working closely with Webb, ably oversaw the Gemini missions, which ended in November 1966. This second phase of Kennedy's going-to-the-moon pledge was a success with regard to its primary goal of conducting manned spaceflight on a

On March 23, 1965, President Lyndon Johnson watches the launch of Gemini 3 space capsule on television in his office at the White House. A framed portrait of John F. Kennedy sits atop the television set.

routine basis. Sixteen different astronauts flew in Project Gemini and spent a collective 1,940 man-hours in space. LBJ publicly expressed his pride in fulfilling JFK's pledge not to throw in the towel on the promise of lunar exploration, following the safe return of *Gemini 12* to Earth. "Ten times in this program of the last twenty months we have placed two men in orbit about the earth in the world's most advanced manned spacecraft," the president said on November 15, 1966, in a written statement. "Ten times we have brought them home. Today's flight was the culmination of a great team effort, stretching back to 1961, and directly involving more than 25,000 people in the National Aeronautics and Space Administration, the Department of Defense, and other government agencies; in the universities and other research centers; and in American industry. Early in 1962, John Glenn made his historic orbital flight and America was in space. Now, nearly five years later, we have completed Gemini and we know that America is in space to stay."

All ten of the manned Gemini missions between 1965 and 1966 took place after Kennedy's death, with Johnson as president. The New Frontier technocrats at NASA were now part of the Great Society. Much of the core leadership group at NASA stayed the same; Robert Gilruth, Hugh Dryden, and Robert Seamans, for example, continued to push the moonshot forward. The indomitable James Webb, rambling on without embarrassment about every aspect of NASA, continued to prod Project Apollo forward during LBJ's presidency. Every day, his steel-trap mind thought about Kennedy, whose picture adorned his office wall.

For Webb the "Kennedy effect" during the Johnson years was the belief that if the United States accomplished the moonshot, then the nation could "do something for grandma with medicine" as well. Only after

January 27, 1967, when a Cape Canaveral launch accident on *Apollo 1* killed astronauts Gus Grissom, Roger Chaffee, and Ed White in a fire that engulfed the cockpit during a training exercise, did Webb get pressure to resign. "We've always known that something like this was going to happen sooner or later," he told the press after the *Apollo 1* debacle. "Who would have thought that the first tragedy would be on the ground?"

Newspaper stories about the *Apollo 1* disaster mentioned how Grissom was an integral part of Kennedy's space corps. Now that Grissom was dead, his *Liberty Bell 7* mission in the summer of 1961 was treated in a more heroic fashion than ever before by a grieving public. Nobody bemoaned the lost Mercury capsule of yesteryear (in 1999 the *Liberty Bell 7* was hauled from the Atlantic Ocean floor at a depth of more than 15,000 feet, courtesy of modern tracking equipment). Just a few days before Grissom died, he completed a first draft of a memoir called *Gemini: A Personal Account of Man's Venture into Space*. In it, he wrote a line that JFK would have treasured: "The conquest of space is worth the risk of life." Grissom was buried with full military honors at Arlington National Cemetery, not far from the eternal flame on President Kennedy's grave.

By October 1968 President Johnson had grown weary of Webb, especially his over-the-top defense of

NASA protocol after the ghastly *Apollo 1* fire. Webb was in the White House with LBJ and opined that his sixty-second birthday was near and that he might retire at that time. Seizing the opportunity to get rid of Webb, LBJ immediately accepted this musing as a firm resignation. And the president insisted that they go to the Press Room and announce it before Webb left the building. A resentful Webb later said he never even talked with his wife about quitting NASA, and it was a real shock. As if to compensate for the rude act, President Johnson presented Webb with the Presidential Medal of Freedom.

History has treated Webb well. When the USSR collapsed in 1991, the declassification of a cache of top secret Kremlin documents proved that the Soviets had *never* stopped plotting their own moonshot (as he had argued). Some NASA cynics mocked Webb during the Johnson years for insisting that the Soviets hadn't thrown in the towel, that the lunar race was real. Webb was right. Only in late 1967, when the Soviets' N1 moon rocket blew up on the pad, killing Russian engineers and designers, did the Kremlin stop competing. And Webb's tireless work ethic and beat-the-Russians drive wasn't forgotten within NASA culture. When the Manned Spacecraft Center in Houston became the Lyndon B. Johnson Space Center in 1973, visitors were greeted by Webb's words displayed on a giant wall: THE WORLD OF SPACE HOLDS VAST PROMISE FOR THE SERVICE OF MAN, AND IT IS A WORLD WE HAVE ONLY BEGUN TO EXPLORE. Webb died that very year and was buried at Arlington National Cemetery, not far from President Kennedy and Gus Grissom.

Former president Dwight Eisenhower continued to squawk that fulfilling Kennedy's moonshot pledge wasn't wise. Always seeking to hold down expenditures, and insisting that to race the Soviets in space was foolhardy, in 1965 he carped to Apollo astronaut Frank Borman that Kennedy's moonshot pledge had been "drastically revised and expanded just after the Bay of Pigs fiasco. . . . It immediately took one single project or experiment out of a thoroughly planned and continuing program involving communication, meteorology, reconnaissance, and future military and scientific benefits and gave the highest priority—unfortunate in my opinion—to a race, in other words, a stunt." Eisenhower died in Washington, DC, on March 28, 1969, of congestive heart failure, never learning that Kennedy's "stunt" became a grand historic reality when Neil Armstrong and Buzz Aldrin walked on the moon that summer.

It wasn't, however, just well-known Republicans like ex-president Dwight Eisenhower and Senator Barry

Goldwater (brigadier general in the air force reserve) who wanted to slash the Project Apollo budget after Kennedy's death. Senator J. William Fulbright of Arkansas, a Democrat, ran a campaign to cut the manned lunar expedition by 10 percent from the 1965 budget. His wholehearted proposal came within four Senate votes of passing, proof that a significant segment of Americans wondered whether the expensive drive to put an astronaut on the moon still made sense. It was noticeable that in states where NASA was spending big dollars (e.g., California, Texas, Florida, Alabama, Mississippi), lawmakers were more pro-moonshot than in places such as Arkansas and Nebraska, which had no particularly large share of the business enterprise pie.

Just five months after Kennedy's death, the New York World's Fair opened in lasting tribute to JFK. Early in his presidency, Kennedy had signed the bill authorizing $17 million for the federal pavilion at the fair whose theme was "Man's Achievement in an Expanding Universe." With Kennedy's moonshot vision fresh on people's minds, the most popular displays at the World's Fair were General Motors' Futurama; New York State's Tent of Tomorrow; Westinghouse's Time Capsule; and the Transportation and Travel Pavilion (with an elaborate tribute to Project Apollo, including a simulated moonscape)—all, in a sense, an homage to JFK's Space Age imagination. Also prominently displayed at the fair was astronaut Scott Carpenter's spacecraft *Aurora 7*, a relic that reminded visitors of JFK's New Frontier heyday. Then there were the Lunar Fountain, the Fountain of Planets, and the Unisphere (an imposing hollow steel globe with longitude and latitude lines circled by three "orbits" symbolic of NASA space satellites) at the Queens, New York, fairgrounds. Clearly, Kennedy had penetrated the psyche of America with his New Frontier technology challenge.

On January 29, 1964, the very Saturn I rocket Kennedy had inspected that past November was successfully launched from Cape Kennedy. Without a hitch, it sent nineteen tons into orbit during this test flight. On a corner of the massive rocket, when guards weren't watching, von Braun had engraved the initials JFK, in honor of his one true political hero. Von Braun had waited for that Saturn success to write Jacqueline Kennedy a letter of condolence on the death of her husband and to stress his commitment to the lunar program. Until Kennedy's moonshot pledge was accomplished, von Braun would politick, coax, cajole, and maneuver Project Apollo to be funded by Congress.

February 1, 1964

Dear Mrs. Kennedy:

In our elation over the successful launch of SA-5 last Wednesday—the fifth in a successful string of launchings of Saturn I rockets, but the first capable of going into orbit—I must tell you how happy and grateful we are that this test came off so well. All of us connected with this undertaking knew only too well how eagerly the late President had been looking forward to this launching, which would at last establish the long-awaited American lead in the capability of orbiting heavy payloads. The trust he had placed in us, and his confidence that we would succeed, offered great encouragement but placed on us an even greater sense of obligation. I am enclosing a picture taken in front of the towering SA-5 rocket at Cape Kennedy on November 16th. The model at the left depicts the upper part of the rocket which is now orbiting the earth once every 94 minutes. The unit in orbit has a length of 83 feet and a weight of 37,800 lbs.

You have been overwhelmed with condolences from all over the world at the tragic death of your beloved husband. Like for so many, the sad news from Dallas was a terrible personal blow to me. We do not know a better way of honoring the late Pres-

ident than to do our very best to make his dream and determination come true that "America must learn to sail on the new ocean of space, and be in a position second to none."

With deepest sympathy—Wernher von Braun

Within a few days von Braun received a handwritten response from JFK's widow, composed on her personal stationery. It read:

February 11, 1964
Dear Dr. von Braun
I so thank you for your letter—about the Saturn—and about my husband.

What a wonderful world it was for a few years—with men like you to help realize his dreams for this country—And you with a President who admired and understood you—so that together you changed the way the world looked at America—and made us proud again.

Please do me one favor—sometimes when you are making an announcement about some spectacular new success—say something about President Kennedy and how he helped turn the tide—so people won't forget.

I hope I am not the only one to feel this way—It

is my only consolation—that at least he was given time to do some great work on this earth, which now seems such a miserable and lonely place without him.

How much more he could have done—but I must not think about that.

I do thank you for your letter.

Sincerely,
Jacqueline Kennedy

Restlessly during the five and a half years between Kennedy's assassination and Neil Armstrong's moon walk, von Braun never forgot Jacqueline Kennedy's request. More than ever, von Braun started going to church regularly (as did Jackie). At public forums, von Braun evoked Kennedy every chance he could, proud they were hitched in history as surely as Thomas Jefferson was to Lewis and Clark. Von Braun was determined to be remembered as Kennedy's space leapfrog avatar, not as the V-2 mastermind who built rockets constructed by forced labor from German concentration camps. For her fortieth birthday, in 1969, the Greek shipping magnate Aristotle Onassis gave Jackie Kennedy, his bride as of October 20, 1968, the perfect gift: a jeweled pair of Apollo 11 globe earrings in memory of Jack.

On November 9, 1967, von Braun's Saturn V moon

rocket made its first successful launch from the pad at the Kennedy Space Center. The spacecraft achieved an altitude of over 11,185 miles and a reentry speed of greater than 7 miles per second. The mission qualified both the Saturn V and the command capsule for further Earth-orbiting missions. If Kennedy had lived, he would have known on this day that the moon was truly in reach by the end of the decade. Von Braun's Huntsville team not only designed the Apollo rockets that brought Americans to the moon but, later, also oversaw the development of the space shuttle propulsion system.

NASA was lucky to have a rocket engineer as talented as von Braun to work on Apollo. But he shouldn't be remembered as an American hero. His direct role in the Nazi concentration camp labor programs, where thousands perished under inhumane conditions, makes him a pariah figure of sorts. As historian Michael J. Neufeld ably summarized in the *German Studies Review*: "Von Braun made a Faustian bargain with the German Army and National Socialist regime in order to pursue his long-term dream of exploring space, and late in World War II found out what that bargain meant. His career, however admirable in many other aspects, serves as an exemplary warning of the dangers of the amoral pursuit of science and technology in the twentieth century—and the twenty-first."

Jackie Onassis smiles at photographer while her husband, Aristotle, admires the *syrtaki* skill of some of his friends dancing (in background) at a party celebrating the former First Lady's fortieth birthday. She is shown wearing the Apollo 11–themed earrings presented to her by her husband two days before.

As the Apollo 11 launch approached in the early summer of 1969, Democrats Bill Moyers (LBJ's former press secretary) and Daniel Patrick Moynihan (future U.S. senator from New York) asked Richard Nixon, who had taken office that January as the nation's thirty-seventh president, to name the spacecraft the *John F. Kennedy*. Presidential assistant Stephen Bull wrote White House advisor H. R. Haldeman about the possibility, noting that it might be "good politics," and interpreted far and wide as an "act of graciousness." Nobody important in the Nixon White House cottoned to the idea. Bryce Harlow, Eisenhower's former White House assistant, fumed about the federal government's having "gone far enough" in "Kennedyizing" NASA space ventures. White House advisor John Ehrlichman warned that if Nixon "fell prey" to naming the rocket after JFK, "the next step will be renaming the moon" after Kennedy "because NBC thinks it would be a good idea." Haldeman decided "positively!!"—as he wrote in the margins of the Bull memo—*against* the Kennedy honorific. Not only did Richard Nixon *not* name the rocket after the man who defeated him in the 1960 presidential election, but he refused to evoke Kennedy in the days before or after the Apollo 11 lunar voyage.

In the hot summertime of 1969 NASA's three-staged Saturn V rocket blasted off from Cape Canaveral. That

July 16, the *Apollo 11* crew headed to the moon, and Kennedy's dream inched even closer to reality. Retreat wasn't an option for there was no turning back. To honor the state of Texas, the three astronauts, Buzz Aldrin, Neil Armstrong, and Michael Collins, had brought a Lone Star flag with them on the mission. Armstrong also brought along a wing fragment extracted from the Wright brothers' famous Kitty Hawk plane as a good luck charm; in doing so he forever linked NACA aviation history with NASA space exploration.

On July 20, a gangly LEM dubbed *Eagle* descended forward on a flat lunar field named Sea of Tranquility. In a moment of high tension radioed live into living rooms on Earth, the astronauts reported that their descent engines were kicking up dust. The first words spoken on the moon were "contact light" from Aldrin. This referred to an *Eagle* sensor that had lit, as anticipated, inside the lander. This was followed by Armstrong saying the iconic "Houston . . . Tranquility Base here. The *Eagle* has landed."

At the NASA Manned Spacecraft Center, all the technicians erupted in spontaneous cheers. Time itself, it seemed, had stopped the second Armstrong had uttered those unforgettable words. The score was the United States had landed, and the USSR had not—

game over. It was as if a new millennium had opened up for the world to embrace with awe and wonder.

While Armstrong and Aldrin walked the lunar surface, Collins had been left piloting the *Columbia* command module around the moon. On his silent trip along the far side, he wrote, "I am absolutely isolated from any known life. I am it. If a count were taken, the score would be three billion plus two over the other side of the moon, and one plus God knows what on this side."

When *Apollo 11* command module *Columbia* returned to Earth on July 24, 1969, having successfully completed its mission, von Braun's life's work had been accomplished and Jacqueline Kennedy's request fulfilled. Watching the televised Apollo event from their home in Cape Cod were Jack's ecstatic elderly parents. It was as if their son had won his moonshot bet with history. The 528 million moon-mad global citizens who watched the historic spectacle on TV delighted in the human achievement. It was as if America's sins in Vietnam had been forgotten for a while. The astronauts wandered only a few hundred feet from the *Eagle.* But they opened up the moon for future travelers. "This is the greatest week in the history of the world since the creation," President Nixon enthused to the astronauts with a broad grin of satisfaction. "As a result of what you've done, the world has never been closer together before."

NASA had beaten by five months President Kennedy's pledge to put a man on the moon by the decade's end. After more than eight days in space, the Apollo astronauts splashed into the Pacific. At Mission Control in Houston, a sentence from JFK's May 25, 1961, special message to Congress flashed on the large headquarters screen: "I believe that this nation should commit itself to achieving the goal, before this decade is out, of landing a man on the moon and returning him safely to the earth." An Apollo 11 logo also appeared on the NASA screen, offering the greatest honor of John F. Kennedy's public career: "Task Accomplished July 1969."

At around that time, an unknown citizen had left a lovely bouquet of flowers on Kennedy's Arlington grave with a thoughtful card that read simply: "Mr. President, the *Eagle* has landed."

The three astronauts chosen for Apollo 11, the first manned lunar landing mission. *From left to right:* Edwin “Buzz” Aldrin, born in Montclair, New Jersey, in 1930; Michael Collins, born in Rome, Italy, in 1931; and Neil Armstrong, born in Wapakoneta, Ohio, in 1930. These space heroes made President John F. Kennedy’s dream a reality.

Acknowledgments

Every fall semester at Rice University I teach two classes: "Twentieth-Century American Presidents" and "The United States in the 1960s and 1970s." My students are always pleased to learn that John F. Kennedy's famous space oration on September 12, 1962, was delivered at Rice's football stadium and that astronaut Neil Armstrong said "Houston. Tranquility Base here," on July 20, 1969, when the lunar module *Eagle* reached the moon. Local pride in all things NASA remains understandably strong in Greater Houston, and especially at Rice, where we have a first-rate Space Institute. One of my undergraduate students, an apprentice rocket scientist, Sam Zorek, helped me collect oral histories and better understand the complicated engineering aspects of space exploration. Immense thanks to Rice colleagues Erin Baezner, Beverly Konzem, Kathleen Canning, Carl Caldwell, David Ruth, Lora Wildenthal, Marie Lynn Miranda, Allan Matusow, and David Leebron. Campus historian Melissa Kean pointed me to a recently discovered batch of NASA documents at Rice's Fondren Library, which I fruitfully mined.

When lecturing at Rice on the American space program, I love to recount the parting gesture of Neil Armstrong and Edwin "Buzz" Aldrin. Just before climbing up the stairs of the *Eagle* to leave the moon, Armstrong asked Aldrin if he had deposited the NASA-sanctioned mementos they planned to leave behind. Aldrin, grateful for the reminder, reached into his shoulder packet, pulled out a package, and placed it on the lifeless lunar surface. Inside the packet were shiny medals honoring two Soviet cosmonauts: Yuri Gagarin (the first human to orbit the Earth, who had died in a 1967 MiG-15 crash) and Vladimir Komarov (killed in 1967, when his *Soyuz 1* parachute didn't open on descent from space). Also left behind by Armstrong and Aldrin was an *Apollo 1* patch commemorating Gus Grissom, Ed White, and Roger Chaffee (who had perished in the *Apollo 1* on-ground accident of 1967) and a gold olive branch pin, symbolic of the peaceful nature of Apollo 11. This NASA satchel still rests there in the lunar dust.

By honoring the deceased cosmonauts, NASA was

encouraging Soviet citizens to proudly participate in the epic American moonshot. And for good reason. Without the prod of Vostok and Voskhod scientists and engineers, Kennedy simply could not have convinced Congress to fund Mercury, Gemini, or Apollo. Between 1969 and 1972, NASA orchestrated five more Apollo missions, with a total of twelve men walking the moon's surface, repeatedly fulfilling Kennedy's dream, even collecting 840 pounds of moon rock. Kennedy's other hope, that the U.S. and Soviet Union could collaborate in space exploration together, had to wait until Richard Nixon left the White House and Gerald Ford was president. On July 15, 1975, Apollo and Soyuz spacecraft launched from Earth within hours of each other. In two days' time they docked. To all appearances, the Cold War space race between the United States and the Soviet Union had ended in partnership.

Reminiscent of NASA during its halcyon years, there was a lot of teamwork in the writing of this book, which took years of research. Most important were the archivists and staff of the John F. Kennedy Presidential Library and Museum (Boston, Massachusetts); Lyndon B. Johnson Presidential Library and Museum (Austin, Texas); John Glenn Center at Ohio State University (Columbus, Ohio); Silicon Valley Archives at Stanford University (Menlo Park, California);

National Archives (Fort Worth, Texas); the National World War II Museum (New Orleans, Louisiana); the Embry-Riddle Aeronautical University (Daytona Beach, Florida); the National Air and Space Museum (Washington, DC); the Marshall Space Flight Center (Huntsville, Alabama); the Canaveral Research Center (Cape Canaveral, Florida); and the NASA History Office (Washington, DC). The first-rate collection of NASA documents at the University of Houston–Clear Lake, in Texas, was especially helpful. Likewise, my friend Mitch Daniels of Purdue University hosted me on campus, where I studied the papers of Gus Grissom and Neil Armstrong in the Special Collections Library.

Beginning in the summer of 1962, Kennedy installed a secret taping system in the White House, presumably to aid in the writing of a White House memoir someday. Up until his death, Kennedy would tape White House conversations, including some on space policy. The Kennedy tapes, safeguarded by the National Archives, were a tremendous boon in writing this book. As were the fine oral histories and space artifacts housed in the JFK Presidential Library. The John F. Kennedy Library Foundation—especially my friend Executive Director Steven M. Rothstein—helped me better understand the intertwined relationship between JFK, computer science, and modern technology.

Throughout my writing and research I spoke about JFK's space policy with members of the president's family. Ethel Kennedy and Congressman Joe Kennedy III had invited me to lecture on the American moonshot in 2015 and 2016 at their summer compound in Hyannis Port, Massachusetts, for a gathering of distinguished guests. In 2017 I co-edited *JFK: A Vision for America* with the former president's nephew Stephen Kennedy Smith, who proofread this book. Other members of the family who helped me include Rory Kennedy, Max Kennedy, Caroline Kennedy, Robert F. Kennedy Jr., Christopher Kennedy, Jean Kennedy Smith, and Kathleen Kennedy Townsend.

Back in 1999 my book *John F. Kennedy and Europe* was published (with an introduction by the legendary speechwriter and counselor to our thirty-fifth president, Theodore Sorensen). At that time Ted suggested that a book about JFK's leadership in space policy was needed. He was right. I began conducting interviews for this book in 2001, starting with Neil Armstrong and John Glenn. Over the years I've learned much from conversations with dozens of participants in the early years of NASA. Special thanks go to Sean O'Keefe (former administrator of NASA and my Sonoma County, California, camping friend) and George Abbey (the former director of Johnson Manned Space Center and a colleague of mine at Rice) for being sounding boards. NASA historian Bruce Odum—based in Huntsville, Alabama—offered me great insight on early rocketry.

John Logsdon, professor emeritus of political science and international affairs at George Washington University's Elliott School of International Affairs, edited an early draft of this book. He was the founder in 1987 and longtime director of GW's Space Policy Institute. His *John F. Kennedy and the Race to the Moon* (2010) and *The Decision to Go to the Moon* (1970) were foundational readings. He is a true gentleman.

I'm deeply indebted to Roger D. Launius—chief NASA historian from 1990 to 2002—for twice proofreading my entire manuscript. I've never known a scholar more generous with his time. Nobody knows more about NASA history than Launius. He is a national treasure. Two of his most recent books—*The Smithsonian History of Space Expedition* (2018) and *Apollo's Legacy* (2019)—were indispensable.

Cold War historian Yanek Mieczkowski, the author of *Eisenhower's Sputnik Moment* (2013), and a first-rate academic historian, gave this manuscript a close proofread. Michael J. Neufeld, author of *Von Braun: Dreamer of Space, Engineer of War* (2007), thoroughly edited my chapters pertaining to the V-1 and V-2 on two different occasions. The National Air and Space

Museum is blessed to have such a fine intellectual as Neufeld serving as senior curator.

The honor roll of friends who helped me out in this project include Brian Lamb, Steve Scully, David Rubenstein, Jamie Kabler, Rodney Krajca, Marie Arana, Emma Juniper, Ian Frederick-Rothwell, Ben Riley, Melissa Schnitzer, Leslie Berlin, Mark Baily, Jon Meacham, Paul Hendrickson, Mark Winkleman, Walter Isaacson, Scott Hubbard, Duvall Osteen, David Gergen, Nate Brostrum, Chip Wiser, Neal Thompson, William Webster, Andrea Lewis, Ben Barnes, John Csepegi, Patt Morrison, Kyle Longley, John Lewis, Clayton Maxwell, Ted Deutch, Geoffrey Cowan, Jill Krastner, Larry Temple, Orly Jaffe, Helen Galen, Linda Forehand, Cynthia Barrett, Meredith Cullen, Irwin Gellman, James Denham, Ted Widmer, Mark Updegrove, Dennis Fabisak, Kathryn Hillhouse, Jessica Lowenthal, Luke Nichter, Kabir Sehgal, and Louis Paul.

It's been a joy working with the New-York Historical Society, established in 1804, on all things related to U.S. presidential history. The director, Louise Mirrer, is a brilliant historian and friend, who is a pleasure to collaborate with on POTUS projects. Others at NYHS who deserve thanks include Dale Gregory, Alexander Kassl, Jennifer Schantz, and Alliy Drago. For nearly two decades Washington's Speakers' Bureau has organized my public lectures. Great thanks to Bernie Swain and Harry Rhoods Jr. for finding ways for me to travel around our nation to lecture on U.S. presidential history and leadership.

I serve on the board of trustees of Brevard College in North Carolina. The president, David Joyce, arranged access for me to visit the Pisgah Astronomical Research Institute, which has a partnership with Brevard. This was a Kennedy-established NASA watch spot from which Gemini and Apollo missions could be tracked. An amazing place to look at the stars high in the Blue Ridge Mountains. It is likewise an honor to serve on the National Council for History Education's (NCHE) Board of Directors. Together we've explored new ways to teach U.S. space history in public schools.

First and foremost, *American Moonshot* is a work of U.S. presidential history (not space studies). In this tradition, I benefited mightily from the fine works of Kennedy scholars such as Robert Dallek, Chris Matthews, Steven Levingston, David Nasaw, Robert Caro, Richard Reeves, Fredrick Kempe, Andrew Cohen, Sally Bedell Smith, Martin W. Sandler, Jeff Shesol, Barbara A. Perry, Philip Nash, David Greenberg, Fredrik Logevall, Thurston Clarke, Nigel Hamilton, and Doris Kearns Goodwin.

Anybody reading my notes can easily ascertain

which books, memoirs, documents, oral histories, and articles I found most useful. Nevertheless, I'd like to give a double thumbs-up to the following space history classics: William E. Burrows, *The New Ocean* (1998); Piers Bizony, *The Man Who Ran the Moon: James E. Webb, NASA, and the Secret History of Project Apollo* (2006); Andrew Chaikin, *A Man on the Moon: The Voyages of Apollo Astronauts* (1994); Paul Dickson, *Sputnik* (2001); Francis French and Colin Burgess, *Into That Silent Sea: Trailblazers in the Space Era, 1961–1965* (2017); Monique Laney, *German Rocketeers in the Heart of Dixie: Making Sense of the Nazi Past During the Civil Rights Era* (2015); Walter A. McDougall, *. . . the Heavens and the Earth: A Political History of the Space Age* (1985); Yanek Mieczkowski, *Eisenhower's Sputnik Moment* (2013); Charles Murray and Catherine Bly Cox, *Apollo* (2004); Asif Siddiqi, *Challenge to Apollo: The Soviet Union and the Space Race, 1945–1974* (2000); William Taubman, *Khrushchev: The Man and His Era* (2003); Neil deGrasse Tyson and Avis Lang, *Accessory to War: The Unspoken Alliance Between Astrophysics and the Military* (2018); and John Noble Wilford, *We Reach the Moon* (1969). All these books are fixtures in my Austin home library.

A number of the Mercury, Gemini, and Apollo astronauts wrote autobiographies. The most useful were John Glenn with Nick Taylor, *John Glenn: A Memoir* (1999); Buzz Aldrin, *Magnificent Desolation: The Long Trip Home from the Moon* (2009); Scott Carpenter with Kris Stoever, *For Spacious Skies* (2003); and Alan Shepard and Deke Slayton, *Moon Shot: The Inside Story of America's Race to the Moon* (1994). The best biography ever written of a NASA astronaut is James R. Hansen's *First Man: The Life of Neil A. Armstrong* (2005). I'm indebted to Professor Hansen for proofreading an early draft of this book.

It is my personal opinion, based on all that I've read, that Wernher von Braun was culpable for war crimes associated with the German Third Reich, using slave labor to build his V-2s during World War II. Too many studies of von Braun try to sugarcoat his questionable Nazi past. While von Braun should be studied and honored within the guided corridors of engineering and space exploration, he should not be treated as a sustainable twentieth-century American hero. I'm grateful that von Braun scholars Dr. Michael J. Neufeld of the National Air and Space Museum and Monique Laney of Auburn University have helped expose the dark side of this enigmatic personality in their first-rate books.

Diane McWhorter, the author of the Pulitzer Prize–winning *Carry Me Home: Birmingham, Alabama:*

The Climactic Battle of the Civil Rights Revolution, proofread the manuscript with laserlike eyes. She is a marvelous person and a walking encyclopedia of Alabama history.

I'm lucky to have Harper as my publisher for nearly a quarter century. President and publisher Jonathan Burnham is one of the most erudite and thoughtful friends I have. There is nobody better to discuss the book world with (and he takes my weekend calls). Doug Jones (publisher), Kate D'Esmond (publicity), Trent Duffy (editorial services), and Katie O'Callaghan (marketing) all deserve thanks. Associate editor Sarah Haugen did an incredible job of preparing this book for publication. Similarly, Matt Hannafin helped me retool and trim the manuscript down to a more reader-friendly size. The photo guru was Lawrence Schiller, who discovered the amazing Kennedy era images within. And then there is Jonathan Jao, my primary editor, counselor, and guide, who never lets me down. His new son, Julian, was born while we were closing this book. May he inherit his father's calmness, wisdom, and work ethic.

Dr. Mark Davidson—my former personal assistant, now chief archivist for Bob Dylan—is an Internet genius. He helped me locate a number of hard-to-find space-related documents and helped me solve primary source problems. My friend Julie Fenster, a marvelous historian, helped me in myriad ways. The book benefited mightily from her amazing sense of Cold War history, editorial instincts, and savvy advice. Sloan Harris, my International Creative Management (ICM) agent, was a huge help on numerous fronts.

My mother died during the writing of this book at age eighty-four (heart attack). During my childhood, she had me watch every Apollo mission on television and took me on a study trip to Cape Canaveral. My ninety-year-old father, Edward Brinkley, has gone on living without her. A great student of American history, he was a marvelous sounding board for this project. I love him dearly.

My personal assistant for the past two years is Erika Bell, a mother of three children in Austin. She grew up in Hubbard, Iowa (population 800), and earned an undergraduate degree from the University of Northern Iowa. She is meticulous, conscientious, and hilarious, and is a warm soul. The Brinkley family adores her. Her husband, Garrick Bell, a longtime law-enforcement officer, is a blessed man.

And, finally, words cannot fully convey my gratitude to my wife, Anne Brinkley, and our three school-age children, Benton, Johnny, and Cassady. Together we visited Florida's Space Coast (Florida), Armstrong

Air and Space Museum (Ohio), Johnson Space Center (Texas), the National Air and Space Museum (Washington, DC), and the John F. Kennedy Presidential Library and Museum (Massachusetts), and many other space-related sites. They also enjoyed the John F. Kennedy Birthplace in Brookline, Massachusetts, run by the National Park Service. Together, we made the American journey to understanding the politics of space exciting and fun.

Notes

Preface: Kennedy's New Ocean

xiv The trophy had been established: "Seven Mercury Astronauts Winners of Collier Trophy," *New York Times*, October 8, 1963, p. 24. Starting in 1957, *Look* magazine had taken over the annual trophy award.

xv "Some of us": "Kennedy Presents the Collier Trophy to Astronauts," *New York Times*, October 11, 1963, p. 19.

xvi Writing the president's obituary: Robert Hotz, "An Indelible Mark," *Aviation Week & Space Technology* 79, no. 23 (December 3, 1963), p. 1.

xx "Oh, I think it's predominantly the responsibility": Douglas Brinkley interview with Neil Armstrong, September 14, 2001, NASA Johnson Space Center Oral History Project Transcript, https://www.nasa.gov/pdf/6228/main_armstrong_oralhistory.pdf. Also with me to interview Neil Armstrong was the historian Stephen E. Ambrose.

xxi "As the clock was ticking": Ibid.

xxi "I believe": *Public Papers of the President of the United States: John F. Kennedy, 1961* (Washington, DC: Government Printing Office, 1962), pp. 403–5.

xxii "You don't run for President": William E. Burrows, *This New Ocean: The Story of the First Space Age* (New York: Random House, 1998), p. 329.

xxiv "talked of the heavens": James Preston quoted in *Congressional Record: Proceedings and Debates of Congress* April 13, 1959, Vol. 105, Pt. 19, Appendix A-2960, 86th Congress, 1st Session. See also Scott Carpenter and Kris Stoever, *For Spacious Skies* (New York: New American Library, 2004), p. 197.

xxvi The technology: Stacey Bredhoff, *Moonshot: JFK and Space Exploration* (Washington, DC: Foundation for the National Archives, 2009), p. 38.

xxvii "Each invested enormous resources": Neil Armstrong, "Introduction," in Alan Shepard and Deke Slayton, with Jay Barbree, *Moon Shot: The Inside Story of America's Apollo Moon Landings* (Atlanta, GA: Turner Publishing, 1994), pp. 8–9.

xxviii "the lack of effort": John M. Logsdon, *John F. Kennedy and the Race to the Moon* (New York: Palgrave Macmillan, 2010), p. 8.

xxix "I think he became convinced": Quoted in ibid., p. 225.

xxx Because NASA worked in tandem: Shelby G. Spires, "The

Teflon Myth and Other Inventions from NASA," *Chicago Tribune*, March 5, 2007.

xxxi "Why, some say, the moon": "Address at Rice University, Houston, Texas, 12 September 1962," Papers of John F. Kennedy, Presidential Papers, President's Office Files, Speech Files, John F. Kennedy Presidential Library, Boston, MA (hereafter "Kennedy Library").

xxxii "he wasn't a technical man": Wernher von Braun, recorded interview by Walter D. Sohier and Eugene M. Emme, March 31, 1964, p. 9, John F. Kennedy Oral History Program, Kennedy Library.

xxxiii "moon shots": Steve Marble, "Former Dodger's Slugger Helped Wally Moon, Whose 'Moon Shots' Helped Team Reach Three World Series Dies at 87," *Los Angeles Times*, February 10, 2018, https://www.latimes.com/local/obituaries/la-me-wally-moon-20180210-html.story.html.

xxxiii As early as Kennedy's Rice: "JFK and NASA" *Houston Press*, September 12, 1962, p. 10; NASA 1960s Vertical Files, Houston Metropolitan Research Center, Houston, TX.

xxxiii "from my earliest boyhood": "News Release: New Exhibition to Celebrate JFK's Love of the Sea," March 27, 2000, https://www.jfklibrary.org/about-us/news-and-press/press-releases/new-exhibit-to-celebrate-jfks-love-of-the-sea.

xxxiv "The eyes of the world": "Address at Rice University, Houston, Texas, 12 September 1962," Papers of John F. Kennedy, Presidential Papers, President's Office Files, Speech Files, Kennedy Library.

xxxv "crawling up on the land": Valerie Neal, Cathleen S.

Lewis, and Franklin Winter, *Spaceflight: A Smithsonian Guide* (New York: Macmillan, 1995), p. 132.

xxxvi "We rejoice": John Noble Wilford, *We Reach the Moon* (New York: Bantam Press, 1969), p. xvii.

xxxvi "We needed the first man": Buzz Aldrin with Ken Abraham, *Magnificent Desolation: The Long Journey Home from the Moon* (New York: Three Rivers Press, 2010), p. 10.

1. Dr. Robert Goddard Meets Buck Rogers

3 Jules Verne published: Burrows, *This New Ocean*, pp. 28–32.

4 "certain narrow-minded people": Jules Verne, *From the Earth to the Moon* (France: Pierre-Jules Hetzel, 1865).

4 "the restless erratic insight": Kurt Vonnegut, *Player Piano* (New York: Charles Scribner's Sons, 1952), pp. 4–5.

6 In the wake of the Wright brothers': Robert H. Goddard, "The Moon Rocket Proposition: Refutation of Some Popular Fallacies," *Scientific American*, February 26, 1921.

8 "supervise and direct": James R. Hansen, *First Man* (New York: Simon and Schuster, 2005), p. 130.

8 new technology's military applications: https://history.nasa.gov/naca/overview.html.

9 NACA established the Langley: Sylvia Doughty Fries, *NASA Engineers and the Age of Apollo* (Washington, DC: NASA, 1992).

10 unveiled his astronautical ideas: Robert H. Goddard, *A*

Method of Reaching Extreme Altitudes, Smithsonian Miscellaneous Collections, vol. 71, no. 2 (Washington, DC: Smithsonian Institution, 1919).

11 "It has often proved true": Robert Goddard, "On Taking Things for Granted," graduation oration, South High School, Worcester, MA, June 24, 1904, in "Frequently Asked Questions About Dr. Robert H. Goddard," Clark University Archives and Special Collections, http://www2.clarku.edu/research/archives/goddard/faq.cfm.

12 "Rocket for Moon": *Colorado Springs Gazette*, January 12, 1920, p. 1.

12 "on nothing that is really impossible": Goddard, *A Method of Reaching Extreme Altitudes*, p. 57.

13 media and public remained uncertain: Paul Dickson, *Sputnik: The Shock of the Century* (New York: Walker, 2001), p. 40.

13 a front-page story: "Believes Rocket Can Reach Moon," *New York Times*, January 12, 1920, p. 1.

13 "need to have something better": "A Severe Strain on Credulity," *New York Times*, Topics of the Times segment, January 13, 1920, p. 12.

13 compared the Worcester rocketeer's: "All Aboard for the Moon," *Philadelphia Inquirer*, January 13, 1920, p. 10.

14 "So much for": Quoted in "Frequently Asked Questions About Dr. Robert H. Goddard."

15 estimate was one hundred thousand dollars: "Nine Want Moon Trip," *St. Albans Messenger*, April 6, 1920, p. 1.

15 "publicity is the worst possible disaster": Paul G. Carter,

"Rockets to the Moon 1919–1944," *American Studies* 15, no. 1 (Spring 1974): 33.

15 nine applications from brave men: "Nine Want Moon Trip," p. 1.

16 "the more practical objects": "To Test Moon-Bound Rocket This Summer," *Philadelphia Inquirer*, January 28, 1921, p. 2.

16 Goddard's solicitations: "Seeks 'Rocket to Moon' Fund," *Kansas City Times*, June 28, 1921, p. 1.

17 constructing a small replica: Dickson, *Sputnik*, p. 47.

18 a description of Goddard's work: Hermann Oberth, *Die Rakete zu den Planetenräumen* [The Rocket into Planetary Space] (München: R. Oldenbourg, 1923).

18 "the American professor Dr. Goddard": Quoted in Asif Siddiqi, "Deep Impact: Dr. Robert Goddard and the Soviet 'Space Fad' of the 1920s," *History and Technology* 20, no. 2 (2004): 99.

19 international team of space travelers: Konstantin Tsiolkovsky, *Outside the Earth* (*Vne Zemli: Nauchno-fantasticheskaya povest*) (Fairfield, CT: Athena Books, 2006); and Burrows, *This New Ocean*, p. 43.

19 "mass fascination with space travel": Siddiqi, "Deep Impact," p. 99.

19 "possibility of cosmic travel": Burrows, *This New Ocean*, p. 43.

20 "I didn't get a watch": Quoted in David Halberstam, *The Fifties* (New York: Ballantine, 1993), p. 609.

20 "The wagon was wholly out of control": Wernher von

Braun, "Space Man: The Story of My Life," *American Weekly*, July 20, 1958, p. 8.

21 "Here was a task": Quoted in Daniel Lang, "A Romantic Urge," *New Yorker*, April 21, 1951, pp. 75–93.

22 "Lunetta" (Little Moon): Bob Ward, *Dr. Space: The Life of Wernher von Braun* (Annapolis, MD: Naval Institute Press, 2015), p. 13.

22 the world was fourfold: Ibid., p. 15.

23 "the first man to walk on the moon": Ernst Stuhlinger, "How It All Began: Memories of an Old-Timer," July 20, 1999, Wernher von Braun Library of Archives, U.S. Space and Rocket Center, Huntsville, AL.

25 he did correspond: "Plans Hop to Moon in a Rocket-Plane," *New York Times*, May 8, 1927, p. 19.

26 ambulances, police cars: Associated Press, "Rocket Plows Skyward Alarming District," *Houston Chronicle*, July 18, 1929, p. 26.

26 residents of Auburn forbade: "Dr. Robert H. Goddard, World Rocket Pioneer," typescript (Washington, DC: NASA), 1960, p. 6.

28 a nervous breakdown: "Halts Moon Rocket Work," *New York Times*, January 17, 1930, p. 3.

28 enormous box-office hit: "Die Frau im Mond/Der Neue Lang-Film," *Vossische Zeitung* (Berlin), October 16, 1929; Heike Langenberg, "Rocket Retrospective," *Nature*, November 8, 2001, p. 152.

30 nine thousand feet in just 22.3 seconds: "Frequently Asked Questions About Dr. Robert H. Goddard."

30 "Morning in the desert": Robert Goddard diary entry (June 2, 1937), quoted in Dave A. Clary, *Rocket Man: Robert H. Goddard and the Birth of the Space Age* (New York: Hyperion, 2003), p. 169.

2: Kennedy, von Braun, and the Crucible of World War II

34 Once, when Jack was a boy: Yanek Mieczkowski, *Eisenhower's Sputnik Moment: The Race for Space and the World* (Ithaca, NY: Cornell University Press, 2013), p. 260.

37 "the *Literary Digest*": Quoted in Nigel Hamilton, *JFK: Reckless Youth* (New York: Random House, 1992), p. 86.

37 "accustomed to the idea": John F. Kennedy to Christian Causs, October 21, 1935, Papers of John F. Kennedy, Personal Papers, Harvard: Harvard Records, Pre-enrollment material: 1935–1936, Kennedy Library.

38 "Jack has rather superior mental ability": Henry Raymont, "Kennedy Data: Years at Harvard," *New York Times*, August 3, 1971, p. 26.

39 accepting nearly every applicant: Harvard was accepting 98.6 percent of its applicants. See Jerome Karabel, *The Chosen: The Hidden History of Admission and Exclusion at Harvard, Yale, and Princeton* (Boston: Houghton Mifflin, 2005), p. 205.

39 the president of Harvard described his method: Karabel, *The Chosen*, p. 174.

39 "I feel that Harvard": John F. Kennedy, application, Papers of John F. Kennedy, Personal Papers, Harvard: Harvard Records, Pre-enrollment material: 1935–1936, Kennedy Library.

41 "Oberth was the first": Hermann Oberth Raumfahrt Museum, http://www.Oberth-museum.org, December 28, 1989.

41 The rocketry concepts developed by Oberth: "Hermann Oberth," NASA, For Educators, September 22, 2010, https://www.nasa.gov/audience/foreducators/rocketry/home/hermann-oberth.html.

43 In 1937, Doolittle went to Germany: Winston Groom, *The Aviators: Eddie Rickenbacker, Jimmy Doolittle, Charles Lindbergh, and the Epic Age of Flight* (Washington, DC: National Geographic, 2013), pp. 182–83.

44 "with the moon only": James H. "Jimmy" Doolittle, *I Could Never Be So Lucky Again* (New York: Bantam, 1994), p. 515.

44 "I naturally cannot turn over": R. H. Goddard to Robert Milikin, September 1, 1936, in Esther Goddard and G. Edward Pendray, eds., *The Papers of Robert H. Goddard, Including the Reports to the Smithsonian Institution and the Daniel and Florence Guggenheim Foundation*, vol. 3 (New York: McGraw-Hill, 1970), pp. 1012–13.

46 amazed by the quality: John F. Kennedy diary, August 18, 1937, Papers of John F. Kennedy, Personal Papers, Early Years, 1928–1940, Kennedy Library.

47 "had the added attraction": John F. Kennedy diary, August 20, 1937, Papers of John F. Kennedy, Personal Papers, Early Years, 1928–1940, Kennedy Library.

47 "The Germans really are too good": John F. Kennedy diary, August 21, 1937, Papers of John F. Kennedy, Personal Papers, Early Years, 1928–1940, Kennedy Library.

47 broad new highways: Billings diary quoted in David Pitts, *Jack and Lem: John F. Kennedy and Lem Billings* (New York: Carroll and Graf, 2009), pp. 52–67.

50 "We were all struck": Michael O'Brien, *John F. Kennedy* (New York: St. Martin's, 2005), p. 94.

50 "For a twenty-two-year-old American": Whalen quoted in ibid., p. 93.

54 "astonishing theoretical knowledge": Ward, *Dr. Space*, p. 18.

54 "If you want more money": Walter R. Dornberger, "The German V-2," *Technology and Culture* 4, no. 4 (Autumn 1963): 397.

55 "arrow stability": Burrows, *This New Ocean*, p. 82. See also Michael J. Neufeld, *The Rocket and the Reich: Peenemünde and the Coming of the Ballistic Missile Era* (Washington, DC: Smithsonian Books, 1995), pp. 23–71.

55 Christened Peenemünde: Dickson, *Sputnik*, p. 52.

56 seventy next-generation Aggregat rockets: Ibid., p. 52.

57 "going to the moon": Dornberger, "The German V-2," p. 399.

57 There is currently a debate: Jared S. Buss, *Willy Ley:*

Prophet of the Space Age (Gainesville: University Press of Florida, 2017).

58 "a liquid-propellant rocket can fly": Quoted in Christopher Potter, *The Earth Gazers: On Seeing Ourselves* (New York: Pegasus, 2018), p. 82.

59 "thousands of major problems": Dornberger, "The German V-2," p. 399.

60 "The slim missile rose slowly": Michael J. Neufeld, *Von Braun: Dreamer of Space, Engineer of War* (New York: Knopf Doubleday, 2007), p. 118.

60 "The shocking German success": O'Brien, *John F. Kennedy*, pp. 99–100.

61 turning his attention away: Michael J. Neufeld, "Hitler, the V-2, and the Battle for Priority, 1939–1943," *The Journal for Military History* 57, no. 3 (July 1993): 528.

63 "revolutionary importance for the conduct of warfare": Quoted in Neufeld, *Von Braun*, p. 128.

3: Surviving a Savage War

68 his most authentic self: Wallace J. Nichols, *Blue Mind: The Surprising Science That Shows How Being Near, in, on, or Under Water Can Make You Happier, Healthier, More Connected, and Better at What You Do* (New York: Little, Brown, 2014), pp. 5–15.

69 "cowards or defeatists": Peter Collier and David Horowitz, *The Kennedys: An American Drama* (New York: Simon and Schuster, 1984), p. 118.

71 put his craft above: Frederick I. Ordway III and Mitchell R. Sharpe, *The Rocket Team* (New York: William Heinemann, 1979), pp. 47–48.

72 was fond of saying: Von Braun quoted in "Wernher von Braun," International Space Hall of Fame at the New Mexico Museum of Space History (1976), http://www.nmspacemuseum.org/halloffame/detail.php?id=29.

73 "was a true ancestor": Dickson, *Sputnik*, p. 52.

73 "In the full glare of the sunlight": Ibid., p. 49.

74 "Europe and the world": Quoted in Halberstam, *The Fifties*, p. 611.

76 "Our big-ship navy": William F. Liebenow, oral history, p. 3, John F. Kennedy Oral History Program, Kennedy Library, https://jfklibrary.org/asset-viewer=archives/JFKOH/Liebenow%2C%20William%20F/JFKOH-WFL-01/JFKOH-WFL-01.

77 "I'm not so crazy": Quoted in Thurston Clarke, *JFK's Last Hundred Days: The Transformation of a Man and the Emergence of a Great President* (New York: Penguin Press, 2013), p. 14.

77 Early on August 2, 1943: John Hersey, "Survival," *The New Yorker*, June 17, 1944, p. 31.

78 "People that haven't been there": Liebenow, oral history, p. 8.

79 "Most of the courage shown": Doris Kearns Goodwin, *The Fitzgeralds and the Kennedys* (New York: St. Martin's Press, 1987), p. 715.

80 mission to find Peenemünde: Neufeld, *Von Braun*, p. 128.

80 "After four weeks of cleanup work": Dornberger, "The German V-2," pp. 404–5.

82 "endurance or death": Leon Jaroff, "The Rocket Man's Dark Side," *Time*, March 26, 2002, content.time.com/time/health/article/0,8599,220201,00.html.

82 Dora: Andre Sellier, *A History of the Dora Camp: The Untold Story of the Nazi Slave Labor Camp* (Chicago, IL: Ivan R. Dee, 2003), p. 5.

82 In later years, a disingenuous von Braun: Dickson, *Sputnik*, p. 53. See also Halberstam, *The Fifties*, pp. 611–12.

83 a *Life* magazine story: John Hersey, "PT Squadron in the South Pacific," *Life*, May 10, 1943, p. 74.

84 He spent most of late winter: John Hellmann, *The Kennedy Obsession: The American Myth of JFK* (New York: Columbia University Press, 1997), p. 43.

84 "They all wait anxiously": Quoted in Edward J. Renehan Jr., *The Kennedys at War, 1937–1945* (New York: Doubleday, 2002), p. 280.

85 Jack always loved being in Florida: Ibid., pp. 214–16.

85 Jack decided to learn how to fly: John F. Kennedy, May 19–June 8, 1944, Flight Log, Shapell Foundation Archive, Herzilya, Israel. Thanks to Lavin Montgomery and Embry-Riddle Aeronautical University for bringing this to my attention.

86 "accelerated schedule": Halberstam, *The Fifties*, p. 608.

89 "a concrete chamber": Graham M. Simons, *Operation LUSTY: The Race for Hitler's Secret Technology* (Barnsley, UK: Pen and Sword Books, 2016), p. 71.

90 "some kind of a gadget": Jacob Neufeld, *The Development of Ballistic Missiles in the United States Air Force, 1945–1960* (Washington, DC: Office of Air Force History, United States Air Force, 1990), p. 35.

90 appeared in *The New Yorker*: Hersey, "Survival."

91 "construction of John F. Kennedy": Hellmann, *The Kennedy Obsession*, p. 43.

92 "I firmly believe": William Doyle, *PT 109: An American Epic of War, Survival, and the Destiny of John F. Kennedy* (New York: William Morrow, 2015), p. xiii.

94 Naval Air Corps pilots: Alan Axelrod, *Lost Destiny: Joe Kennedy Jr. and the Doomed WWII Mission to Save London* (New York: Palgrave Macmillan, 2015), p. 160.

95 incrementally pulling out: Simons, *Operation LUSTY*, p. 72.

96 Third Infantry Division captured: Steven J. Zaloga, *German V-Weapon Sites, 1943–45* (London: Osprey, 2008), p. 14.

97 "an oasis of love": Edward M. Kennedy, *True Compass* (New York: Twelve, 2009), p. 63.

97 "While he's singing": Kate Thom Kelley, oral history, p. 14, National Archives and Records Administration, Office of Presidential Libraries, Kennedy Library.

4: Who's Afraid of the V-2?

99 "Britons pondered the possibility": "Ten-Ton Robots Called Possible as Flying Bombs Batter Britain," *New York Times*, July 18, 1944, p. 6.

100 missile test: Neufeld, *The Rocket and the Reich*, pp. 158–90.

101 installed SS general Hans Kammler: Sellier, *A History of the Dora Camp*, p. 103; and Heinrich Himmler, "Orders to Hans Kammler Concerning Construction Work for the V2 Missile Program," http://nuremberg.law.harvard.edu/documents/4279-orders-to-hans-kammler?q=heinrich+himmler.

102 This rocket landed near: Kenneth Lipartito and Orville Butler, *A History of the Kennedy Space Center* (Gainesville: University Press of Florida, 2007), p. 33.

103 launch two V-2s at London: Norman Longmate, *Hitler's Rockets: The Story of the V-2s* (London: Hutchinson, 1985), p. x.

104 "There wasn't a mark": Clare Heal, "The Day Hitler's Silent Killer Came Falling on Chiswick," (London) *Express*, September 7, 2014.

104 "Von Braun was completely devastated": Quoted in Neufeld, *Von Braun*, p. 184.

105 "When the first V-2 hit London": Ibid., p. 185.

107 "With a satisfied eye he witnessed": Pierre Boulle, *The Bridge over the River Kwai* (New York: Presidio Press, 1982), p. 96.

109 "Once the rockets are up": Tom Lehrer, "Wernher von Braun," Album/CD *That Was the Year That Was* (Reprise/Warner), recorded in San Francisco, July 1965. http://www.protestsonglyrics.net/Humorous_Songs/Wernher-Von-Braun.phtml.

110 "his new missiles until November 8": Winston Churchill, *The Second World War*, vol. 6: *Triumph and Tragedy* (Boston: Houghton Mifflin, 1953), p. 47.

111 "V-2 type rocket appears to be": Robert Goddard, *The Papers of Robert H. Goddard* (New York: McGraw-Hill, 1970), 9:1548.

111 the V-2 had been copied directly: Frank H. Winter, "Did the Germans Learn from Goddard? An Examination of Whether the Rocketry of R. H. Goddard Influenced German Pre–World War II Missile Development," *History of Rocketry and Astronautics* (San Diego: American Astronautical Society, 2016), pp. 106–11.

115 almost 700 V-2s per month: Walter A. McDougall, *. . . the Heavens and the Earth: A Political History of the Space Age* (New York: Basic Books, 1985), p. 42.

116 "but its role in history": Potter, *The Earth Gazers*, p. 115.

118 "We despise the French": Quoted in Ordway III and Sharpe, *The Rocket Team*, p. 274.

119 "I think you're nuts": Dickson, *Sputnik*, p. 59.

119 Toftoy had U.S. Army troops race: McDougall, *. . . the Heavens and the Earth*, p. 44.

120 "We defeated Nazi armies": Quoted in G. A. Tokady, "Soviet Rocket Technology," *Technology and Culture* 4 (Fall 1963): 523.

121 plenty of machine and rocket: Eugene Reichl, *Project Mercury* (Atglen, PA: Schiffer, 2016), p. 4.

123 "Imagine . . . finding": Lindbergh quoted in Winston Groom, *The Aviators: Eddie Rickenbacker, Jimmy Doolit-*

tle, Charles Lindbergh, and the Epic Age of Flight (Washington, DC: National Geographic, 2013), pp. 437–38.

123 shipped to the Annapolis Experiment Station: Milton Lehman, *This High Man: The Life of Robert H. Goddard* (New York: Farrar, Straus and Giroux, 1963), p. 387.

123 agreement "was a sham": McDougall, *. . . the Heavens and the Earth*, pp. 44–45.

127 "Dropping the bombs": Harry S. Truman to James L. Cate, December 6, 1952. Truman Papers, President's Secretary's File (Atomic Bomb), Harry S. Truman Presidential Library, Independence, Missouri.

129 "A screaming comes across the sky": Thomas Pynchon, *Gravity's Rainbow* (New York: Penguin, 1973), p. 3.

130 Considered wards of the army: Ward, *Dr. Space*, p. 63.

131 dubbed himself a POP: Ibid.

131 "The GIs sized me up": Wernher von Braun, "Why I Chose America," *American Magazine*, July 1952.

131 the Soviets would appropriate: Doran Baker, "The University, Electrical Engineering and Space Travel," Paper 77, no. 75, in *USU Faculty Honor Lectures* (Logan: Utah State University Press, 1979).

132 But some of the imported Germans were sent back: Brian Crim, *Our Germans: Operation Paperclip and the National Security State* (Baltimore: Johns Hopkins University Press, 2018).

132 Arthur Rudolph, a close colleague: Dennis Piszkiewicz, *The Nazi Rocketeers: Dreams of Space and Crimes of War* (Mechanicsburg, PA: Stackpole, 2007), p. 225.

132 Hubertus Strughold: Colin A. Ross, *The CIA Doctors* (Richardson, TX: Manitou Communications, 2006).

133 he managed to design his first trajectories: Ward, *Dr. Space*, p. 65.

135 "If we may assume": Quoted in Sean Kalic, *U.S. Presidents and the Militarization of Space, 1946–1967* (College Station: Texas A&M University Press, 2012), pp. 17–18.

137 "fresh-faced, charming young war hero": David Nasaw, *The Patriarch: The Remarkable Life and Turbulent Times of Joseph P. Kennedy* (New York: Penguin, 2012), p. 603. See also "Articles: by John F. Kennedy in the Hearst Newspapers, 1945," Kennedy Library, www.jfklibrary.org/Asset-Viewer/Archives/JFKPOF-129-003.aspx.

138 "Jack Kennedy was the only pol in Boston": Quoted in John T. Shaw, *JFK in the Senate: Pathway to the Presidency* (New York: Palgrave Macmillan, 2013), p. 26.

139 the performance of President Truman: Gary A. Donaldson, *Truman Defeats Dewey* (Lexington: University of Kentucky Press, 2015), p. 6.

140 Democrats scrambled: Donald R. McCoy, "Harry S. Truman: Personality, Politics, and Presidency," *Presidential Studies Quarterly* 12, no. 2 (Spring 1982): 223.

142 "I told them that Soviet Russia today": John F. Kennedy, "The Time Has Come: Radio Speech on Russia," October 1946, Papers of John F. Kennedy, Pre-Presidential Papers, Kennedy Library.

144 "The war made us get serious": Quoted in Doyle, *PT 109*, p. xiii.

144 "What we do now will": John F. Kennedy, congressional campaign radio broadcast, 1946, quoted in *John F. Kennedy in His Own Words*, ed. Eric Freedman and Edward Hoffman (New York: Citadel Press, 2005), p. 140.

146 "It seems to be a law of nature": Potter, *The Earth Gazers*, p. 144; Ward, *Dr. Space*, p. 68.

5: Spooked into the Space Race

152 Having spent a third of a trillion: Harry S. Truman, "Speech to Congress, March 12, 1947," in *Debating the Origins of the Cold War: American and Russian Perspectives*, ed. Ralph B. Levering, Vladimir O. Pechatnov, Verena Botzenhart-Viehe, and C. Earl Edmondson (Lanham, MD: Rowman and Littlefield, 2002), p. 83.

153 "an almost impenetrable veil of secrecy": Arthur C. Clarke, *The Making of a Moon* (New York: Harper and Brothers, 1957), pp. 18–19.

154 technological leaps: Boris Chertok, *Rockets and People*, ed. Asif A. Siddiqi, vol. 3: *Hot Days of the Cold War* (Washington, DC: Government Printing Office, 2009), p. 284.

154 "no ballistic missiles worth mentioning": Mieczkowski, *Eisenhower's Sputnik Moment*, p. 40.

154 a new National Military Establishment: "About the Department of Defense," U.S. Department of Defense, www.defense.gov/About/.

155 Major General Curtis LeMay: "General Curtis Emerson LeMay," U.S. Air Force, http://www.af.mil/About-Us/Bi

ographies/Display/Article/106462/general-curtis-emerson-lemay/.

156 "Whose imagination is not fired": RAND Corporation, Preliminary Design of an Experimental World-Circling Spaceship (SM-11827), May 2, 1946, www.rand.org/content/dam/rand/pubs/special_memoranda/2006/SM11827part1.pdf.

157 rocket ultimately named the Viking: Mike Gruntman, *Blazing the Trail: The Early History of Spacecraft and Rocketry* (Reston, VA: American Institute of Aeronautics and Astronautics, 2004), p. 216.

157 "The U.S. Navy wanted no part": Baker, "The University, Electrical Engineering and Space Travel."

159 The U.S. Army was continuing: Neufeld, *The Rocket and the Reich*, pp. 270–71.

161 Electronic navigation and anti-icing: Winston Groom, *The Aviators*, 434.

163 "one of the laziest men": Quoted in Robert Caro, *The Passage of Power: The Years of Lyndon Johnson* (New York: Alfred A. Knopf, 1982), p. 29.

163 "the most important thing in history": James Mahaffey, *Atomic Awakening: A New Look at the Future History of Nuclear Power* (New York: Pegasus, 2009), pp. viii–ix.

163 stripping this authority from the military: Clark Clifford, *Counsel to the President* (New York: Anchor, 1992), p. 277.

164 the USSR lagged behind: Mieczkowski, *Eisenhower's Sputnik Moment*, p. 13.

165 a thrust yield of at least eight hundred thousand pounds:

Erik Bergaust, *Rocket City U.S.A.* (New York: Macmillan, 1963), p. 14.

165 Soviet Union had detonated: Robert C. Albright, "Grave Senate Hears Soviet A-bomb News," *Washington Post*, September 24, 1949, p. 1.

166 "months and even years": United Press, "Indifference to Civil Defense Cited," *Washington Post*, October 10, 1949, p. 11.

170 "onrushing tide of communism": John F. Kennedy, *John Fitzgerald Kennedy: A Compilation of Statements and Speeches Made During His Service in the United States Senate and House of Representatives* (Washington, DC: Government Printing Office, 1964), pp. 41–42.

170 "I never had the feeling": Quoted in Shaw, *JFK in the Senate*, p. 26.

172 "current emotional heat wave": Genevieve Reynolds, "Capitol Has Own Heat Wave, Attributed to John Kennedy," *Washington Post*, May 22, 1949, p. 57.

172 "I knew Jack": John D. Lane, oral history interviews, October 12 and December 6, 2006, Senate Historical Office, no. 2, p. 49, https://www.senate.gov/artandhistory/history/resources/pdf/OralHistory_LaneJohn.pdf (hereafter "Lane oral history interviews").

174 Addison's disease: Richard Lacayo, "How Sick Was J.F.K.?" *Time*, November 24, 2002, http://content.time.com/time/magazine/article/0,9171,393754,00.html.

174 "Sometimes you read that [Jack] was a reluctant figure": Quoted in Thomas Oliphant and Curtis Wilke, *The Road* to Camelot: Inside JFK's Five-Year Campaign (New York: Simon and Schuster, 2017), p. 8.

176 "at a tempo for peace": Quoted in Ward, *Dr. Space*, p. 74.

176 being "too fantastic": Halberstam, *The Fifties*, pp. 607–8.

178 "McMahon was home ill": Lane oral history interviews, no. 2, p. 54.

179 "the hardest campaigner": Quoted in Caro, *The Passage of Power*, p. 30.

181 "Kennedy told me later": Lane oral history interviews, no. 1, p. 29.

182 courted by both political parties: Robert H. Ferrell, "Eisenhower Was a Democrat," *Kansas History* 13, no. 3 (Autumn 1990): 135.

183 "to demand the abolition": Robert C. Byrd, *The Senate 1789–1989: Addresses on the History of the United States Senate* (Washington, DC: Government Printing Office, 1988), 1:606.

184 Saint Lawrence Seaway: Caro, *The Passage of Power*, 30.

185 a senator or journalist: Paul B. Fay, recorded interview with James A. Oesterle, November 11, 1970, p. 214, John F. Kennedy Library Oral History Program, Kennedy Library.

186 "Every gun that is made": Dwight Eisenhower, "The Chance for Peace" speech, April 16, 1953, https://www.eisenhower.archives.gov/all_about_ike/speeches/chance_for_peace.pdf.

187 Eisenhower, in his postpresidential memoir: Dwight D. Eisenhower, *Waging Peace, 1956–1961; The White House Years* (New York: Doubleday, 1965), pp. 207–9.

187 To put this into perspective: Mieczkowski, *Eisenhower's Sputnik Moment*, p. 40.

187 the Redstone would go on: Eugene M. Emme, *Aeronautics and Astronautics, 1915–1960* (Columbus, OH: BiblioGov, 2012), p. 72.

189 "You should know how advertising": Potter, *The Earth Gazers*, pp. 146–47.

191 "At the present time our engineering efforts": Quoted in Roger D. Launius, *The U.S. Space Program and American Society* (Carlisle, MA: Discovery Enterprises, 1998), p. 16.

192 "In addition to the cogent scientific arguments": Alan Dulles to Charles Erwin Wilson, January 29, 1955, CIA Library Reading Room, https://www.cia.gov/library/readingroom/docs/DOC_0006513734.pdf.

193 Kennedy's stature in the Senate: Oliphant and Wilke, *The Road to Camelot*, p. 2.

194 "This was the first time": Caro, *The Passage of Power*, p. 50.

195 Soviet domination in ICBM development: Philip Nash, "Bear Any Burden? John F. Kennedy and Nuclear Weapons," in *Cold War Statesmen Confront the Bomb: Nuclear Diplomacy Since 1945*, ed. John Lewis Gaddis et al. (New York: Oxford University Press, 1999), p. 122.

200 Sixty-seven countries: Manu Saadia, "Is America Facing Another Sputnik Moment?" *New Yorker*, October 4, 2017.

202 "specifically ordered to forget": "Space: Reach for the Stars," *Time*, February 17, 1958, pp. 21–25.

6: *Sputnik* Revolution

206 "Senator and Mrs. Kennedy": Wernher von Braun, recorded interview by Walter D. Sohier and Eugene M. Emme, March 31, 1964, p. 1, John F. Kennedy Oral History Program, Kennedy Library.

207 "the accident occurred with an obsolescent type": Ibid.

208 "The senator pointed at the close relationship": Ibid., p. 2.

209 "I would raise a question": Ibid.

210 "When I published my first space novel": Neil McAleer, *Arthur C. Clarke: An Authorized Biography* (Chicago: Contemporary Books, 1992), p. 215.

211 "The cost of continuing": S. Everett Gleason, "Discussion at the 329th Meeting of the National Security Council, Wednesday, July 3, 1957," NSC Records, DDE Presidential Papers, Dwight D. Eisenhower Presidential Library, Abilene, Kansas, July 5, 1957, p. 2.

211 the night of October 4, 1957: Dickson, *Sputnik*, p. 1.

212 Kremlin confirmed the stunning report: William J. Jorden, "Soviets Fire Earth Satellite into Space: 560 Miles High," *New York Times*, October 5, 1957, p. 1.

212 three boldface lines: Ibid.

213 American Radio Relay League: Dickson, *Sputnik*, p. 13.

213 "second in importance": Hannah Arendt, *The Human Condition* (Chicago: University of Chicago Press, 1958), p. 5.

215 While *Sputnik* was newsworthy: Responses of twenty-two-year-old female and forty-year-old white male, File

87: Correspondence, Rhoda Metraux Papers, Library of Congress Manuscript Division, Washington, DC.

216 "In the pre-Sputnik days": Rachel Carson to Dorothy Freeman, February 1, 1958, in *Rachel Carson: Silent Spring and Other Writings on the Environment,* ed. Sandra Steingraber (New York: Library of America, 2018), p. 374.

217 "Oh Little *Sputnik*": Quoted in Burrows, *This New Ocean,* p. 192.

218 "In the Open West": Lyndon Baines Johnson, *Vantage Point: Perspectives of the Presidency, 1963–69* (New York: Holt, Rinehart and Winston, 1971), p. 272.

218 George Reedy, a high-powered Democratic strategist: George E. Reedy to Lyndon B. Johnson, October 17, 1957, Lyndon B. Johnson Presidential Library, Austin, Texas.

220 "The Roman Empire controlled": Ibid., pp. 275–76.

221 he rejected the idea: Craig Ryan, *Sonic Wind: The Story of John Paul Stapp and How a Renegade Doctor Became the Fastest Man on Earth* (New York: Liveright, 2015), p. 257.

221 the launch of *Sputnik* "spooked": Neil deGrasse Tyson, *Space Chronicles: Facing the Ultimate Frontier* (New York: W. W. Norton, 2012), p. 5.

222 an official investigation to gauge: Kevin J. Fernlund, *Lyndon B. Johnson and Modern America* (Norman: University of Oklahoma Press, 2009), p. 151.

223 "the age of *Sputnik*": "Blue Key Banquet" (speech), University of Florida, Gainesville, Florida, October 18, 1957,

item JFKSEN-0898-018, Papers of John F. Kennedy, Pre-Presidential Papers, Senate Files: Speeches and the Press, Speech Files: 1953–1960, Kennedy Library.

224 "losing the satellite and missile race": Christopher A. Preble, "Who Ever Believed in the 'Missile Gap'? John F. Kennedy and the Politics of National Security," *Presidential Studies Quarterly* 33, no. 4 (2003): 801–26.

224 British reporter informed von Braun: Neufeld, *Von Braun*, p. 311.

224 "suddenly been vaccinated": Ibid., pp. 311–12.

224 "words tumbled over one another": Thomas M. Coffey, *Iron Eagle: The Turbulent Life of General Curtis LeMay* (New York: Crown, 1986), p. 349.

224 "useless hunk" . . . "basketball game": Wilson and Adams quoted in Halberstam, *The Fifties*, pp. 624–25.

225 "a brown star racing northward": Bob Kealing, *Kerouac in Florida: Where the Road Ends* (Arbiter Press, 2011), p. 30.

226 a weightless environment: Dr. David Whitehouse, "First Dog in Space Died Within Hours," BBC News, October 28, 2002.

226 "outstripped the leading capitalist country": William Taubman, *Khrushchev: The Man and His Era* (New York: W. W. Norton, 2003), p. 378.

227 "a battle more important": Robert A. Divine, *The Sputnik Challenge: Eisenhower's Response to the Soviet State* (New York: Oxford University Press, 1993), pp. xv–xvi.

227 "a total politician": Kenneth P. O'Donnell and David F. Powers, with Joe McCarthy, "*Johnny, We Hardly Knew*

Ye": Memories of John Fitzgerald Kennedy (Boston: Little, Brown, 1972), pp. 127–28.

229 "the solemn consequences": "Kansas Democratic Club Banquet" (speech), Topeka, Kansas, November 6, 1957, item JFKSEN-0898-027, Papers of John F. Kennedy, Pre-Presidential Papers, Senate Files: Speeches and the Press, Speech Files: 1953–1960, Kennedy Library.

229 "It is now apparent": Ibid.

230 "stood in so critical a position": Associated Press, "Kennedy Assails U.S. Missile Lag," *New York Times*, November 7, 1959, p. 16.

230 "go frantic": Dwight Eisenhower, President's New Conference, March 26, 1958, Dwight D. Eisenhower Presidential Library, Abilene, Kansas (hereafter "Eisenhower Library").

231 "the race for advantage": Hal Willard, "Kennedy Calls for Federal School Aid," *Washington Post*, October 11, 1957, p. 1.

231 comparing American and Soviet performance: "Crisis in Education," *Life*, March 24, 1958, p. 26.

231 "It's the Americans' turn": "U.S. Turn Now, Bulganin Says," *New York Times*, November 12, 1957, p. 27.

233 "It would have been better": Philip Nash, *The Other Missiles of October: Eisenhower, Kennedy, and the Jupiters, 1957–1963* (Chapel Hill: University of North Carolina Press, 1999), p. 3.

233 "in a race" with the Kremlin: "John Foster Dulles to James C. Hagerty, October 8, 1957: 'Draft Statements on the Soviet Satellite,'" October 5, 1957, John Foster Dulles Papers, Eisenhower Library, https://history.nasa.gov/sputnik/15.html. See also *Dwight D. Eisenhower, The White House Years,* vol. 2: *Waging Peace: 1956–1961* (New York: Doubleday, 1965).

234 "Vanguard will never make it": Quoted in Neufeld, *Von Braun*, p. 312.

234 reserved a late-January launch date: "Reach for the Stars," *Time*, February 17, 1958, http://content.time.com/time/subscriber/article/0,33009,862899-7,00.html.

235 "a confusing aura": Norman Mailer, *Of a Fire on the Moon* (Boston: Little, Brown, 1970), p. 70.

236 von Braun adorned the covers: *Time*, February 17, 1958; *Life*, November 18, 1957.

236 "I cannot share": "Letters," *Time*, March 3, 1958, http://content.time.com/time/magazine/article/0,9171,893837,00.html.

237 "I wasn't truly aware": Wernher von Braun to Alan Fox, January 22, 1971, Correspondence file, Wernher von Braun Library and Archives, U.S. Space and Rocket Center, Huntsville, Alabama.

238 "small satellite spheres": Constance McLaughlin Green and Milton Lomask, *Project Vanguard* (Mineola, NY: Dover, 2009), p. 198.

239 Kennedy regarded the "mismanagement": Ibid., p. 202.

239 "duplicating each other's efforts": "Kansas Democratic Club Banquet."

240 "unannounced but unabashed run": "Man Out Front," *Time*, December 2, 1957, p. 19.

244 Kennedy had not written: Herbert S. Parmet, *JFK: The Presidency of John F. Kennedy* (New York: Dial Press, 1983).

244 fought the accusation relentlessly: Klaus P. Fischer, *America in White, Black and Gray* (New York: Continuum, 2006), p. 96n10.

245 "we must work as though": "Johnson's Talk to Democratic Senators," *New York Times*, January 8, 1958, p. B3.

246 "where the initiative": "Legislative Origins of the National Aeronautics and Space Act of 1958: Proceedings of an Oral History Workshop," Conducted April 3, 1992, Moderated by John M. Logsdon, *Monographs in Space History*, no. 8 (Washington, DC: NASA History Office, 1998), https://history.nasa.gov/40thann/legorgns.pdf.

246 "the flaunting of the Soviets": "Women's Club of Richmond" (speech), Richmond, Virginia, January 20, 1958, item JFKSEN-0899-005, Papers of John F. Kennedy, Pre-Presidential Papers, Senate Files: Speeches and the Press, Speech Files: 1953–1960, Kennedy Library.

248 "lap up publicity and attention": Quoted in Michelle L. Evans, *The X-15 Rocket Plane: Flying the First Wings into Space* (Lincoln: University of Nebraska Press, 2013), p. 70.

249 "God gave man a fixed number": Quoted in Milton O. Thompson, *At the Edge of Space: The X-15 Flight Pro-*

gram (Washington, DC: Smithsonian Institution Press, 1992), pp. 9–41.

250 "had a mind that absorbed": Quoted in Richard R. Truly, "Neil A. Armstrong," in *Memorial Tributes: National Academy of Engineering* (Washington, DC: National Academies Press, 2013).

250 "Neil was probably the most": Thompson, *At the Edge of Space*, pp. 9–41.

251 "All in all": Hansen, *First Man*, p. 53.

7: Missile Gaps and the Creation of NASA

256 "When the elevator": "Harvard Club" (speech), Boston, Massachusetts, March 21, 1958, item JFKSEN-0900-012, Papers of John F. Kennedy, Pre-Presidential Papers, Senate Files: Speeches and the Press, Speech Files: 1953–1960, Kennedy Library.

257 "cautioned Dr. von Braun": John Medaris, *Countdown for Decision* (New York: G. P. Putnam, 1960), pp. 143–73.

258 "It makes us feel that we paid": Ibid.

258 "I sure feel a lot": Mieczkowski, *Eisenhower's Sputnik Moment*, p. 129.

258 "not make too big": Dickson, *Sputnik*, p. 175.

259 "This is the beginning": Wernher von Braun quoted in "America in Space, January 31, 2008, 50th Anniversary," *Huntsville Times* Commemorative Edition, February 1, 2008, https://www.kozmiclazershow.com/Huntsville-Times%20Gala%20Edition.pdf.

259 it proved a trouper: Green and Lomask, *Project Vanguard*, p. 187.

260 Advanced Research Projects Agency: Lloyd Norman, "G.E. Executive Chosen to Head Space Agency," *Chicago Tribune*, February 8, 1958, p. 2.

260 "It doesn't look so screwball": Jack Manno, *Arming the Heavens: The Hidden Military Agenda for Space, 1945–1995* (New York: Dodd, Mead, 1984), p. 52.

261 Such a nuclear arsenal: Halberstam, *The Fifties*, p. 702.

262 "Dear Jack, don't buy": Larry Sabato, *The Kennedy Half-Century: The Presidency, Assassination, and Lasting Legacy of John F. Kennedy* (New York: Bloomsbury, 2013), p. 46.

263 Select Committee on Astronautics: Mieczkowski, *Eisenhower's Sputnik Moment*, p. 141.

263 "LBJ was eager": Quoted in Dickson, *Sputnik*, pp. 151–52.

264 the National Defense Education Act: Lawrence J. McAndrews, *Broken Ground: John F. Kennedy and the Politics of Education* (New York: Routledge, 2012), p. 10.

265 Teller's group concluded: Lieutenant General Donald L. Putt, USAF deputy chief of staff development, to Hugh L. Dryden, NACA director, January 31, 1958, NASA Historical Reference Collection, NASA HQ.

268 earned the Distinguished Flying Cross: Thompson, *At the Edge of Space*, pp. 9–41.

268 On his inaugural X-15 flight: Richard Branson, *Reach for the Skies: Ballooning, Birdmen, and Blasting into Space* (New York: Penguin, 2011), p. 257.

270 "The highest priority should go": Quoted in George Robinson, "Space Law, Space War, and Space Exploration," *Journal of Social and Political Studies* 5 (Fall 1980): 165.

271 "exercising control over aeronautical": The full text of the National Aeronautics and Space Act of 1958 can be found on the NASA website, https://history.nasa.gov/spaceact.html.

271 "General Donald Putt recently called": Dr. W. H. Pickering to Dr. W. V. Houston (president of the Rice Institute), April 19, 1958, Fondren Library, NASA Files, Rice University, Houston, Texas.

272 "While we [the NACA] knew": Doolittle, quoted in Dik Alan Daso, *Doolittle: Aerospace Visionary* (Dulles, VA: Potomac Books, 2003), pp. 105–6.

272 "the most capable element in the Nation": Albon B. Hailey, "Army Fights Proposal to Transfer Its Space Experts to Civilian Agency," *Washington Post*, October 23, 1958, p. A1.

273 "Americans were no longer": "Eighth Annual Pittsburgh World Affairs Forum" (speech), Pittsburgh, Pennsylvania, April 18, 1958, item JFKSEN-0900-021, Papers of John F. Kennedy, Pre-Presidential Papers, Senate Files: Speeches and the Press, Speech Files: 1953–1960, Kennedy Library.

274 offer Eisenhower advice: John C. Donovan, *The Cold Warrior: A Policy Making Elite* (Lexington, MA: D. C. Heath, 1974), p. 134.

274 "unexpected Soviet development": Greg Herken, *Counsels of War* (New York: Knopf, 1985), p. 113.

275 "Our nation could have afforded": Quoted in Preble, "Who Ever Believed in the 'Missile Gap'?," pp. 801–26.

276 U-2 photographs proved: Glenn Hastedt, "Reconnaissance Satellites, Intelligence, and National Security," in Steven J. Dick and Roger D. Launius, eds., *Societal Impact of Spaceflight* (Washington, DC: Government Printing Office, 2007), p. 369.

277 control of outer space: Alexander McDonald, *The Long Space Age: The Economic Origins of Space Exploration from Colonial America to the Cold War* (New Haven, CT: Yale University Press, 2017), pp. 165–66.

277 "As sure as anything": Ben Price, "Medaris Has to Be out of This World," *Washington Post*, February 2, 1958, p. E3.

277 most top army brass: Hailey, "Army Fights Proposal to Transfer Its Space Experts to Civilian Agency," p. A1.

278 "discoveries that have military value": John M. Logsdon et al., eds., *Exploring the Unknown: Selected Documents in the History of the U.S. Civil Space Program*, vol. 1: *Organizing for Exploration* (Washington, DC: NASA, 1995), pp. 334–45.

278 were all incorporated into NASA: Burrows, *This New Ocean*, pp. 213–16; McDougall, *. . . the Heavens and the Earth*, pp. 170–76.

279 George C. Marshall Space Flight Center: "Dr. Wernher von Braun; First Center Director, July 1, 1960–Jan. 27, 1970," MSFC History Office, Marshall Space Flight Center, https://history.msfc.nasa.gov/vonbraun/bio.html.

279 "No doubt this mighty rocket system": *Public Papers of*

Dwight D. Eisenhower, 1960–1961 (Washington, DC: Government Printing Office, 1961), p. 690.

279 "Employees had been reassured": James R. Hansen, *Spaceflight Revolution: NASA Langley Research Center from Sputnik to Apollo* (Toronto: ChiZine Publications, 2017), p. 2.

280 NASA launched *Pioneer 1*: NASA Langley Research Center from Sputnik to Apollo, https://nssdc.gsfc.nasa.gov/nmc/spacecraft/display.action?id=1958-007A.

280 Doolittle refused: William D. Putnam and Eugene M. Emme, "I Was There: The Tremendous Potential of Rocketry," *Air and Space Magazine*, September 2012, https://www.airspacemag.com/space/i-was-there-the-tremendous-potential-of-rocketry-18946468/.

281 "the same technical value": U.S. House of Representatives Select Committee on Astronautics and Space Exploration, "Authorizing Construction for NASA," 85th Congress, science advisor James Killian (Washington, DC, 1958), pp. 9–12.

281 From 1950 to 1952: Glennan quoted in Mieczkowski, *Eisenhower's Sputnik Moment*, p. 285.

282 "One purpose of Eisenhower's strategic posture": McDougall, *. . . the Heavens and the Earth*, p. 200.

283 "the balance sheet of a year": Hanson W. Baldwin, "The *Sputnik* Era—Where the U.S. and Soviet Union Stand," *New York Times*, October 5, 1958, p. E6.

288 a "non-entity": Frank Van Riper, *Glenn: The Astronaut*

Who Would Be President (New York: Empire Books, 1983), p. 103.

289 morning of July 16, 1957: George C. Larson, "John Glenn's Project Bullet," *Air and Space Magazine*, July 2009, https://www.airspacemag.com/history-of-flight/john-glenns-project-bullet-138177585/.

289 "pretty good position": Van Riper, *Glenn*, p. 123.

292 "I have never seen anybody": George Smathers, Oral History, Lyndon Baines Johnson Presidential Library, Austin, Texas (hereafter "Johnson Library").

294 "match the Russians missile for missile": "Remarks of Senator John F. Kennedy, Jefferson-Jackson Day Dinner, Detroit, Michigan, May 23, 1959," Papers of John F. Kennedy, Pre-Presidential Papers, Senate Files: Speeches and the Press, Speech Files: 1953–1960, Kennedy Library, https://www.jfklibrary.org/Research/Research-Aids/JFK-Speeches/Detroit-MI_19590523.aspx.

295 "outer space is fast becoming the heart and soul": U.S. Congress, House Select Committee on Astronautics and Space Exploration, *The United States and Outer Space*, H.R. Rep. No. 2710, 85th Congress, 2nd Session (1959), p. 6.

296 "second-place status": Mieczkowski, *Eisenhower's Sputnik Moment*, p. 244.

8: Mercury Seven to the Rescue

300 a "mystical lure of the unknown": Ray Allen Billington, *America's Frontier Heritage* (Albuquerque: University of New Mexico Press, 1963), p. 26.

302 "It is my pleasure": Logsdon and Launius, eds., *Exploring the Unknown*, vol. 7: *Human Spaceflight: Mercury, Gemini, and Apollo* (Washington, DC: Government Printing Office, 2008).

303 "We didn't know what": Quoted in DeGroot, *Dark Side of the Moon*, p. 107.

303 The press gushed enthusiasms: James Reston quoted in Carpenter and Stoever, *For Spacious Skies*, p. 197.

304 "poked, prodded": Michael Collins, *Carrying the Fire: An Astronaut's Journey* (New York: Farrar, Straus, and Giroux, 1974), p. 27.

304 Psychologists also administered: DeGroot, *Dark Side of the Moon*, p. 105.

305 "I'd go so far as to say": Wally Schirra with Richard N. Billings, *Schirra's Space* (Annapolis, MD: Naval Institute Press, 1988), pp. 65–66.

306 "Scientist alone": Allen Ginsberg, "Poem Rocket," in *Kaddish and Other Poems* (San Francisco: City Lights Books, 1961), p. 38.

307 "opened the door": Lehman, *This High Man*, p. 46.

307 "Man is still the best computer": Quoted in Charles R. Pellegrino and Joshua Stoff, *Chariots for Apollo: The Untold*

Story Behind the Race to the Moon (New York: Avon, 1999), p. 15.

307 What von Braun envisioned: "IBM Commemorates NASA's 50th Anniversary of First U.S. Manned Space Flight and the IBMers Who Supported It," May 5, 2011, https://www-03.ibm.com/press/us/en/pressrelease/34449.wss.

308 "Kennedy identifies enthusiastically": Van Riper, *Glenn*, p. 36.

310 biggest story of the day: Max Frankel, "Soviet Rocket Hits Moon After 35 Hours; Arrival Is Calculated Within 84 Seconds; Signals Received Till Moment of Impact," *New York Times*, September 14, 1959, p. 1.

311 Kennedy essentially agreed with a snarky: Swenson, Grimwood, and Alexander, *This New Ocean*, pp. 281–88.

312 "first but, first and": Michael R. Beschloss, *The Crisis Years: Kennedy and Khrushchev, 1960–1963* (New York: HarperCollins, 1991), p. 28.

9: Kennedy for President

316 "formative stage": "U.S. Aeronautics and Space Activities," Report to Congress from the President of the United States, January 31, 1962, https://history.nasa.gov/presrep1961.pdf.

316 "a serious man on a serious mission": Quoted in Clarke, *JFK's Last Hundred Days*, p. 343.

319 "Jack was always out": Shaw, *JFK in the Senate*, p. 183.

320 a "pathetic" congressman: "Reminiscences of President Lyndon Baines Johnson," August 19, 1969, Oral History Collection, Johnson Library.

320 referred to as "the boy": O'Brien, *John F. Kennedy*, p. 430.

320 "sort of Irishman": Elizabeth Hardwick, *The Collected Essays of Elizabeth Hardwick* (New York: New York Review of Books, 2017), p. 85.

321 "I had not realized": Glennan quoted in Potter, *The Earth Gazers*, p. 171.

322 "intrinsic merit": Michael R. Beschloss, "Kennedy and the Decision to Go to the Moon," in Roger D. Launius and Howard E. McCurdy, eds., *Spaceflight and the Myth of Presidential Leadership* (Urbana: University of Illinois Press, 1997), p. 60.

322 "high-thrust space vehicles": Roger E. Bilstein, *Stages to Saturn: A Technological History of the Apollo/Saturn Launch Vehicles* (Washington, DC: Government Printing Office, 1980), p. 50.

322 "I am profoundly worried": Wernher von Braun, "Space," *Washington Post*, March 20, 1960 (reprinted in *Current News*).

322 "space is the greatest new frontier": T. Keith Glennan, comments on "Report from Outer Space," *World Wide 60*, NBC, May 14, 1960. See also Susan Landrum Magnus, "Conestoga Wagons to the Moon: The Frontier, The American Space Program, and National Identity" (PhD dissertation, Ohio State University, 1999), p. 72.

324 "undeviated Republicanism": Logsdon, *John F. Kennedy and the Race to the Moon*, p. 7.

324 "Whatever the scale and pace": Kalic, *US Presidents and the Militarization of Space, 1946–1967*, pp. 62–63.

328 "its cherished Jupiter missile": Erik Bergaust, *Wernher von Braun* (Lanham, MD: Stackpole, 1976), p. 406.

328 "the image of the god Apollo": Quoted in Courtney G. Brooks, James M. Grimwood, and Loyd S. Swenson Jr., *Chariot for Apollo: The NASA History of Manned Lunar Spacecraft to 1969* (Washington, DC: NASA History Series SP-4205, Washington, DC: Government Printing Office, 1979), p. 15.

328 "the next spacecraft beyond Mercury": Hugh C. Dryden, "NASA Mission and Long-Range Plan," in *NASA–Industry Program Plans Conference* (Washington, DC: Government Printing Office, 1960), p. 8.

329 lambasted the Johnson campaign: Larry Tye, *Bobby Kennedy: The Making of a Liberal Icon* (New York: Random House, 2016), p. 111.

329 "uneasy and joyless marriage": Richard Nixon, *RN: The Memoirs of Richard Nixon* (New York: Grosset and Dunlap, 1978), p. 215.

330 "The New Frontier of which I speak": John F. Kennedy, "Address of Senator John F. Kennedy Accepting the Democratic Party Nomination for the Presidency of the United States—Memorial Coliseum, Los Angeles," July 15, 1960, at John Woolley and Gerhard Peters, The American Presidency Project, https://www.presidency.ucsb.edu

/documents/address-senator-john-f-kennedy-accepting -the-democratic-party-nomination-for-the.

331 giant aerospace corporations: Rachel Reeves, "Aerospace: The Industry That Built South Bay" *Easy Reader News*, October 17, 2013. https://easyreadernews.com/aerospace -chronicles-industry-built-south-bay/.

331 "Even before Kennedy took office": William E. Leuchtenburg, *The American President: From Teddy Roosevelt to Bill Clinton* (New York: Oxford University Press, 2015), p. 387.

332 United States had lost its lead: "Split Over Space Issue," *New York Times*, October 26, 1960.

332 advocate for accelerated deployment: Mieczkowski, *Eisenhower's Sputnik Moment*, pp. 245–46.

333 "I could expose that phony": Quoted in Gary Donaldson, *The First Modern Campaign: Kennedy, Nixon and the Election of 1960* (Lanham, MD: Rowman and Littlefield, 2007), p. 128.

333 "This is year three": Allan C. Fisher Jr., "Exploring Tomorrow with the Space Agency," *National Geographic* 118, no. 1 (July 1960).

333 Kennedy would sometimes retell: Author interview with Ted Sorensen, July 18, 2004, Boston, Massachusetts.

334 "The people of the world": "Speech of Senator John F. Kennedy, Multnomah Hotel, Portland, OR," September 7, 1960, The American Presidency Project, http://www.pres idency.ucsb.edu/ws/index.php?pid=25675.

335 apparently spotted *Echo 1*: Chertok, *Rockets and People*, p. 47.

335 Strelka gave birth: Allison Gee, "Pushinka: A Cold War Puppy the Kennedys Loved," BBC News Magazine, January 6, 2014, www.bbc.com/news/magazine-24837199.

336 "with too many slums": "Address to the Greater Houston Ministerial Association" (video), September 12, 1960, Kennedy Library, https://www.jfklibrary.org/Asset-Viewer/ALL6YEBJMEKYGMCntnSCvg.aspx.

336 "I am tired of reading": John F. Kennedy, "Speech by Senator John F. Kennedy at a Democratic Fund-Raising Dinner in Syracuse, N.Y.," September 29, 1960, Kennedy Library, https://www.jfklibrary.org/asset-viewer/archives/JFKSEN/0912/JFKSEN-0912-021.

336 "We have been repeatedly reassured": Jonathan Croyle, "John F. Kennedy Campaigns in Syracuse in 1960," *Syracuse Post-Standard*, September 29, 2016, p. 1.

337 "I look up and see": "'Face-to-Face, Nixon-Kennedy' Vice President Richard M. Nixon and Senator John F. Kennedy Fourth Joint Television-Radio Broadcast, October 21, 1960," Kennedy Library, https://www.jfklibrary.org/Research/Research-Aids/JFK-Speeches/4th-Nixon-Kennedy-Debate_19601021.aspx.

337 "You may be ahead of us in rocket thrust": Ibid.

338 "The Republican presidential candidate": John W. Finn, "Johnson Assails U.S. Space Delay," *New York Times*, October 31, 1960.

10: Skyward with James Webb

344 "Let the word go forth": John F. Kennedy: "Inaugural Address," January 20, 1961, Kennedy Library, https://www.jfklibrary.org/learn/about-jfk/historic-speeches/inaugural-address.

345 "I was so proud of Jack": Ralph G. Martin, *A Hero for Our Time: An Intimate Story of the Kennedy Years* (New York: Macmillan, 1983), p. 12.

346 Each of these sites: Neal, Lewis, and Winter, *Spaceflight: A Smithsonian Guide*, pp. 58–59.

347 a three-stage Minuteman: T. A. Heppenheimer, *Countdown: A History of Space Flight* (New York: John Wiley and Sons, 1999), p. 146.

348 some Republicans faux-congratulated: McDougall, . . . *the Heavens and the Earth*, p. 328.

348 "Who ever believed": Mieszkowski, *Eisenhower's Sputnik Moment*, p. 246.

350 "neither fair nor carefully prepared": Bilstein, *Stages to Saturn*, p. 54.

351 "met with complete failure": T. Keith Glennan, *The Birth of NASA: The Diary of T. Keith Glennan* (Washington, DC: NASA History Office, 1993), p. 304.

351 NASA's primate program: Burrows, *This New Ocean*, p. 317.

353 seventeen candidates: Piers Bizony, *The Man Who Ran the Moon: James Webb, NASA, and the Secret History of*

Project Apollo (New York: Thunder's Mouth Press, 2006), p. 15.

354 "technological anticommunism": McDougall, . . . *the Heavens and the Earth*, pp. 344–45.

356 "blabbermouth": Bizony, *The Man Who Ran the Moon*, p. 18.

357 advice on the best person: W. Henry Lambright, *Powering Apollo: James E. Webb of NASA* (Baltimore: Johns Hopkins University Press, 1995), pp. 82–83.

358 "I want you because": Ibid., p. 84.

358 "President Kennedy said": Nola Taylor Redd, "James Webb: Early NASA Visionary," Space.com, November 21, 2017.

358 failed missions and dead astronauts: McDougall, . . . *the Heavens and the Earth*, p. 309.

360 "one of the ablest": Wolfe, *The Right Stuff*, pp. 226–27.

360 reassuring them of his commitment: Roscoe Drummond, "NASA Now Could Come into Own," *Washington Post*, February 15, 1961, p. E5.

361 "John Kennedy, perhaps for the first": Logsdon, *John F. Kennedy and the Race to the Moon*, p. 64.

361 *Vostok 1* cosmonaut: "Soviet Lands Man After Orbit of World," *Washington Post*, April 13, 1961, p. A1.

361 "We feel there is no better means": "Administrator's Presentation to the President," March 21, 1961, NASA History Office, Washington, DC.

362 JFK had a tireless advocate: McDougall, . . . *the Heavens and the Earth*, p. 317.

11: Yuri Gagarin and Alan Shepard

366 Gagarin completed a single low orbit: Gruntman, *Blazing the Trail*, p. 345.

367 "Modest; embarrasses": Asif Siddiqi, *Challenge to Apollo: The Soviet Union and the Space Race, 1945–1974* (Washington, DC: Government Printing Office, 2000), pp. 261–62.

368 "The road to the stars": Roger D. Launius, *The Smithsonian History of Space Exploration: From the Ancient World to the Extraterrestrial Future* (Washington, DC: Smithsonian Books, 2018), p. 7.

369 weighing 5 tons: Roger D. Launius, *Frontiers of Rocket Exploration*, 2nd ed. (Santa Barbara, CA: Greenwood, 2004), p. 99.

371 "walking on thin ice": Neal Thompson, *Light This Candle: The Life and Times of Alan Shepard, America's First Spaceman* (New York: Crown, 2004), p. 250.

371 "However tired anybody may be": John F. Kennedy, "The President's News Conference," April 12, 1961, at John Woolley and Gerhard Peters, The American Presidency Project, https://www.presidency.ucsb.edu/documents/the-presidents-news-conference-211.

371 "had no real grasp": Theodore Sorensen, *Kennedy* (New York: Harper and Row, 1965), p. 345.

372 the president's disposition: Hugh Sidey, *John F. Kennedy, President* (New York: Atheneum, 1964), p. 119.

373 "our two major organizational concepts": James Webb

to Keith Glennan, April 14, 1961, NASA History Office, https://history.nasa.gov/SP-4105.pdf.

374 When Shepard first learned: Francis French and Colin Burgess, *Into That Silent Sea: Trailblazers of the Space Era, 1961–1965* (Lincoln: University of Nebraska Press, 2009), p. 57.

374 "the most expensive funeral": Quoted in ibid., p. 57.

374 "I didn't like it worth a damn": Christopher Kraft, *Flight: My Life in Mission Control* (New York: Dutton, 2001), p. 132.

375 "Of course, we tried to derive": Nikita Khrushchev, *Khrushchev Remembers: The Last Testament* (Boston: Little, Brown, 1974), p. 53.

375 "the Americans talked a lot": Hugh Sidey, "How the News Hit Washington—with Some Reactions Overseas," *Life*, April 21, 1961, pp. 26–27.

375 "trying to break the chains of imperialism": McDougall, *. . . the Heavens and the Earth*, pp. 245–46.

377 "Is there any place": Hamish Lindsay, *Tracking Apollo to the Moon* (London: Springer-Verlag, 2001), p. 21.

380 "Jefferson's expeditions": Julie M. Fenster, *Jefferson's America: The President, the Purchase, and the Explorers Who Transformed a Nation* (New York: Crown, 2016), p. 367.

380 "Kennedy began to really get": Logsdon, *John F. Kennedy and the Race to the Moon*, p. 77.

382 "anguished and fatigued": Quoted in John Noble Wilford, "Race to Space, Through Lens of Time," *New York*

Times, May 23, 2011, https://www.nytimes.com/2011/05/24/science/space/24space.html.

382 a gala at the White House: Godfrey Hodgson, *JFK and LBJ* (New Haven, CT: Yale University Press, 2015), p. 109.

383 "All you bright fellows": John Noble Wilford, "Race to Space, Through the Lens of Time," *New York Times*, May 23, 2011.

384 "Do we have a chance": Quoted in Richard W. Orloff and David M. Harland, *Apollo: The Definitive Sourcebook* (New York: Springer Science and Business Media, 2006), p. 12.

385 "dramatic accomplishments in space": Lyndon B. Johnson to John F. Kennedy, April 28, 1961, LBJ Library, http://www.lbjlibrary.org/assets/uploads/news/LBJ-response-to-JFK.pdf

386 "Manned exploration of the moon": Vice President Lyndon B. Johnson to President John F. Kennedy, April 28, 1961, memorandum, LBJ Library, http://www.lbjlibrary.org/assets/uploads/news/LBJ-response-to-JFK.pdf.

387 "a performance jump by a factor 10": Wernher von Braun to Lyndon Johnson, April 29, 1961, NASA Historical Reference Collection, NASA Headquarters, Washington, DC, https://history.msfc.nasa.gov/vonbraun/documents/vp_ljohnson.pdf.

388 "We were being rushed": Quoted in Bergaust, *Wernher von Braun*, pp. 406–7.

389 secretly arranging a summit meeting: Günter Bischof,

Stefan Karner, and Barbara Stelzl-Marx, eds., "John F. Kennedy and His European Summitry in Early June 1961," in *The Vienna Summit and Its Importance in International History* (Lanham, MD: Lexington, 2014), p. 113.

390 one final test: Ernest Barcella, "Balloonist's Widow Gets Kennedy Call," *Washington Post*, May 7, 1961, p. A6.

392 "He was hard to get": Quoted in French and Burgess, *Into That Silent Sea*, p. 45.

393 "I'm cooler than you are": Quoted in John Noble Wilford, "Alan B. Shepard Jr. Is Dead at 74; First American to Travel in Space," *New York Times*, July 23, 1998, p. 1.

395 "Boy, what a ride!": quoted in Richard Witkin, "U.S. Hurls Man 115 Miles into Space," *New York Times*, May 6, 1961, p. 1.

395 "My name is José Jimenez": Thompson, *Light This Candle*, p. 259.

395 "Hello, commander": Ibid., p. 306.

396 "the greatest 'suspense drama'": Howard Stentz, "Space Shot Top Television Thriller," *Houston Chronicle*, May 5, 1961, p. 16.

396 "a symbol of the twentieth century": Quoted in Logsdon, *John F. Kennedy and the Race to the Moon*, p. 225.

397 "We had a big laugh": Quoted in French and Burgess, *Into That Silent Sea*, p. 74.

398 "even more thrilled": Alan B. Shepard Oral History, June 12, 1964, pp. 1–2, John F. Kennedy Oral History Program, Kennedy Library.

398 "Thus Alan Shepard": Steven Watts, *JFK and the Masculine Mystique: Sex and Power on the New Frontier* (New York: Thomas Dunne/St. Martin's, 2016), p. 342.

399 TASS's rote criticism: Von Hardesty and Gene Eisman, *Epic Rivalry: The Inside Story of the Soviet and American Space Race* (Washington, DC: National Geographic Society, 2007), p. 125.

399 "incentive to everyone": News Conference 11, May 5, 1961. Kennedy Library, https://www.jfklibrary.org/archives/other.resources/john-f-kennedy-press-conference-11.

400 "In choosing the lunar landing mission": Logsdon, *The Decision to Go to the Moon*, p. 124.

400 the point of Project Apollo: McDonald, *The Long Space Age*, pp. 166–67.

401 "He talked to them": "Posthumous DFC Given to Prather," *Washington Post*, May 21, 1961, p. B4.

12: "Going to the Moon": Washington, DC, May 25, 1961

405 The White House had billed: W. H. Lawrence, "Kennedy Asks 1.8 Billion This Year to Accelerate Space Exploration, Add Foreign Aid, Bolster Defense," *New York Times*, May 26, 1961, p. 1.

406 "As far as President Kennedy": Quoted in Deborah Hart Strober and Gerald S. Strober, *The Kennedy Presidency: An Oral History of the Era* (Washington, DC: Brassey's, 2003), pp. 248–49.

406 consult with a wide array: Theodore Sorensen, *Counselor: A Life at the Edge of History* (New York: HarperCollins, 2008), p. 336.

407 "You're the people": Hugh Dryden, interview by Walter D. Sohier, Arnold Frutkin, and Eugene M. Emme, March 26, 1964, p. 18, John F. Kennedy Oral History Program, Kennedy Library.

407 moon was starting to win out: Wolfe, *The Right Stuff*, pp. 218–19.

408 "JFK was obviously prepping": Paul Haney, foreword to French and Burgess, *Into That Silent Sea*, p. xviii.

408 "public discovery": Daniel Boorstin, "The Rise of Public Discovery," in John M. Logsdon et al., *Apollo in Historical Context* (Washington, DC: Space Policy Institute/George Washington University, 1990), p. 21.

410 "Well, I'm glad they got": Quoted in John Glenn with Nick Taylor, *John Glenn: A Memoir* (New York: Bantam/Random House, 1999), p. 275.

411 Congress had already signaled: "Dollars for Man to Go to Moon," *Business Week*, June 3, 1961, p. 18.

412 "What we had in mind": James E. Webb, interview with H. George Frederickson, Henry J. Anna, and Barry Kelmachter, May 15, 1969, NASA Historical Reference Collection, NASA Headquarters, Washington, DC.

412 "I'm a relatively cautious person": Lambright, *Powering Apollo*, p. 95.

414 "Every time the Air Force": Paul B. Stares, *Space Weapons and US Strategy* (London: Croom Helm, 1985), p. 64.

417 no "man-on-the-moon" requirement: U.S. Congress, Senate Committee on Armed Services, Military Procurement Authorization for Fiscal Year 1964, Hearing, 88th Cong., 1st sess. (1963), p. 152.

419 Kennedy would be pushing: W. H. Lawrence, "President to Ask an Urgent Effort to Land on Moon," *New York Times*, May 23, 1961, p. 1.

420 to buy NASA more time: Mike Wall, "The Moon and the Man at 50: Why JFK's Space Exploration Speech Still Resonates," Space.com, May 25, 2011.

420 by ad-libbing: The John F. Kennedy Presidential Library has various versions of the speech. The original reading copy of the speech, with Kennedy's edits, can be located in Papers of John F. Kennedy, Presidential Papers, President's Office Files, Box 34, Kennedy Library.

421 "our skills and our capital": John F. Kennedy, "Special Message to the Congress on Urgent National Needs," May 25, 1961, at John Woolley and Gerhard Peters, The American Presidency Project, https://www.presidency.ucsb.edu/documents/special-message-the-congress-urgent-national-needs.

421 "If we are to win the battle": Ibid.

422 "We go into space": Ibid.

423 "I believe this nation should commit": Ibid.

423 "his audience was skeptical": Theodore Sorensen, *Counselor: A Life at the Edge of History* (New York: HarperCollins, 2008), p. 336.

424 "stunned doubt and disbelief": Logsdon, *John F. Kennedy and the Race to the Moon*, p. 115.

424 Republican leaders scribbled notes: Alvin Shuster, "Congress Wars on Cost, But Likes Kennedy's Goals," *New York Times*, May 26, 1961, p. 13.

425 "The president is ahead": Robert C. Albright, "President's Arms, Space Aims Get Full Backing by Congress," *Washington Post*, May 27, 1961, p. A6.

426 "doubting the value": Robert Seamans, *Aiming at Targets* (Washington, DC: Government Printing Office, 1997), p. 88.

426 "our debt may reach": "Public Debt Limit, Hearing Before the Committee on Finance," U.S. Senate, 87th Cong., 1st sess., June 27, 1961 (Washington, DC: Government Printing Office, 1961).

427 "Damn it, I taught": Quoted in Beschloss, *The Crisis Years*, p. 166.

427 "make the so-called race": *Washington* (DC) *Sunday Star*, June 13, 1971.

428 "anybody who would spend $40 billion": Mieczkowski, *Eisenhower's Sputnik Moment*, p. 269.

428 "Space technology will eventually": 107 Cong. Rec. S917496 (June 8, 1961).

428 "It will cost thirty-five billion": Quoted in Mieczkowski, *Eisenhower's Sputnik Moment*, p. 265.

429 "My head seemed to fill with fog": Kraft, *Flight*, p. 143.

429 "We've only put Shepard": Ibid.

429 Robert Gilruth: Charles Murray and Catherine Bly Cox, *Apollo* (New York: Simon and Schuster, 1989), p. 4.

431 "Of course, the moon": Neufeld, *Von Braun*, p. 354.

431 "What Kennedy did with the moon program": Quoted in Strober and Strober, *The Kennedy Presidency*, pp. 251–52.

433 "what would propel us to the moon": Quoted in Neil McAleer, *Arthur C. Clarke* (Chicago: Contemporary Books, 1992), p. 216.

433 "power to persuade": Richard E. Neustadt, *Presidential Power: The Politics of Leadership* (New York: Wiley, 1960), p. 10.

433 "greatest open-ended peacetime commitment": McDougall, *. . . the Heavens and the Earth*, p. 305.

433 NASA's annual operating budget: James L. Kauffman, *Selling Outer Space: Kennedy, the Media, and Funding for Project Apollo, 1961–1963* (Tuscaloosa: University of Alabama Press, 1994), p. 2.

433 "spirit of discovery": Sorensen, *Kennedy*, p. 525.

13: Searching for Moonlight in Tulsa and Vienna

436 "seething with excitement": "The Moon by 1967 or Bust," *Business Week*, June 3, 1961, p. 18.

436 "deals with the very heart": John F. Kennedy: "Remarks by Telephone to the Conference on Peaceful Uses of Space Meeting in Tulsa," May 26, 1961. Online at Gerhard Peters and John T. Woolley, The American Presidency Project, https://www.presidency.ucsb.edu/documents/remarks-telephone-the-conference-peaceful-uses-space-meeting-tulsa.

438 "Man has progressed": United Press International, "Moon Landing Sure, Top Scientists Say," May 26, 1961.

441 "If you want me in the landings": A. M. Sperber, *Murrow: His Life and Times* (New York: Fordham University Press, 1998), p. 624.

442 "Landlords will not rent": Edward R. Murrow, May 24, 1961, National Press Club Luncheon Speakers, Library of Congress Recorded Sound Research Center, Library of Congress, Washington, DC, https://www.loc.gov/rr/record/pressclub/murrow.html.

444 "The first colored man": Edward R. Murrow to President John F. Kennedy, April 23, 1962, "Memorandum for the President," Kennedy Library.

445 "The Russians are now graduating": Quoted in "An American on the Moon—A $20 Billion Boondoggle?" *U.S. News & World Report*, August 20, 1962, p. 59.

446 "I am quite sure": "Thirring: Sending a Man in Space Is Nonsense," *U.S. News & World Report*, May 22, 1961, p. 49.

446 "really rather a nuisance": "Man Is a Nuisance in Space," *U.S. News & World Report*, August 20, 1962, p. 56.

446 "I'm not saying that it's unwise": Ibid., p. 57.

447 dark side of the moon program: Elizabeth Gibney, "The Quest to Crystallize Time," *Nature* 543 (March 9, 2017): 165.

449 "What was difficult for us": Eric Berger, "JFK's Speech Today Would Be Hard to Believe," *Houston Chronicle,* September 12, 2012, p. 1.

450 the best general design: Anthony Young, *The Saturn V F-1 Engine: Powering Apollo into History* (New York: Springer Science and Business Media, 2008), p. 18.

450 secure a massive government contract: John W. Finney, "Capital Worried by Lags in Plans on Race to Moon," *New York Times,* August 13, 1961, p. 1.

453 "We thought it was too risky": Quoted in James R. Hansen, "Enchanted Rendezvous: John C. Houbolt and the Genesis of the Lunar-Orbit Rendezvous Concept," *Monograph in Aerospace History,* ser. 4 (January 25, 1999), p. 4, https://nasa.gov.archive/nasa/casi.ntrs.nasa.gov/1996 0014824.pdf.

453 "Houston's first reaction": Quoted in Bergaust, *Wernher von Braun,* p. 408.

454 "No, Jim, I cannot bring": Keith Glennan to James Webb, July 21, 1961, Glennan Personal Papers, Archives, Kevin Smith Library, Case Western Reserve University, Cleveland, Ohio.

455 five times the speed of sound: "X-15 Rocket Plane Flies 3,370 M.P.H.," *Chicago Tribune,* May 26, 1961, p. 2.

456 American commitment to space: "Remarks at the Dedication of the Aerospace Medical Health Center, San Antonio, Texas, November 21, 1963," *Public Papers of the Presidents of the United States: John F. Kennedy, 1963* (Washington, DC: Government Printing Office, 1964), p.

882, https://www.jfklibrary.org/Research/Research-Aids/JFK-Speeches/San-Antonio-TX_19631121.aspx.

457 "With respect to the possibility": The President to Chairman Khrushchev, Vienna, June 3, 1961, Memorandum of Conversation, p. 2, Office of the Historian, https://history.state.gov/historicaldocuments/frus1961-63v24/d107.

458 "At first he said no": Ibid.

458 "if we cooperate": Quoted in James Schefter, *The Race* (New York: Doubleday, 1999), p. 145.

458 categories of "competition" and "cooperation": W. D. Kay, "John F. Kennedy and the Two Faces of the U.S. Space Program," *Presidential Studies Quarterly* 28, no. 3 (Summer 1998): 576–79.

459 "Mr. Kennedy's reaction": James Reston, "Kennedy Is Firm on Defense Aims," *New York Times*, June 6, 1961, p. 1.

14: Moon Momentum with Television and Gus Grissom

463 "You know, Lyndon": Robert Dallek, "Johnson, Project Apollo, and the Politics of Space Program Planning," in Launius and McCurdy, eds., *Spaceflight and the Myth of Presidential Leadership*, p. 72. See also Newton Minow, interview, March 19, 1971 (oral history), Johnson Library.

464 "Space was the platform": Johnson, *Vantage Point*, p. 285.

465 NASA's proposed Manned Spacecraft Center: McDougall, . . . *the Heavens and the Earth*, pp. 302–3. See also Minow interview.

465 "Many friends of Lyndon Johnson": Shepard, Slayton, with Barbree, *Moon Shot,* p. 165.

467 "Maybe I watch more newscasts": Don Mahan, *ETC: A Review of General Semantics* 27, no. 1 (March, 1964): 114–15.

469 "Kennedy felt very strongly": Pierre E. G. Salinger, recorded interview by Theodore H. White, August 10, 1965 (no. 2), p. 108, John F. Kennedy Library Oral History Program, Kennedy Library.

469 every sixteen days: Ibid., p. 109.

470 "reveal his innate characteristics": Virginia Kelly, "Will TV Audience Get Too Much JFK?" *Long Beach Independent Press-Telegram,* February 19, 1961, p. 14.

471 "the real factor in all of this": "Dangerous White House Procedure," *Shreveport* (LA) *Times,* January 31, 1961, p. 6.

471 "undisguised exercises": "The Skill of President Kennedy," *Los Angeles Times,* February 3, 1961, p. 41.

471 "Whatever the doubts": Ibid.

475 "while the eyes": Walter Cronkite, quoted in DeGroot, *Dark Side of the Moon,* p. 253.

475 hosting the network's live coverage: Lyle Johnston, *"Good Night, Chet"* (Jefferson, NC: McFarland, 2003), p. 83.

476 "an open policy" of space information: John Krige, "NASA as an Instrument of U.S. Foreign Policy," in Dick and Launius, eds., *Societal Impact of Spaceflight,* pp. 210–11.

476 "competition without war": McDougall, *. . . the Heavens and the Earth,* p. 241.

476 "between freedom and tyranny": John F. Kennedy, "Special Message to the Congress on Urgent National Needs," May 25, 1961, at John Woolley and Gerhard Peters, The American Presidency Project, https://www.presidency.ucsb.edu/documents/special-message-the-congress-urgent-national-needs.

478 acquired the nickname "Gus": Betty Grissom and Henry Still, *Starfall* (New York: Thomas Y. Crowell, 1974), pp. 8–12.

478 "I usually flew wing position": Quoted in John Darrell Sherwood, *Officers in Flight Suits: The Story of American Air Force Fighter Pilots in the Korean War* (New York: New York University Press, 1996), pp. 11–13.

479 "a little bear": Gordon Cooper with Bruce Henderson, *Leap of Faith: An Astronaut's Journey into the Unknown* (New York: HarperCollins, 2000), p. 22.

480 "I can see the coast": "Mercury-Redstone 4 Mission Journal, Friday, July 21, 1961," *John Pfannerstill's Space Chronicle*, NASA History Office, Washington, DC, https://history.nasa.gov/40thmerc7/MR-4.html.

481 "Well, I was scared": Mary C. White, "Detailed Biographies of Apollo I Crew—Gus Grissom," NASA History Office, https://history.nasa.gov/Apollo204/zorn/grissom.htm.

482 "the great Cape Canaveral tape caper": See, for instance, *Harrisburg* (IL) *Daily Register*, July 26, 1961.

482 "It won't happen again": United Press International, "U.S. Orbit Shot Next Year to Avoid Tape 'Booboo,'" *Cumberland* (MD) *Evening Times*, July 26, 1961.

483 "That was the last thing he wanted": Colin Burgess, *Lib-*

erty Bell 7: The Suborbital Mercury Flight of Virgil I. Grissom (Cham, Switzerland: Springer, 2014), p. 149.

483 "was angry about being blamed": Quoted in Francis French and Colin Burgess, *In the Shadow of the Moon* (Lincoln: University of Nebraska Press, 2007), p. 1.

483 "Once again we have demonstrated": "Space Bill Is Approved by Kennedy," *Spokane Daily Chronicle*, July 21, 1961, p. 11.

484 the *Republic of Technology*: Quoted in McDougall, . . . *the Heavens and the Earth*, p. 452.

484 "Ever since Kennedy declared": Quoted in Neufeld, *Von Braun*, pp. 364–65.

484 "big-spending and self-promoting ways": Ibid.

487 "As planning for Apollo began": Robert Seamans, introduction to 1975 edition of Robert Cortright, ed., *Apollo Expeditions to the Moon: The NASA History* (Mineola, NY: Dover, 2009), p. xii.

488 "a sophisticated senior official": James Webb, "A Perspective on Apollo," in Cortright, ed., *Apollo Expeditions to the Moon*, p. 12.

488 "My answer was just as direct": Ibid.

489 "The fact that every part": Alan Shepard quoted in Gene Kranz, *Failure Is Not an Option: Mission Control from Mercury to Apollo 13 and Beyond* (New York: Simon and Schuster, 2000), p. 201.

490 the geographical advantages: Howard Simons, "Canaveral Selected for Moon Shots," *Washington Post*, August 25, 1961, p. A4.

491 buying eighty thousand acres: United Press International, "Expansion for Moon Shots Finds Canaveral Area Set," *Washington Post*, August 27, 1961, p. A6.

491 "The American test site": Walter Cronkite, *A Reporter's Life* (New York: Random House, 1996), p. 272.

492 difference in the two reactions: "A City with Growing Pains," *Palm Beach Post*, July 11, 1965, p. 66.

493 an important role in raising consciousness: W. Henry Lambright, "NASA and the Environment: Science in a Political Context," in Dick and Launius, eds., *Societal Impact of Spaceflight*, pp. 313–30.

494 Manned Spacecraft Center to Houston: Eric Berger, "A Worthy Endeavor: How Albert Thomas Won Houston NASA's Flagship Center," *Houston Chronicle*, September 14, 2003.

497 "like a divine right monarch": Jack Valenti, *This Time, This Place* (New York: Crown, 2007), p. 30.

497 "The key to the selection": Quoted in William D. Angel Jr., "The Politics of Space: NASA's Decision to Locate the Manned Spacecraft Center in Houston," *Houston Review* 6, no. 2 (1984): 66.

498 waves of astromania swept: Kevin M. Brady, "NASA Launches Houston into Orbit: The Economic and Social Impact of the Space Agency on South Texas, 1961–1969," in Dick and Launius, eds., *Societal Impact of Spaceflight*, pp. 451–65, https://history.nasa.gov/sp4801-chapter23.pdf.

499 to create a planned community: Ibid., pp. 452–53. The NBA franchise the Rockets was originally founded in San

Diego in 1967 because the Atlas rocket was manufactured in Southern California.

15: Godspeed, John Glenn

504 countdown had begun: "U.S. Follows Launching," *New York Times*, August 7, 1961, p. 7.

505 17.5 orbits of Earth: Details of the *Vostok 2* can be found at Anatoly Zak, RussiaSpaceWeb.com, https://spaceflight.nasa.gov/outreach/SignificantIncidents/assets/vostok-2-mission.pdf.

507 boosted Kremlin military spending: Arthur M. Schlesinger Jr., *A Thousand Days: John F. Kennedy in the White House* (Boston: Houghton Mifflin, 1965), p. 385.

507 ratcheted up his rhetoric: Bischof, Karner, and Stelzl-Marx, "Introduction," in *The Vienna Summit and Its Importance in International History*, p. 24.

508 the *Pittsburgh Press* fretted: "East Germans Fleeing Reds One a Minute," *Pittsburgh Press*, August 6, 1961, p. 1.

508 "looked about the size of a marble": Associated Press, "Satellite Is Seen in S.C.," *Greenwood* (SC) *Index-Journal*, August 7, 1961, p. 5.

509 "the lag can never be made up": "Soviet Shot Viewed by World as Big Advance in Space Race," *New York Times*, August 7, 1961, p. 7.

509 ninety-minute monologue broadcast: *Pittsburgh Press*, August 7, 1961, p. 1.

511 "antifascist protection rampart": Tye, *Bobby Kennedy,* p. 246.

511 "It's not a very nice solution": John Lewis Gaddis, *The Cold War: A New History* (London: Penguin Press, 2005), p. 115.

512 "permanently kill[ed] Mercury-Redstone 5": Donald K. Slayton and Michael Cassutt, *Deke! U.S. Manned Space Flight: From Mercury to Shuttle* (New York: Tom Doherty Associates, 1994), p. 104.

513 "The next flight": Donald K. Slayton, *Deke! U.S. Manned Space: From Mercury to the Shuttle* (New York: Forge Books, 1995).

513 Mercury capsule was fired off: Burrows, *This New Ocean,* p. 340.

514 Holmes established a third NASA project: Lambright, *Powering Apollo,* pp. 108–11.

515 "The chimpanzee who is flying": Theodore C. Sorensen, *"Let the Word Go Forth": The Speeches, Statements and Writings of John F. Kennedy, 1947 to 1963* (New York: Delacorte, 1988), p. 174.

516 "less than a 50-50 chance": John Troan, "January Still Target of Moon Shot; 2 Test Failures May Not Delay On-Spot Probe," *Pittsburgh Press,* November 21, 1961, p. 24.

516 "The Russians will be able": Henry A. Berry Jr., "Anti-Satellite Weapon for Soviets Seen," *Shreveport* (LA) *Times,* December 1, 1961, p. 1.

517 "Never before has a major scientific venture": George J.

Feldman, "Spacemen and the Law—Or Lack of It," *New York Times*, October 8, 1961, p. M17.

518 was suddenly delayed: Drew Pearson, "Washington Merry-Go-Round," *San Mateo* (CA) *Times*, February 24, 1962, p. 14.

518 "hurry-up plans": Jerry T. Baulch, "U.S. Delays Manned Space Shot," *Troy* (NY) *Record*, December 7, 1961, p. 1.

521 "Gemini's a Corvette": Quoted in Walter Cunningham, foreword to French and Burgess, *In the Shadow of the Moon*, p. 2.

522 "This nation belongs among": "Transcript of the President's Address to Congress on Domestic and World Affairs," *New York Times*, January 12, 1962, p. 12.

524 "one human being": John H. Glenn Jr., recorded interview by Walter D. Sohier, June 12, 1964, pp. 1–2, John F. Kennedy Library Oral History Program, Kennedy Library.

524 the astronaut offered to come back: Glenn and Taylor, *John Glenn*, p. 394.

525 "John tries to behave": John Dille, introduction to William E. Burrows et al., *We Seven: By the Astronauts Themselves* (New York: Simon and Schuster, 1962), p. 13.

526 "Don't be scared": Glenn and Taylor, *John Glenn*, p. 258.

527 "[M]ay the good Lord": Ibid., p. 344.

527 businesses locked their doors: "He's Off!" *Dover* (OH) *Daily Reporter*, February 20, 1962, p. 18.

527 a bank robber got away: "Two Enter Guilty Plea in Holdup," *Allentown* (PA) *Morning Call*, February 23, 1962, p. 9.

528 Michigan Bell Telephone: "Time Almost Stood Still in

Michigan," *Traverse City* (MI) *Record-Eagle*, February 21, 1962, p. 9.

528 Casey Stengel: "Everything Halts for Glenn's Hop," *Titusville* (FL) *Herald*, February 21, 1962, p. 1.

528 On site at Cape Canaveral: "The Man in the Street," *Fremont* (OH) *News-Messenger*, February 28, 1962, p. 28; Seymour Beubis, "'Go, Baby, Go,' Shouted at Atlas," *Fort Lauderdale News*, February 20, 1962, p. 10.

529 "Wonderful as man-made art may be": Glenn and Taylor, *John Glenn*, p. 350.

529 for ten straight hours: Douglas Brinkley, *Cronkite* (New York: Harper, 2012), p. 233.

529 "united the nation and the world": Jack Gould, "Radio, TV Networks Convey Drama of Glenn Feat," *New York Times*, February 21, 1962, p. 91.

530 "I knew that if the shield": Quoted in Colin Burgess, *Friendship 7: The Epic Orbital Flight of John H. Glenn, Jr.* (New York: Springer, 2015), p. 138.

531 "How do you feel": Thompson, *Light This Candle*, p. 275.

532 outpouring of love and excitement: Michael J. Neufeld, "Mercury Capsule *Friendship 7*," in Michael J. Neufeld, ed., *Milestones of Space: Eleven Iconic Objects from the Smithsonian National Air and Space Museum* (Minneapolis: Zenith Press, 2014), pp. 16–19.

532 "The best moment": Walter Cronkite, "Outstanding Moments During TV Coverage of John Glenn," *New York Herald Tribune*, April 29, 1962.

532 "The distance to the moon": Shepard and Slayton, with Barbree, *Moon Shot,* p. 152.

532 "Orbit Day": Bob Wells, "A World Symbol for Space Age," *Long Beach Independent*, February 21, 1962, p. 3.

533 "vastly impressed by John Glenn": C. L. Sulzberger, *The Last of the Giants* (New York: Macmillan, 1970), p. 915.

534 "Here I am in Lucerne": Quoted in Ward, *Dr. Space*, p. 131.

535 "We were on the plane," Glenn and Taylor, *John Glenn*, pp. 372–73.

535 "I still get a hard to define feeling inside": Glenn quoted in DeGroot, *Dark Side of the Moon*, p. 159.

535 "We have a long way": Quoted in Ward, *Dr. Space*, p. 131.

536 "risk putting him back in space again": Bob Jacobs, "NASA Remembers American Legend John Glenn," Washington, DC, NASA Headquarters, December 8, 2016, https://www.nasa.gov/press-release/nasa-remembers-american-legend-john-glenn.

536 "When I came back": John Glenn, oral history interview, June 12, 1964, pp. 4–5, Kennedy Library, www.jfklibrary.org/sites/default/files/archives/JFKOH/Glenn%2C%20John%20H/JFKOH-JHG-01/JFKOH-JHG-01-TR.pdf.

538 "I think early in the program": Ibid.

539 "His vision set an inspiring example": Glenn and Taylor, *John Glenn*, p. 373.

540 "It was so wonderful": Author interview with Ethel Kennedy, March 14, 2018, Palm Beach, Florida.

541 "May I humbly offer": Martin W. Sandler, ed., *The Letters of John F. Kennedy* (New York: Bloomsbury, 2013), p. 150.

541 "If our countries pooled": Chairman Khrushchev to President Kennedy, February 21, 1962, Moscow, Historical Documents, *Foreign Relations of the United States, 1961–1963*, vol. 6, document 35, Office of the Historian, Department of State, Washington, DC, https://history.state.gov/historicaldocuments/frus1961-63v06/d35. See also Preston Grover, "K Calls for Joint U.S.-Soviet Space Effort," *Washington Post*, February 22, 1962, p. A16.

542 "I am replying to his message": "Text of President Kennedy's Conference with the Press: Response to Khrushchev," *Washington Post*, February 22, 1962, p. A10.

543 commemorative stamps: Cathleen S. Lewis, "The Birth of the Soviet Space Museum: Creating the Earthbound Experience of Space Flight During the Golden Years of the Soviet Space Programme, 1957–68," in Martin Collins and Douglas Millard, eds., *Studies in the History of Science and Technology*, Artefacts Series (East Lansing: Michigan State University Press, 2005).

543 "a well-thought-out scientific": Teasal Muir-Harmony, "*Friendship* 7's 'Fourth Orbit,'" in Neufeld, ed., *Milestones of Space*, pp. 16–17.

544 "100,000 Foot Club": Alan J. Stein, "Astronaut John Glenn Visits the Seattle World's Fair on May 10, 1962," www.historylink.org/File/3697.

16: Scott Carpenter, *Telstar,* and Presidential Space Touring

548 "He has yet to provoke": George Gallup, "Critics of Kennedy Lack Unity of Aims," (Phoenix) *Arizona Republic,* April 2, 1962, p. 15.

548 "acquire billions' worth": Bergaust, *Wernher von Braun,* p. 411.

549 "from muskrats to moon ships": "NASA's Michoud Assembly Facility," NASA, www.nasa.gov/centers/marshall/michoud/maf_history.html.

549 Two hundred times heavier: John Noble Wilford, "Wernher von Braun, Rocket Pioneer, Dies," *New York Times,* June 18, 1977, pp. 1, 19.

550 all three networks broadcast: Barbara A. Perry, *Jacqueline Kennedy: First Lady of the New Frontier* (Lawrence: University Press of Kansas, 2004), p. xiv.

550 three out of four TVs: Mary Ann Watson, "A Tour of the White House: Mystique and Tradition," *Presidential Studies Quarterly* 18, no. 1 (Winter 1988): 95.

551 The army was the most cooperative branch: Stephen B. Johnson, "The History and Historiography of National Security Space," in Steven J. Dick and Roger D. Launius, eds., *Critical Issues in the History of Spaceflight* (Washington, DC: NASA, 2006), p. 536.

553 "could render no greater service": John Kennedy to Nikita Khrushchev, March 7, 1962, item JFKWHSFPS-010-010, Papers of John F. Kennedy, Presidential Papers, White

House Staff: Files of Pierre Salinger, Subject Files: 1961–1964, Khrushchev/Kennedy letters: 7 March 1962–20 January 1963, Kennedy Library.

555 "Let the atom be a worker": Quoted in United Press International, "U.S. Soviet Plan Exchange on Atom Ideas," *Tyrone* (PA) *Daily Herald*, May 21, 1963, p. 2.

556 "Ever since the longbow": Hugh Sidey, introduction to John F. Kennedy, *Leadership: The European Diary of John F. Kennedy: Summer 1945* (Washington, DC: Regnery, 1995), pp. xix–xxi.

556 authorized a leading Southern California think tank: Stares, *Space Weapons and US Strategy*, p. 74.

557 Strategic Defense Initiative: Sharon Begley, "A Safety Net Full of Holes," *Newsweek*, March 22, 1992; William J. Broad, "Technical Failures Bedevil Star Wars," *New York Times*, September 18, 1990; Kevin Crowley, "The Strategic Defense Initiative (SDI): Star Wars," The Cold War Museum, Warrenton, VA, www.coldwar.org/articles/80s/SDI-StarWars.asp.

558 "[The president] made it clear": Glenn T. Seaborg, oral history, June 11, 1964; June 18, 1964; June 25, 1964; June 27, 1964; July 1, 1964 (no. 1), p. 85, John F. Kennedy Oral History Program, Kennedy Library, https://www.jfklibrary.org/Asset-Viewer/Archives/JFKOH-GTS-01.aspx.

558 "On April 23": Ibid., p. 94.

559 discuss aerospace technology: "Wilton Talk Slated by Defense Official," *Bridgeport* (CT) *Post*, May 7, 1962, p. 47.

559 "long-standing proposal": Richard Witkin, "Pentagon to

Push Space Plan Study," *New York Times*, May 13, 1962, p. 56.

560 "The furor that greeted": Stares, *Space Weapons and US Strategy*, p. 77.

561 "The reports from [our] representatives": Seaborg, oral history interview, 1965 (no. 1), p. 99, Kennedy Library; Carroll Kirkpatrick, "Kennedy Hails Flight in Talk with Carpenter," *Washington Post*, May 25, 1962, p. A8.

564 "He was completely ignoring": Kraft, *Flight*, p. 170.

564 In preparation for reentry: Carpenter and Stoever, *For Spacious Skies*, pp. 294–95.

566 "President Kennedy bent down": Ibid., pp. 304–5.

566 "One might argue that Carpenter": Wolfe, *The Right Stuff*, p. 315.

569 "I know there has been disturbance": John F. Kennedy, "The President's News Conference," May 9, 1962, Kennedy Library, http://www.presidency.ucsb.edu/ws/index.php?pid=8642.

570 "the most intricate instrument": John F. Kennedy, "Address at Rice University in Houston on the Nation's Space Effort," September 12, 1962, at Woolley and Peters, The American Presidency Project, www.presidency.ucsb.edu/ws/index.php?pid=8862.

572 "Telstar was the first true communications satellite": Neufeld, "Mercury Capsule *Friendship 7*," in *Milestones of Space*, p. 25.

575 Rockefeller was leading Goldwater: "Romney Gaining in

Popularity as G.O.P. Presidential Choice," *St. Louis Post-Dispatch*, July 17, 1962, p. 23.

575 "The clock has already run too long": "Barry Calls for All-Out Military Space Program," *Tucson Daily Citizen*, July 17, 1962, p. 6.

575 "The armed forces should already": United Press International, "Space Warfare Plans Urged," *Pittsburgh Press*, July 17, 1962, p. 10.

576 the Senate opened a debate: John G. Norris, "Senate Hears Demands for Building Strong Military Capability," *Washington Post*, August 21, 1962, p. A1.

577 "The United States believes": "Kennedy to Tour Space Facilities," *New York Times*, September 6, 1962, p. 16.

578 selection of Hancock County: Robert Dallek, *An Unfinished Life: John F. Kennedy, 1917–1963* (Boston: Little, Brown, 2003), p. 515.

578 "This is the vehicle designed": "JFK Sees Saturn Test at Huntsville Center," *Montgomery Advertiser*, September 12, 1962, p. 1.

579 "Just as the last echoes": Bergaust, *Wernher von Braun*, p. 207.

579 "I understand that Dr. Wiesner": Bilstein, *Stages to Saturn*, p. 67; Logsdon, *John F. Kennedy and the Race to the Moon*, p. 144.

580 "Look at von Braun": Quoted in Bergaust, *Wernher von Braun*, p. 207.

581 "Jerry's going to lose it": Quoted in Charles Murray and

Catherine Bly Cox, *Apollo: Race to the Moon* (New York: Simon and Schuster, 1989), p. 143.

581 "Who said John Glenn": "JFK Cape Briefing 'Brief,'" *Orlando Evening Star*, September 2, 1962, p. 2.

581 "We shall be first": Quoted in Ward, *Dr. Space*, p. 132.

581 the two Mercury astronauts: Jay Barbree, *"Live from Cape Canaveral": Covering the Space Race, From Sputnik to Today* (New York: Smithsonian/Collins, 2007), p. 98.

17: "We Choose to Go to the Moon": Rice University, September 12, 1962

588 "I do not know whether": John F. Kennedy, quoted in *Houston Press*, September 12, 1962 (Rice University Special Collections).

588 "Some from Texas might disagree": Quoted in John Lewis Gaddis, *On Grand Strategy* (New York: Penguin Press, 2018), pp. 311–12.

589 "The people here realize": "Kennedy's Visit," *Houston Chronicle*, September 12, 1962, sec. 1, p. 20.

589 "May God continue to guard": Forrest Fischer, "Kennedy Puts U.S. in Orbit . . . Hails Heat of Space Effort," *Houston Press*, September 12, 1962.

589 "Sixty firms have moved": "JFK and NASA," p. 10.

590 "everyone perspired" in the "roaster": Marie Dauplaise, "Dignitaries Sat Close to JFK—But Never Heard a Word," *Houston Press*, September 12, 1962, p. 2.

591 Pitzer committed the university: Jessica P. Cannon, "Owls

in Space: Rice University's Connections to NASA Johnson Space Center," *Houston in History* 6, no. 1 (1963): 33.

592 "educational pilot plant": "JFK and NASA," p. 10.

592 "I can remember it clearly today": Eric Berger, "JFK's Speech Today Would Be Hard to Believe," *Houston Chronicle*, September 12, 2012.

593 "We meet at a college noted for knowledge": Kennedy, "Address at Rice University in Houston on the Nation's Space Effort," https://www.jfklibrary.org/asset-viewer/archives/JFKPOF/040/JFKPOF-040-001.

593 "Surely the opening vistas": Ibid.

594 "The exploration of space": Ibid.

594 "politically uncommon fiscal candor": Neil deGrasse Tyson, *Space Chronicles* (New York: W. W. Norton, 2002), pp. 1–5.

594 "This year's space budget": Kennedy, "Address at Rice University in Houston on the Nation's Space Effort."

595 "We choose to go to the moon": Ibid.

596 "British explorer George Mallory": "Address at Rice University, Houston, Texas, 12 September 1962," p. 3, Papers of John F. Kennedy, Presidential Papers, President's Office Files, Speech Files, Kennedy Library.

596 the stadium erupted in applause: "46 Keel Over from Heat During President's Speech," *Houston Chronicle*, September 12, 1962, sec. 1, p. 2.

596 "I remember the times": Mark Carreau, "The Quest Begins: At Rice, President Kennedy Inspired a Nation to Look at the Stars," *Houston Chronicle*, October 10, 2002, p. 46.

597 "Now you guys do the details!": Ted Sorensen told this story on *Focus on Youth*, a syndicated radio show hosted by Sandy Kenyon of ABC News. Thanks to documentary filmmaker Rory Kennedy—daughter of RFK and close friend—for bringing this to my attention.

597 "I certainly remember it": Douglas Brinkley, interview with Neil Armstrong, September 19, 2001, in Clear Lake City, Texas.

598 "encapsulates all of recorded history": Jade Boyd, "JFK's 1962 Moon Speech Still Appeals 50 Years Later," Rice University News & Media, August 30, 2008, http://news.rice.edu/2012/08/30/jfks-1962-moon-speech-still-appeals-50-years-later/.

598 "blew me away" . . . "I came away": Ibid.

598 The astronauts escorted the president: Cannon, "Owls in Space," p. 33.

600 within the decade: Warren Burkett, "Kennedy Pleased with Houston Space Briefing," *Houston Chronicle*, September 13, 1962.

600 "To talk of placing": John F. Kennedy, "Remarks at the NASA Manned Spacecraft Center in Houston," September 12, 1962, Gerhard Peters and John T. Woolley, The American Presidency Project, https://www.presidency.ucsb.edu/documents/remarks-the-nasa-manned-spacecraft-center-houston.

601 "By all means, we must carry on": Dwight D. Eisenhower, "Are We Headed in the Wrong Direction," *Saturday Evening Post*, August 11, 1962, p. 24.

603 Kennedy needed Eisenhower's . . . support: E. W. Kenworthy, "Eisenhower Tells Kennedy of Tour," *New York Times*, September 11, 1962, p. 1.

604 "Every citizen of this country": Tom Yarborough, "Kennedy Visit," *St. Louis Post-Dispatch*, September 13, 1962, p. 5.

605 contractor on the Apollo lunar module: Richard Thruelsen, *The Grumman Story* (New York: Praeger, 1976).

605 "By forceful implication": Alvin Spivak, "Pledge to Beat Russia to Moon," *St. Louis Post-Dispatch*, September 13, 1962, p. 1.

606 "all of us are so committed to the sea": John F. Kennedy, "Remarks on Australian Ambassador's Dinner for America's Cup Race," Newport, Rhode Island, September 14, 1962, item JFKPOF-040-005, Papers of John F. Kennedy, Presidential Papers, President's Office Files, Speech Files, Kennedy Library.

608 "If at any time the Communist build-up": Press Conference, September 13, 1962, item JFKPOF-057-012, Papers of John F. Kennedy, Presidential Papers, President's Office Files, Kennedy Library.

18: Gemini Nine and Wally Schirra

612 Gemini's launch schedule: Project Gemini: A Chronology, Part 1 (B), "Concept and Design," NASA Goddard Space Flight Center, Greenbelt, Maryland, https://history.nasa.gov/SP-4002/p1b.htm.

612 "NASA was not looking for": Colin Burgess, *Moon Bound: Choosing and Preparing NASA's Lunar Astronauts* (New York: Springer Praxis Books, 2013), p. 54.

613 included six civilians: "2d Generation of Spacemen Make World Debut in Houston," *Houston Chronicle*, September 17, 1962, p. 1.

613 about two years younger: "Space Voyagers Rarin' to Orbit," *Life*, April 20, 1959, p. 22.

613 "born under the second law": Neil Armstrong, "The Engineered Century," March 1, 2000, National Academy of Engineering, www.nae.edu/publications/bridge/thevertiginousmarchoftechnology/theengineeredcentury.aspx.

614 Slayton instructed Armstrong to report to Houston: Hansen, *First Man*, p. 202.

615 "In the opinion of individuals": Ibid., pp. 207–8.

615 The Gemini Nine were the lucky test pilots: Grissom quoted in French and Burgess, *In the Shadow of the Moon*, p. 2.

615 from $530 million to $745 million: Eugene Reichl, *Project Gemini* (Atglen, PA: Schiffer, 2013), p. 18.

617 "Many women are employed": O. B. Lloyd Jr. to Susan Marie Scott, June 18, 1962, University of Houston—Clear Lake, NASA Archive Collection, No. 2018-0001, Records of NASA Johnson Space Center, Library and Archives of the University of Houston, Clear Lake, Clear Lake City, Texas.

617 "NASA did not state gender": French and Burgess, *Into That Silent Sea*, p. 202.

619 They graduated with flying colors: Dianna Wray, "The Real Story of NASA's First Female Astronauts," *Miami New Times*, September 19, 2017, www.miaminewtimes.com/content/printView/9681122. See also Margaret Weitekamp, *Right Stuff, Wrong Sex: America's First Women in Space Program* (Baltimore: Johns Hopkins University Press, 2004).

620 Some of the choice headlines: Roger Launius to Douglas Brinkley, November 5, 2018.

620 "I think this gets back to the way": Quoted in Martha Ackmann, *The Mercury 13: The True Story of Thirteen Women and the Dream of Space Flight* (New York: Random House, 2004), p. 168.

622 $5,147, to $3,283: Andrew Cohen, *Two Days in June: John F. Kennedy and the 48 Hours That Made History* (Toronto: McClelland and Stewart, 2014), p. 62.

622 The idea of an African American: Roger D. Launius, *The Smithsonian History of Space Exploration: From the Ancient World to the Extraterrestrial Future* (Washington, DC: Smithsonian Books, 2018), p. 142.

625 "My rambunctious approach": Quoted in French and Burgess, *Into That Silent Sea*, p. 232.

625 "Not a fancy name": Ibid.

626 "In mission control, I winked": Kraft, *Flight*, p. 178.

626 "The President was always extremely interested": Richard Witkin, "Schirra Orbits Earth Six Times, Landing Near Carrier in Pacific After Almost Flawless Flight," *New York Times*, October 4, 1962, p. 24.

627 "I ate and I wasn't hungry": "'Astronauts Will Eat, Sleep by Numbers'—Schirra," *Houston Chronicle*, October 8, 1962, clipping, Fondrun Library, Rice University.

627 "would have turned a robot green": Shepard and Slayton, with Barbree, *Moon Shot*, p. 157.

627 still paled in comparison: Witkin, "Schirra Orbits Earth Six Times," p. 1.

628 At a press conference: Transcript of Press Conference for Walter Schirra, Rice University, October 8, 1962, Fondrum Library, NASA Archives, Rice University, Houston.

628 At 9:25 a.m.: Ernest May and Philip R. Zelikow, eds., *The Kennedy Tapes* (New York: William Morrow, 2002), p. 31.

628 "I know who you are": Jerry T. Baulch, "It's Back to Work Now," *Pittsburgh Post-Gazette*, October 17, 1962, p. 2.

629 "Why would the Soviets": May and Zelikow, *The Kennedy Tapes*, p. 49.

630 hundred-dollar-a-seat fund-raiser: George Tagge, "Kennedy Plugs for Yates," *Chicago Tribune*, October 20, 1962, p. 1.

631 "the space program was understandably preoccupied": Krantz, *Failure Is Not an Option*, p. 94.

632 "how much bad advice": Sheldon M. Stern, *The Week the World Stood Still: Inside the Secret Cuban Missile Crisis* (Palo Alto, CA: Stanford University Press, 2005), p. 217.

634 put the problem succinctly: "Space Goals Put Strain on Budget," *New York Times*, November 5, 1962, pp. 1, 6.

635 "keep me thinking of the taxpayers' money": "Reaching for the Moon," *Time*, August 10, 1962, p. 54.

635 "one which we intend to win": "Address at Rice University, Houston, Texas, 12 September 1962," p. 3.

637 "programs for scientific investigations in space": W. D. Kay, *Defining NASA: The Historical Debate over the Agency's Mission* (Albany: State University of New York Press, 2012), p. 82.

638 "No, sir, I do not": "Transcript of Presidential Meeting in the Cabinet Room of the White House: Supplemental Appropriations for the National Aeronautics and Space Administration (NASA), November 21, 1962," tape 63, Papers of John F. Kennedy, Presidential Papers, President's Office Files, Presidential Recordings collections, Kennedy Library.

640 "The people that are going to furnish": Ibid.

641 "I would certainly not favor": Ibid.

643 "We ought to get it": Ibid.

644 "In Berlin you spent": Ibid.

19: State of Space Exploration

649 "an extraordinary technical accomplishment": "Remarks on Presentation of Mariner II Model," January 17, 1963, JFKPOF042-024-p0001, Papers of John F. Kennedy, Presidential Papers, President's Office Files, Speech Files, Kennedy Library.

650 The overall earmark: Michio Kaku, *The Future of Humanity: Terraforming Mars, Interstellar Travel, Immortality, and Our Destiny Beyond Earth* (New York: Doubleday, 2018), p. 30.

650 New Frontier budget as "austere": Howard Simons, "Increase of $2 Billion Asked in Space Funds," *Washington Post*, January 18, 1963, p. A12.

651 "maintain a position of world leadership": The Budget Message of the President, January 1963, p. 18, item JFKPOF-071-006, Papers of John F. Kennedy, Presidential Papers, President's Office Files, Legislative Files, Kennedy Library.

651 "Efforts are being concentrated": Ibid., p. 19.

652 Former president Eisenhower: Mieczkowski, *Eisenhower's Sputnik Moment*, p. 268.

652 He even went so far: Arthur Krock, "Mr. Kennedy's Management of the News," *Fortune*, March 1963, pp. 198–202.

652 "We have had limited success": Associated Press, "News Tinted by Kennedy Says Krock," *Des Moines Register*, February 24, 1963, p. 5. See also Krock, "Mr. Kennedy's Management of the News," p. 82.

653 "The official [White House] release of information": Arthur Krock, "Mr. Kennedy's Management of the News," *Fortune*, March 1963.

654 "The eyes of all ages": Arthur C. Clarke, "Space Flight and the Spirit of Man," *Reader's Digest*, February 1962.

655 "a race of education and research": Hal Willard, "Kennedy Calls for Federal School Aid," *Washington Post*, October 11, 1957, p. 1.

655 confident that advanced computer technology: Leslie Berlin, *Troublemakers: Silicon Valley's Coming of Age* (New York: Simon and Schuster, 2017), p. 11.

656 "Remember when NASA": Ibid., p. 23.

657 conduct a thorough review: John F. Kennedy to Lyndon B. Johnson, April 9, 1963, and Lyndon B. Johnson to John F. Kennedy, May 13, 1963, John F. Kennedy Presidential Files, NASA Historical Reference Collection, NASA Headquarters, Washington, DC.

657 BENEFITS TO NATIONAL ECONOMY: Lyndon B. Johnson to the President, May 13, 1963, with attached report, John F. Kennedy Presidential Files, NASA Historical Reference Collection, NASA Headquarters, Washington, DC.

660 "That's just wonderful": Quoted in Ward, *Dr. Space*, p. 132.

661 from his White House bedroom: "President Watches," *Philadelphia Inquirer*, May 16, 1963, p. 3.

662 "I had to initiate retrofire": L. Gordon Cooper Jr., "Astronaut's Summary Flight Report," in Walter C. Williams, Kenneth S. Kleinknecht, William M. Bland Jr., and James E. Bost, eds., *Mercury Project Summary*, NASA Publication SP-45 (Washington, DC: NASA, 1963), p. 356, https://history.nasa.gov/SP-45/contents.htm.

662 "I know that a good many people": John F. Kennedy, "Remarks Upon Presenting the NASA Distinguished Service Medal to Astronaut L. Gordon Cooper," May 21, 1963, at John Woolley and Gerhard Peters, The American Presidency Project, https://www.jfklibrary.org/asset-viewer/archives/JFKWHF/WHN15/JFKWHF-WHN15/JFKWHF-WHN15.

663 "One of the things which warmed us": Ibid.

664 "Help us in our future space endeavors": Carroll Kilpatrick, "250,000 Give Cooper Hero's Welcome," *Washington Post*, May 22, 1963, p. 1.

665 once back in Houston: Carpenter and Stoever, *For Spacious Skies*, pp. 310–11.

665 "other aspects of human needs": "Critics Urge Slowdown in U.S. Moon Program," *Asbury Park* (NJ) *Press*, July 1, 1963, p. 6.

666 "To allow the Soviet Union": Ibid.

666 "the United States space program is receiving": Howard Simons, "Space Program Being Scrutinized by Budget-Minded Congress," *Washington Post*, July 1, 1963, p. A2.

667 Webb warned that cuts: Roulhac Hamilton, "House Group Cuts NASA Drastically," *Orlando Sentinel*, July 10, 1963, p. 2.

667 "will lead a great America": No. 3 Cong. Rec. H13906 (August 1, 1963) (statement of Rep. Fulton).

668 "We cannot say definitely": "The Soviet Space Program," Central Intelligence Agency, National Intelligence Estimate 11-1-62, December 5, 1962, p. 6.

670 "in a state of perfect inertia": Max Frankel, "Test-Ban Hopes Linger," *New York Times*, May 19, 1963, p. E4.

671 "it would provide insurance": Paul B. Stares, *The Militarization of Space* (Ithaca, NY: Cornell University Press, 1985), p. 88.

672 "examine our attitude toward peace itself": John F. Kennedy, "Commencement Address at American University in

Washington: June 10, 1963," at Woolley and Peters, The American Presidency Project, https://www.presidency.ucsb.edu/documents/commencement-address-american-university-washington.

673 "the best speech by any president since Roosevelt": Taubman, *Khrushchev: The Man and His Era*, p. 602.

674 "freedom is indivisible": "Ich bin ein Berliner" speech, "Remarks of President John F. Kennedy at the Rudolph Wilde Platz, Berlin, June 26, 1963," Kennedy Library, https://www.jfklibrary.org/Research/Research-Aids/JFK-Speeches/Berlin-W-Germany-Rudolph-Wilde-Platz_19630626.aspx.

20: "The Space Effort Must Go On"

678 At the Marshall Space Flight Center: Richard Witkin, "Kennedy Pushes Project Apollo," *New York Times*, July 8, 1962.

679 "the antithesis of fiscal soundness": Dwight D. Eisenhower to Charles Halleck, May 26, 1963, White House Presidents, NASA History Office, Washington, DC.

680 no "substantial military value": Bizony, *The Man Who Ran the Moon*, pp. 101–2.

680 "The man-in-space program": Richard Witkin, "Lunar Program in Crisis," *New York Times*, July 11, 1963.

680 When a top-tier NASA engineer: "Lift-Off! The U.S. Space Program," Kennedy Library, https://www.jfklibrary.org/Exhibits/Permanent-Exhibits/Lift-off-The-US-Space-Program.aspx.

681 "The point of the matter": "News Conference 58," July 17, 1963, Papers of John F. Kennedy, Presidential Papers, President's Office Files, Kennedy Library.

681 "He told me many times": Interview with Sergei N. Khrushchev, *Red Files*, PBS, www.pbs.org/redfiles/moon/deep/interv/m_int_sergei_khrushchev.htm.

682 "interested in an international program": Howard Simons, "JFK Probed Kremlin on Joint Moon Trip," *Washington Post*, September 22, 1963, p. A1.

684 "The bombers were stopped": Hugh Sidey quoted in Andrew Cohen, *Two Days in June: John F. Kennedy and the 48 Hours That Made History* (Toronto: McClelland and Stewart, 2014), pp. 22–23.

686 "It is rarely possible": John F. Kennedy, "Special Message to the Senate on the Nuclear Test Ban Treaty," August 8, 1963, Gerhard Peters and John T. Woolley, The American Presidency Project, https://www.presidency.ucsb.edu/documents/special-message-the-senate-the-nuclear-test-ban-treaty.

687 At a White House meeting: Steven Levingston, *Kennedy and King: The President, the Pastor, and the Battle over Civil Rights* (New York: Hachette Books, 2017), p. 426.

688 "a ray of light appears": Carroll Kilpatrick, "Kennedy's Aides, 6 Other Scientists Join in Endorsing Test-Ban Treaty," *New York Times*, August 25, 1963, p. A1.

690 "We'll have worked to fly by": John F. Kennedy, meeting with James Webb (audiotape), September 18, 1963, Kennedy Tapes, Kennedy Library. See also Ted Widmer, *Lis-*

tening In: The Secret White House Recording of John F. Kennedy (New York: Hyperion, 2012).

692 "this can be an asset": Carolyn Y. Johnson, "JFK Had Doubts About Moon Landing," *Boston Globe*, May 25, 2011.

693 "if we cooperate": John Noble Wilford, "Race to Space, Through the Lens of Time," *New York Times*, May 24, 2011, p. D1.

693 short- and medium-range missiles: "Treaty on Principles Governing the Activities of States in the Exploration and Use of Outer Space, Including the Moon and Other Celestial Bodies," January 27, 1967, Bureau of Arms Control, Verification and Compliance, U.S. State Department, www.state.gov/t/isn/5181.htm.

696 "In a field where": John F. Kennedy, "Address Before the 18th General Assembly of the United Nations," September 20, 1963, at John Woolley and Gerhard Peters, Kennedy Library, https://www.jfklibrary.org/archives/other-resources/john-f-kennedy-speeches/united-nations-19630920.

696 "were just exercises in image-building": Wilford, "Race to Space Through the Lens of Time," p. D1.

697 "We have received no response": John F. Kennedy Press Conference, October 9, 1963, Kennedy Library, https://www.jfklibrary.org/Asset-Viewer/Archives/JFKPOF-061-003.aspx.

699 that Khrushchev was "unquestionably planning": Bureau of Scientific Intelligence, CIA, "A Brief Look at the Soviet

Space Program," October 1, 1963, National Security File, Box 308, Kennedy Library.

700 Unbeknownst to the CIA: Logsdon, *John F. Kennedy and the Race to the Moon*, p. 179.

700 "If the United States ever experiences": Arthur Krock, "The Intra-Administration War in Vietnam," *New York Times*, October 3, 1963, p. 34.

701 "I don't know of any technical problem": John W. Finney, "Apollo Capsules Now Being Built," *New York Times*, October 13, 1963, p. 80.

703 "What could be better": Theodore Shabad, "Russians Report Launching Craft That Shifts Orbit," *New York Times*, November 2, 1963, p. 1.

703 "an homage": Douglas Brinkley, *Dean Acheson: The Cold War Years, 1953–71* (New York: Yale University Press, 1992), p. 172.

704 create proposals on accommodating: National Security Action Memorandum No. 271, November 12, 1963, Papers of John F. Kennedy, Presidential Papers, National Security Files, Meetings and Memoranda Series: National Security Action Memoranda, Kennedy Library.

21: Cape Kennedy

708 biggest payload that any nation: Richard Witkin, "First U.S. Hydrogen-Fueled Rocket Is Orbited," *New York Times*, November 28, 1963, p. 9.

708 "began to realize the dimensions": John M. Logsdon, "An-

alyzing the New Kennedy Tape," *Space Review*, May 31, 2011, www.thespacereview.com/article/1856/1.

709 "the myths of the future": Clarke, "Space Flight and the Spirit of Man," p. 78.

709 "We gave him a first-class": Gordon Cooper with Bruce Henderson, *Leap of Faith: An Astronaut's Journey into the Unknown* (New York: HarperCollins, 2000), p. 131.

710 "Now, be sure": Robert Seamans, *Aiming at Targets: The Autobiography of Robert C. Seamans, Jr.* (Charleston, SC: CreateSpace Independent Publishing Platform, 2012), p. 115.

712 "Many Americans": Kennedy, "Remarks at Aero-Space Medical Health Center Dedication, San Antonio, Texas, 21 November 1963," Papers of John F. Kennedy, Presidential Papers, President's Office Files, Speech Files, Kennedy Library.

712 The term *space medicine*: Since 1976, NASA has annually published *Spinoff*, a handsome publication featuring technological innovations from space research. I read all copies to glean the technologies that were most viable. There is a profusion of mythology regarding the medical advances NASA did or didn't innovate.

713 "Examinations of the astronauts' physical": Kennedy, "Remarks at Aero-Space Medical Health Center Dedication, San Antonio, Texas, 21 November, 1963."

714 "how, as a boy": "Remarks at Aero-Space Medical Health Center Dedication, San Antonio, Texas, 21 November 1963."

715 "We were at Brooks Air Force Base": Cooper, with Henderson, *Leap of Faith*, p. 159.

716 "It would be a disgrace": "Harris GOP Chief Urges Cordiality," *Houston Chronicle*, November 21, 1963, p. 9.

717 the NASA connection had lured corporations: Cody C. Stanley, "Albert Thomas: Space in the Bayou," MA thesis, January, 2016, Stephen F. Austin State University, Nacogdoches, Texas, https://library.sfasu.edu/find/SummonRecord/FETCH-proquest_journals_18346716083.

718 By the time the Manned Spacecraft Center: Kevin M. Brady, "NASA Launches Houston into Orbit: The Economic and Social Impact of the Space Agency on Southeast Texas, 1961–1969," in Dick and Launius, eds., *Societal Impact of Spaceflight*, pp. 451–65.

718 "He has helped steer this country": "Remarks at Representative Albert Thomas Dinner, Houston Coliseum, Texas, 21 November 1963," Papers of John F. Kennedy, Presidential Papers, President's Office Files, Speech Files, Kennedy Library.

719 "[Kennedy] and I had a big argument": Quoted in Dallek, "Johnson, Project Apollo, and the Politics of Space Program Planning," p. 73.

720 "We have regained the initiative": "Undelivered Remarks for Dallas Citizens Council, Trade Mart, Dallas, Texas, 22 November 1963," Papers of John F. Kennedy, Presidential Papers, President's Office Files, Speech Files, Kennedy Library.

721 "a mecca for those who see": Saul Friedman, "Dallas

Frame of Mind Has Led to Outbreaks of Violence Before," *Houston Chronicle*, November 22, 1963, p. 6.

722 "He wanted to go to the Cape": Olin E. Teague to NASA historian Eugene Emme, January 24, 1979, NASA Historical Reference Collection, NASA Headquarters, Washington, DC. See also Clarke, *JFK's Last Hundred Days*, p. 34.

722 On a more somber note: Dickson, *Sputnik*, 55.

723 "That was a bad day": Dr. Robert Gilruth, oral history interview with the National Air and Space Museum, February 27, 1987, Washington, DC, https://airandspace.si.edu/research/projects/oral-histories/TRANSCPT/GILRUTH5.HTM.

723 "I called Bobby and Ethel": Glenn and Taylor, *John Glenn*, p. 395.

724 "We loved John": Douglas Brinkley interview with Ethel Kennedy, December 14, 2018.

725 Over the decades: Author interview with John Glenn, November 1, 2004, Columbus, Ohio.

726 "He was devastated": Thompson, *Light This Candle*, pp. 294–96.

726 "Before the assassination": Carpenter and Stoever, *For Spacious Skies*, pp. 317–18.

728 "everything will be different": Taubman, *Khrushchev: The Man and His Era*, pp. 603–5.

728 from Pensacola to Houston: Hansen, *First Man*, pp. 223–24.

728 a "50-50 chance": Paul Gallagher, "Neil Armstrong Last Interview: Rare Glimpse of Man and Moon Mission,"

Guardian, August 25, 2012, https://www.guardian.com/science/2012/aug/25/neil-armstrong-last-interview.

729 "that's going to be forgotten": Jacqueline Kennedy Onassis, oral history interview, January 11, 1974, pp. 7–8, John F. Kennedy Oral History Program, Kennedy Library.

729 would be renamed the John F. Kennedy Space Center: Cabell Phillips, "Canaveral Space Center Renamed Cape Kennedy," *New York Times*, November 29, 1963, p. 1.

730 "The loss of John Kennedy": Cooper, with Henderson, *Leap of Faith*, p. 131.

731 "make sure Mr. Webb": Bonnie Holmes/Wernher von Braun office diary, December 9, 1963, in private collection.

731 the "one time I ever saw": "Behind the Scenes with von Braun," *Huntsville Times*, September 8, 1994, p. 1.

732 achieved a major milestone: Witkin, "First U.S. Hydrogen-Fueled Rocket."

epilogue: the triumph of apollo 11

736 "We cannot be the first": Lyndon B. Johnson speech, October 25, 1964, Atlantic University, Boca Raton, Florida, www.fau.edu/fiftieth/speech.php.

736 Johnson evoked the martyred JFK: Roger D. Launius "What Are Turning Points in History, and What Were They for the Space Age?" in Dick and Launius, eds., *Societal Impacts of Spaceflight*, pp. 34–35.

737 "Ten times in this program": "Statement to the President Following the Completion of the Final Flight in the Gemini

Program," November 15, 1966, Public Papers of Lyndon B. Johnson.

738 "do something for grandma with medicine": Roger D. Launius, "Interpreting the Moon Landings: Project Apollo and the Historians," *History and Technology* 22, no. 3 (September 2006): 225–55.

739 "We've always known": Webb quoted in Dickson, *Sputnik*, p. 219.

739 Newspaper stories about the *Apollo 1*: Burgess, *Liberty Bell 7: The Suborbital Mercury Flight of Virgil I. Grissom*, p. 219.

739 By October 1968 President Johnson had grown weary: W. Henry Lambright, *Powering Apollo: James E. Webb of NASA* (Baltimore: Johns Hopkins University Press, 1995).

740 Webb was right: Rod Pyle, *Destination Moon: The Apollo Missions in the Astronauts' Own Words* (New York: HarperCollins/Smithsonian, 2005), 8.

741 "The world of space holds vast promise": "James E. Webb" (biography), International Space Hall of Fame, New Mexico Museum of Space History, Alamogordo, New Mexico, www.nmspacemuseum.org/halloffame/detail.php?id=122.

741 "drastically revised and expanded": Dwight D. Eisenhower to Frank Borman, June 18, 1965. Dwight D. Eisenhower Presidential Library, Abilene, Kansas. Also in Yanek Mieczkowski, *Eisenhower's Sputnik Moment: The Race for Space and World Peace*, p. 268.

742 the New York World's Fair: McDougall, . . . *the Heavens and the Earth*, p. 399.

743 engraved the initials JFK: Chris Matthews, *Jack Kennedy: Elusive Hero* (New York: Simon and Schuster, 2011), p. 402.

746 "In our elation" . . . "I so thank you": Wernher von Braun to Jacqueline Kennedy, February 1, 1964, and Jacqueline Kennedy to Wernher von Braun, February 11, 1964, both in Correspondence file, Wernher von Braun Library and Archives, U.S. Space and Rocket Center, Huntsville, Alabama.

747 Von Braun's Huntsville team: Andrew Duna and Stephen Waring, *Power to Explore: History of Marshall Space Flight Center 1960–1990* (Washington, DC: NASA, 2018), p. 3.

747 "Von Braun made a Faustian bargain": Michael J. Neufeld, "Wernher von Braun, the SS, and Concentration Camp Labor: Questions of Moral, Political, and Criminal Responsibility," *German Studies Review* 25, no. 1 (February 2002). The quote is from a published letter he wrote defending his interpretation on December 2002, which appeared in "Wernher von Braun and Concentration Camp Labor: An Exchange," *German Studies Review* 26, no. 1 (2003).

749 an "act of graciousness": Stephen Bull to H. R. Haldeman, June 13, 1969, Memorandum, Richard Nixon Presidential Library, Yorba Linda, California. See also John Logsdon, *After Apollo?: Richard Nixon and the American Space Program* (New York: Palgrave Macmillan, 2015).

749 refused to evoke: John Logsdon, ed., *The Penguin Book* *of Outer Space Exploration* (New York: Penguin, 2018), p. 236.

750 "Houston . . .": "One Small Step . . . A Giant Leap," *Houston Magazine*, September 1969, p. 18.

751 "I am absolutely isolated": Collins, *Carrying the Fire*, p. 402.

751 The 528 million moon-mad global citizens: Wilford, *We Reach the Moon*, p. xvii.

752 "Task Accomplished July 1969": Logsdon, *John F. Kennedy and the Race to the Moon*, p. 223.

Bibliography

Abbas, Khwaja Ahmad. *Till We Reach the Stars: The Story of Yuri Gagarin.* New York: Asia Publishing House, 1961.

Ackmann, Martha. *The Mercury 13: The Untold Story of Thirteen American Women and the Dream of Space Flight.* New York: Random House, 2003.

Aldrin, Buzz, with Ken Abraham. *Magnificent Desolation: The Long Journey Home from the Moon* (New York: Three Rivers Press, 2010).

Aldrin, Buzz, and Malcolm McConnell. *Men from Earth.* New York: Bantam, 1989.

Arendt, Hannah. *The Human Condition.* Chicago: University of Chicago Press, 1958.

The Astronauts Book. Norwich, UK: Panther, 1966.

Armstrong, Neil A., et al. *First on the Moon: A Voyage with Neil Armstrong, Michael Collins, and Edwin E. Aldrin, Jr.* With Gene Farmer and Dora Jane Hamblin. Boston: Little, Brown, 1970.

Axelrod, Alan. *Lost Destiny: Joe Kennedy Jr. and the Doomed WWII Mission to Save London.* New York: Palgrave Macmillan, 2015.

Baker, David. *The History of Manned Spaceflight.* New York: Crown, 1981.

Bartos, Adam, and Svetlana Boym. *Kosmos: A Portrait of the Russian Space Age.* New York: Princeton Architectural + PHS, 2001.

Bell, Joseph N. *Seven into Space: The Story of the Mercury Astronauts.* Chicago: Popular Mechanics, 1960.

Benson, Charles, and William B. Faherty. *Gateway to the Moon: Building the Kennedy Space Center Launch Complex.* Gainesville: University Press of Florida, 2001.

Bergaust, Erik. *Wernher von Braun.* Lanham, MD: Stackpole, 1976.

Beschloss, Michael R. *The Crisis Years: Kennedy and Khrushchev, 1960–1963.* New York: HarperCollins, 1991.

———. *Mayday: Eisenhower, Khrushchev, and the U-2 Affair.* New York: Harper and Row, 1986.

Bilstein, Roger E. *Stages to Saturn: A Technological History of the Apollo/Saturn Launch Vehicles.* NASA SP-4206. Washington, DC: Government Printing Office, 1980.

Bizony, Piers. *The Man Who Ran the Moon: James E. Webb, NASA, and the Secret History of Project Apollo.* New York: Thunder's Mouth Press, 2006.

Boomhower, Ray E. *Gus Grissom: The Lost Astronaut.* Indianapolis: Indiana Historical Society Press, 2004.

Borisenko, Ivan, and Alexander Romanov. *Where All Roads into Space Begin.* Moscow: Progress Publishers, 1982.

Boulle, Pierre. *The Bridge over the River Kwai.* New York: Presidio Press, 1982.

Boyne, Walter J. *Beyond the Wild Blue: A History of the U.S. Air Force, 1947–1997.* New York: St. Martin's, 1997.

Bredhoff, Stacey. *Moonshot: JFK and Space Exploration.* Washington, DC: Foundation for the National Archives, 2009.

Brinkley, Douglas. *Cronkite.* New York: HarperCollins, 2012.

Brooks, Courtney G., James M. Grimwood, and Loyd S. Swenson Jr., *Chariots for Apollo: The*

NASA History of Manned Lunar Spacecraft to 1969. NASA SP-4205. Washington, DC: Government Printing Office, 1979.

Burchett, Wilfred, and Anthony Purdy. *Gagarin: First Man in Space.* London: Panther, 1961.

———. *Gherman Titov's Flight into Space.* London: Panther, 1962.

Burgess, Colin. *Liberty Bell 7: The Suborbital Mercury Flight of Virgil I. Grissom.* Cham, Switzerland: Springer, 2014.

Burgess, Colin, Kate Doolan, and Bert Vis. *Fallen Astronauts: Heroes Who Died Reaching for the Moon.* Lincoln: University of Nebraska Press, 2003.

Burrows, William E. *This New Ocean: The Story of the First Space Age.* New York: Random House, 1999.

Burrows, William E., Gordon Cooper, John Glenn, Virgil Grissom, Walter Schirra, Alan Shepard, and Donald Slayton. *We Seven: By the Astronauts Themselves.* New York: Simon and Schuster, 1962.

Buss, Jared. *Willy Ley: Prophet of the Space Age.* Gainesville: University Press of Florida, 2017.

Cadbury, Deborah. *Space Race: The Epic Battle Between America and the Soviet Union for the Domination of Space.* New York: HarperPerennial, 2007.

Caidin, Martin. *The Astronauts: The Story of Project Mercury.* New York: E. P. Dutton, 1961.

Caro, Robert. *The Passage of Power: The Years of Lyndon Johnson.* New York: Alfred A. Knopf, 1982.

———. *Man into Space.* New York: Pyramid, 1961.

———. *Rendezvous in Space.* New York: E. P. Dutton, 1962.

Carpenter, Scott, and Kristen Stoever. *For Spacious Skies.* Waterville, ME: Thorndike, 2002.

Chaikin, Andrew. *A Man on the Moon.* New York: Penguin, 1994.

Chappell, Carl L. *Seven Minus One.* Madison, IN: New Frontier, 1968.

Clark, Phillip. *The Soviet Manned Space Program.* New York: Salamander, 1988.

Clarke, Thurston. *JFK's Last Hundred Days.* London: Penguin, 2014.

Clifford, Clark. *Counsel to the President.* New York: Anchor, 1992.

Cobb, Jerrie. *Solo Pilot.* Sun City Center, FL: Jerrie Cobb Foundation, 1997.

Cobb, Jerrie, and Jane Rieker. *Woman into Space: The*

Jerrie Cobb Story. Englewood Cliffs, NJ: Prentice Hall, 1963.

Cochran, Jacqueline. *The Stars at Noon.* Boston: Little, Brown, 1954.

Cochran, Jacqueline, and Maryann Bucknum Brinley. *Jackie Cochran: The Story of the Greatest Woman Pilot in Aviation History.* New York: Bantam, 1987.

Coffey, Thomas M. *Iron Eagle: The Turbulent Life of General Curtis LeMay.* New York: Crown, 1986.

Cohen, Andrew. *Two Days in June: John F. Kennedy and the 48 Hours That Made History.* Toronto: McClelland and Stewart, 2014.

Collins, Michael. *Carrying the Fire: An Astronaut's Journeys.* New York: Farrar, Straus and Giroux, 1974.

———. *Liftoff: The Story of America's Adventure in Space.* New York: Grove, 1988.

Compton, William David. *Where No Man Has Gone Before: A History of Apollo Lunar Exploration Missions.* NASA SP-4214. Washington, DC: Government Printing Office, 1989.

Cooper, Gordon, with Bruce Henderson. *Leap of Faith: An Astronaut's Journey into the Unknown.* New York: HarperCollins, 2000.

Cormier, Zeke, Barrett Tillman, Wally Schirra, and

Phil Wood. *Wildcats to Tomcats: The Tailhook Navy.* St. Paul, MN: Phalanx, 1995.

Crim, Brian. *Our Germans: Operation Paperclip and the National Security State.* Baltimore: Johns Hopkins University Press, 2018.

Cronkite, Walter. *A Reporter's Life.* New York: Random House, 1996.

Cunningham, Walter. *The All-American Boys: An Insider's Look at the U.S. Space Program.* New York: Macmillan, 1977.

Dallek, Robert. *Camelot's Court: Inside the Kennedy White House.* New York: Harper, 2013.

Daniels, Patricia. *Astronauts: Past, Present, Future.* Washington, DC: National Geographic, 2017.

———. *An Unfinished Life: John F. Kennedy, 1917–1963.* Boston: Little, Brown, 2003.

Day, Dwayne A., John M. Logsdon, and Brian Latell, eds. *Eye in the Sky: The Story of the CORONA Spy Satellites.* Washington, DC: Smithsonian Institution Press, 1998.

Dethloff, Henry C. *Suddenly, Tomorrow Came: The NASA History of the Johnson Space Center.* NASA SP-4307. Washington, DC: Government Printing Office, 1993.

Diamond, Edwin. *The Rise and Fall of the Space Age.* Garden City, NY: Doubleday, 1964.

Dick, Steven J., and Roger D. Launius, eds. *Societal Impact of Spaceflight.* NASA SP2007-4801. Washington, DC: Government Printing Office, 2007.

Dickson, Paul. *Sputnik: The Shock of the Century.* New York: Walker, 2001.

Divine, Robert A. *The Sputnik Challenge: Eisenhower's Response to the Soviet Satellite.* New York: Oxford University Press, 1993.

Doolittle, James H. *I Could Never Be So Lucky Again.* New York: Bantam, 1994.

Doran, Jamie, and Piers Bizony. *Starman: The Truth Behind the Legend of Yuri Gagarin.* London: Bloomsbury, 1998.

Doyle, William. *PT-109: An American Epic of War, Survival, and the Destiny of John F. Kennedy.* New York: William Morrow, 2015.

Emme, Eugene M. *Aeronautics and Astronautics, 1915–1960.* Columbus, OH: BiblioGov, 2012.

Fallaci, Oriana. *If the Sun Dies.* Trans. Pamela Swinglehurst. New York: Atheneum, 1966.

Fernlund, Kevin J. *Lyndon B. Johnson and Modern America.* Norman: University of Oklahoma Press, 2009.

Freni, Pamela. *Space for Women: A History of Women*

with the Right Stuff. Santa Ana, CA: Seven Locks Press, 2002.

Fries, Sylvia Doughty. *NASA Engineers and the Age of Apollo.* Washington, DC: NASA, 1992.

Furniss, Tim. *Manned Spaceflight Log.* London: Jane's, 1983.

Fursenko, Aleksandr, and Timothy Naftali. *One Hell of a Gamble: Khrushchev, Castro, and Kennedy, 1958–1964.* New York: W. W. Norton, 1997.

Gaddis, John Lewis. *On Grand Strategy.* New York: Penguin, 2018.

Gagarin, Valentin. *My Brother Yuri.* Moscow: Progress Publishers, 1973.

Gagarin, Yuri. *Road to the Stars.* Moscow: Foreign Language Publishing, n.d.

Gagarin, Yuri, and Vladimir Lebedev. *Survival in Space.* New York: Praeger, 1969.

Gibson, Edward G., ed. *The Greatest Adventure: Apollo 13 and Other Space Adventures.* Sydney: C. Pierson, 1994.

Glenn, John, and Nick Taylor. *John Glenn: A Memoir.* New York: Bantam/Random House, 1999.

Glennan, T. Keith. *The Birth of NASA: The Diary of T. Keith Glennan.* NASA SP-4105. Washington, DC: Government Printing Office, 1993.

Goddard, Robert. *The Papers of Robert H. Goddard.* New York: McGraw-Hill, 1970.

Golovanov, Yaroslav. *Korolev: Fakty i Mify* [Korolev: Facts and Myths]. Moscow: Nauka, 1994.

Goodwin, Doris Kearns. *The Fitzgeralds and the Kennedys: An American Saga.* New York: Simon & Schuster, 1987.

Greenberg, David. *Republic of Spin: An Inside History of the American Presidency.* New York: W. W. Norton, 2016.

Griffith, Alison. *The National Aeronautics and Space Act: A Study of the Development of Public Policy.* Washington, DC: Public Affairs Press, 1962.

Grissom, Betty, and Henry Still. *Starfall.* New York: Thomas Y. Crowell, 1974.

Grissom, Virgil. *Gemini! A Personal Account of Man's Venture into Space.* London: Macmillan, 1968.

Groom, Winston. *The Aviators: Eddie Rickenbacker, Jimmy Doolittle, Charles Lindbergh, and the Epic Age of Flight.* Washington, DC: National Geographic, 2013.

Gubarev, Vladimir. *The Man from a Legend.* Moscow: Progress, 1988.

Gurney, Gene, and Clare Gurney. *The Soviet Manned*

Space Programme: Cosmonauts in Orbit. New York: Franklin Watts, 1972.

Halberstam, David. *The Fifties.* New York: Villard, 1993.

Hall, Rex, and David J. Shayler. *The Rocket Men: Vostok and Voskhod, the First Soviet Manned Spaceflights.* Chichester, UK: Praxis, 2001.

Hamish, Lindsay. *Tracking Apollo to the Moon.* London: Springer-Verlag, 2001.

Harford, James. *Korolev: How One Man Masterminded the Soviet Drive to Beat America to the Moon.* New York: John Wiley, 1997.

Harland, David M. *Exploring the Moon: The Apollo Expeditions.* New York: Springer, 2008.

Hartmann, William, Ron Miller, Andrei Sokolov, and Vitaly Myagkov, eds. *In the Stream of Stars: The Soviet/American Space Art Book.* New York: Workman, 1990.

Harvey, Dodd, and Linda Ciccoritti. *U.S.-Soviet Cooperation in Space.* Washington, DC: Center for International Studies/University of Miami, 1974.

Hawthorne, Douglas. *Men and Women of Space.* San Diego: Univelt, 1992.

Heiken, Grant, and Eric Jones. *On the Moon: The Apollo Journals.* New York: Springer, 2007.

Hellman, John. *The Kennedy Obsession: The American Myth of JFK*. New York: Columbia University Press, 1997.

Hilsman, Roger. *To Move a Nation: The Politics of Foreign Policy in the Administration of John F. Kennedy.* Garden City, NY: Doubleday, 1967.

Hodgson, Godfrey. *JFK and LBJ: The Last Two Great Presidents.* New Haven, CT: Yale University Press, 2015.

Holden, Henry M., and Captain Lori Griffith. *Ladybirds II: The Continuing Story of American Women in Aviation.* Mount Freedom, NJ: Black Hawk, 1993.

Holmes, Jay. *Americans on the Moon: The Enterprise of the '60s.* Philadelphia: Lippincott, 1962.

Holt, Nathalia. *Rise of the Rocket Girls: The Women Who Propelled Us, from Missiles to the Moon to Mars.* New York: Little, Brown and Company, 2016.

Hooper, Gordon. *The Soviet Cosmonaut Team.* Vol. 2: *Cosmonaut Biographies.* Suffolk, UK: GRH Publications, 1990.

Johnson, Lyndon Baines. *The Vantage Point: Perspectives of the Presidency, 1963–1969.* New York: Holt, Rinehart and Winston, 1971.

Johnson, Nicholas. *The Soviet Reach for the Moon.* New York: Cosmos Books, 1995.

Kamanin, Nikolai. *The Hidden Cosmos.* Moscow: Infortekst, 1995.

———. *"I Feel Sorry for Our Guys": General N. Kamanin's Space Diaries.* NASA Publication No. TT-21658. Washington, DC: NASA, 1993.

Kelley, Kevin W., ed. *The Home Planet.* Reading, MA: Addison-Wesley; and Moscow: Mir Publishers, 1988.

Kelly, Dr. Fred. *America's Astronauts and Their Indestructible Spirit.* Blue Ridge Summit, PA: Aero, 1986.

Kelly, Thomas J. *Moon Lander: How We Developed the Lunar Module.* Washington, DC: Smithsonian Institution Press, 2001.

Kennedy, Edward M. *True Compass: A Memoir.* New York: Twelve, 2009.

Kevles, Bettyann Holtzmann. *Almost Heaven: The Story of Women in Space.* New York: Basic Books, 2003.

Killian, James R., Jr. *Sputnik Scientists and Eisenhower: A Memoir of the First Special Assistant to the President for Science and Technology.* Cambridge, MA: MIT Press, 1977.

Kistiakowsky, George B. *A Scientist at the White*

House: The Private Diary of President Eisenhower's Special Assistant for Science and Technology. Cambridge, MA: Harvard University Press, 1976.

Kraft, Christopher. *Flight: My Life in Mission Control.* New York: Dutton, 2001.

Kranz, Gene. *Failure Is Not an Option: Mission Control from Mercury to Apollo 13 and Beyond.* New York: Simon and Schuster, 2000.

Krug, Linda. *Presidential Perspectives on Space Exploration: Guiding Metaphors from Eisenhower to Bush.* New York: Praeger, 1991.

Lambright, W. Henry. *Powering Apollo: James E. Webb of NASA.* Baltimore: Johns Hopkins University Press, 1995.

Laney, Monique. *German Rocketeers in the Heart of Dixie: Making Sense of the Nazi Past During the Civil Rights Era.* New Haven, CT: Yale University Press, 2015.

Launius, Roger D. *After Apollo: The Legacy of the American Moon Landings.* New York: Oxford University Press, 2013.

———. *Frontiers of Rocket Exploration.* Santa Barbara, CA: Greenwood, 2004.

———. *The Smithsonian History of Space Exploration from the Ancient World to the Extraterrestrial*

Future. Washington, DC: Smithsonian Books, 2018.

Launius, Roger D., John M. Logsdon, and Robert W. Smith, eds. *Reconsidering Sputnik: Forty Years Since the Soviet Satellite.* Amsterdam: Harwood Academic Publishers, 2000.

Launius, Roger D., and Howard E. McCurdy, eds. *Spaceflight and the Myth of Presidential Leadership.* Urbana: University of Illinois Press, 1997.

Lebedev, L., B. Lukyanov, and A. Romanov. *Sons of the Blue Planet.* New Delhi: NASA/New Delhi Amerind Publishing, 1973.

Lehman, Milton. *This High Man: The Life of Robert H. Goddard.* New York: American Book–Stratford Press, 1963.

Leonov, Alexei. *I Walk in Space.* Moscow: Malysh Publishers, 1980.

———. *The Sun's Wind.* Moscow: Progress, 1977.

Leonov, Alexei, and Vladimir Lebedev. *Space and Time Perception by the Cosmonaut.* Moscow: Mir Publishers, 1971.

Leonov, Alexei, and Andrei Sokolov. *Life Among Stars.* Moscow, 1981.

———. *The Stars Are Awaiting Us.* Moscow: Publisher Unknown, 1967.

Lewis, Richard S. *Appointment on the Moon.* New York: Viking, 1968.

Lincoln, Evelyn. *My Twelve Years with John F. Kennedy.* New York: David McKay, 1965.

Linn, Brian McAllister. *Elvis's Arm: Cold War GIs and the Atomic Battlefield.* Cambridge, MA: Harvard University Press, 2016.

Lipartito, Kenneth, and Orville Butler. *A History of the Kennedy Space Center.* Gainesville: University Press of Florida, 2007.

Logevall, Fredrik. *Choosing War: The Lost Chance for Peace and the Escalation of War in Vietnam.* Berkeley: University of California Press, 1999.

Logsdon, John M. *After Apollo?: Richard Nixon and the American Space Program.* New York: Palgrave Macmillan, 2015.

———. *The Decision to Go to the Moon: The Apollo Project and the National Interest.* Cambridge, MA: MIT Press, 1970.

———. *John F. Kennedy and the Race to the Moon.* New York: Palgrave Macmillan, 2010.

Logsdon, John M., Dwayne Day, and Roger Launius, eds. *Exploring the Unknown: Selected Documents in the History of the U.S. Civil Space Program.* Vol. 1: *Organizing for Exploration.*

NASA SP-4407. Washington, DC: Government Printing Office, 1995.

———. *Exploring the Unknown: Selected Documents in the History of the U.S. Civil Space Program.* Vol. 2: *External Relationships.* NASA SP-4407. Washington, DC: Government Printing Office, 1996.

———. *Exploring the Unknown: Selected Documents in the History of the U.S. Civil Space Program.* Vol. 7: *Human Spaceflight: Mercury, Gemini, and Apollo.* NASA SP-2008-4407. Washington, DC: Government Printing Office, 2008.

Longmate, Norman. *Hitler's Rockets: The Story of the V-2s.* London: Hutchinson, 1985.

Lothian, Lady Antonella. *Valentina, First Woman in Space: Conversations with A. Lothian.* Edinburgh: Pentland Press, 1993.

Mahaffey, James. *Atomic Awakening: A New Look at the History and Future of Nuclear Power.* New York: Pegasus, 2009.

Mailer, Norman. *Of a Fire on the Moon.* Boston: Little, Brown, 1970.

Maranin, I. A., S. Shamsutdinov, and A. Glushko. *Soviet and Russian Cosmonauts, 1960–2000.* Moscow: Novosti Kosmonautika Publishers, 2001.

Matthews, Chris. *Jack Kennedy: Elusive Hero.* New York: Simon and Schuster, 2011.

McAleer, Neil. *Arthur C. Clarke: An Authorized Biography.* Chicago: Contemporary Books, 1992.

McCray, W. Patrick. *Keep Watching the Skies! The Story of Operation Moonwatch and the Dawn of the Space Age.* Princeton, NJ: Princeton University Press, 2008.

McCullough, David. *The Wright Brothers.* New York: Simon and Schuster, 2015.

McCurdy, Howard E. *Inside NASA: High Technology and Organizational Change in the U.S. Space Program.* Baltimore: Johns Hopkins University Press, 1993.

———. *Space and the American Imagination.* Washington, DC: Smithsonian Instution Press, 1997.

McCutchan, Ann. *Where's the Moon?* College Station: Texas A&M University Press, 2016.

McDonnell, Virginia B. *Dee O'Hara, Astronauts' Nurse.* New York: Thomas Nelson, 1965.

McDougall, Walter, A. *. . . the Heavens and the Earth: A Political History of the Space Age.* New York: Basic Books, 1985.

Medvedev, Roy, and Zhores A. Medvedev.

Khrushchev: The Years in Power. New York: W. W. Norton, 1978.

Mieczkowski, Yanek. *Eisenhower's Sputnik Moment: The Race for Space and World Prestige.* Ithaca, NY: Cornell University Press, 2013.

Miller, Merle. *Lyndon: An Oral Biography.* New York: G. P. Putnam's Sons, 1980.

Mindell, David A. *Digital Apollo: Human and Machine in Spaceflight.* Cambridge, MA: MIT Press, 2008.

Murphy, Thomas P. *Science, Geopolitics, and the Federal Spending.* Lexington, MA: D. C. Heath, 1971.

Murray, Charles, and Catherine Bly Cox. *Apollo: Race to the Moon.* New York: Simon and Schuster, 1989.

Nasaw, David. *The Patriarch: The Remarkable Life and Turbulent Times of Joseph P. Kennedy.* New York: Penguin Press, 2012.

Neal, Valerie, Cathleen Lewis, and Franklin Winter. *Spaceflight: A Smithsonian Guide.* New York: Macmillan, 1995.

Nelson, Craig. *Rocket Men: The Epic Story of the First Men on the Moon.* London: John Murray Publishers, 2009.

Neufeld, Jacob. *The Development of Ballistic Missiles*

in the United States Air Force, 1945–1960. Washington, DC: Office of Air Force History, United States Air Force, 1990.

Neufeld, Michael J. *Reich: The Rocket and the Peenemünde and the Coming of the Ballistic Missile Era.* Washington, DC: Smithsonian Books, 1995.

———. *Von Braun: Dreamer of Space, Engineer of War.* New York: Alfred A. Knopf, 2007.

Neufeld, Michael J., ed. *Milestones of Space: Eleven Iconic Objects from the Smithsonian National Air and Space Museum.* Minneapolis: Zenith Press, 2014.

Neustadt, Richard E. *Presidential Power: The Politics of Leadership.* New York: Wiley, 1960.

Newport, Curt. *Lost Spacecraft: The Search for Liberty Bell 7.* Burlington, ON: Apogee, 2002.

Nichols, Wallace. *Blue Mind: The Surprising Science That Shows How Being Near, in, on, or Under Water Can Make You Happier, Healthier, More Connected, and Better at What You Do.* New York: Little, Brown, 2014.

Nolen, Stephanie. *Promised the Moon: The Untold Story of the First Women in the Space Race.* New York: Four Walls Eight Windows, 2003.

Oberg, James. *Red Star in Orbit.* New York: Random House, 1967.

O'Brien, Michael. *John F. Kennedy.* New York: St. Martin's, 2005.

O'Donnell, Kenneth P., and David F. Powers with Joe McCarthy. *"Johnny, We Hardly Knew Ye": Memories of John Fitzgerald Kennedy.* Boston: Little, Brown, 1972.

O'Leary, Brian. *The Making of an Ex-Astronaut.* Boston: Houghton Mifflin, 1970.

Oliphant, Thomas, and Curtis Wilkie. *The Road to Camelot: Inside JFK's Five-Year Campaign.* New York: Simon and Schuster, 2017.

Ordway, Frederick, III, and Mitchell R. Sharpe. *The Rocket Team.* New York: William Heinemann, 1979.

Outer Space and Man. Moscow: Mir Publishers, 1967.

Parmet, Herbert S. *JFK: The Presidency of John F. Kennedy.* New York: Dial Press, 1983.

Perret, Geoffrey. *Jack: A Life Like No Other.* New York: Random House, 2001.

Pogue, William. *Space Trivia.* Burlington, ON: Apogee, 2003.

Ponomareva, Valentina. *The Female Face of the Cosmos.* Moscow: Gelios Publications, 2002.

Project Mercury. New York: Time Life, 1964.

Pushkin, Alexander. *Eugene Onegin*. London: Penguin Classics, 1977.

———. *The Queen of Spades and Other Stories*. London: Penguin Classics, 1962.

Pynchon, Thomas. *Gravity's Rainbow*. New York: Penguin, 1973.

Reeves, Richard. *President Kennedy: Profile of Power*. New York: Simon and Schuster, 1993.

Renehan, Edward J., Jr. *The Kennedys at War, 1937–1945*. New York: Doubleday, 2002.

Results of the Second U.S. Manned Suborbital Flight. Washington, DC: NASA History Office, 1961.

Riabchikov, Evgeny. *Russians in Space*. New York: Doubleday, 1971.

Romanov, A. *Spacecraft Designer: The Story of Sergei Korolev*. Moscow: Novosti Press, 1976.

Rosholt, Robert L. *An Administrative History of NASA, 1958–1963*. Washington, DC: Government Printing Office, 1966.

Rusk, Dean. *As I Saw It*. New York: W. W. Norton, 1990.

Ryan, Craig. *Sonic Wind*. New York: Liveright/ W. W. Norton, 2015.

Sacknoff, Scott, ed. *In Their Own Words: Conversations with the Astronauts and the*

Men Who Led America's Journey into Space. Bethesda, MD: Space Publications, 2003.

Salinger, Pierre. *With Kennedy.* Garden City, NY: Doubleday, 1996.

Salter, James. *Burning the Days.* New York: Vintage, 1997.

———. *The Hunters.* New York: Vintage, 1999.

Sandler, Martin W., ed. *The Letters of John F. Kennedy.* New York: Bloomsbury, 2013.

Schefter, James. *The Race: The Uncensored Story of How America Beat Russia to the Moon.* New York: Random House, 1999.

Schirra, Wally, and Richard Billings. *Schirra's Space.* Boston: Quinlan Press, 1988.

Schlesinger, Arthur M., Jr. *A Thousand Days: John F. Kennedy in the White House.* Boston: Houghton Mifflin, 1965.

Scott, David, and Alexei Leonov. *Two Sides of the Moon.* New York: Simon and Schuster, 2004.

Seaborg, Glenn T., with the assistance of Benjamin S. Loeb. *Kennedy, Khrushchev, and the Test Ban.* Berkeley: University of California Press, 1981.

Seamans, Robert C., Jr. *Aiming at Targets: The Autobiography of Robert C. Seamans Jr.* NASA SP-4106. Washington, DC: Government Printing Office, 1996.

———. *Project Apollo: The Tough Decisions.* NASA Monographs in Aerospace History no. 31, SP-2005-4537. Washington, DC: Government Printing Office, 2005.

Sellier, André. *A History of the Dora Camp: The Untold Story of the Nazi Slave Labor Camp That Secretly Manufactured V-2 Rockets.* Chicago: Ivan R. Dee, 2003.

Sharpe, Mitchell R. *It Is I, Seagull.* New York: Thomas Y. Crowell, 1975.

Shatalov, Vladimir, and Mikhail Rebrov. *Cosmonauts of the USSR.* Moscow: Prosveshcheniye, 1980.

Shaw, John. *JFK in the Senate: Pathway to the Presidency.* New York: St. Martin's Press, 2013.

Sheehan, Neil. *A Fiery Peace in the Cold War: Bernard Schriever and the Ultimate Weapon.* New York: Random House, 2009.

Shelton, William R. *Man's Conquest of Space.* Washington, DC: National Geographic Society, 1968.

Shepard, Alan, and Deke Slayton, with Jay Barbree. *Moon Shot: The Inside Story of America's Race to the Moon.* Atlanta: Turner Publishing, 1994.

Sherwood, John Darrell. *Officers in Flight Suits: The Story of American Air Force Fighter Pilots in the*

Korean War. New York: New York University Press, 1996.

Schick, Ron, and Julia Van Haaften. *The View from Space: American Astronaut Photography, 1962–1972*. New York: Clarkson Potter, 1988.

Siddiqi, Asif A. *Challenge to Apollo: The Soviet Union and the Space Race, 1945–1974*. NASA SP-2000-4408. Washington, DC: Government Printing Office, 2000.

———. *The Soviet Space Race with Apollo*. Gainesville: University Press of Florida, 2003.

———. *Sputnik and the Soviet Space Challenge*. Gainesville: University Press of Florida, 2003.

Sidey, Hugh. *John F. Kennedy, President*. New York: Atheneum, 1964.

Silverberg, Robert. *First American into Space*. Derby, CT: Monarch, 1961.

Simons, Graham M. *Operation LUSTY: The Race for Hitler's Secret Technology*. Barnsley, UK: Pen and Sword Books, 2016.

Skolnikoff, Eugene B. *Science, Technology, and American Foreign Policy*. Cambridge, MA: MIT Press, 1967.

Slayton, Donald K., and Michael Cassutt. *Deke! U.S. Manned Space: From Mercury to the Shuttle*. New York: Forge Books, 1994.

Smaus, Jewel Spangler, and Charles B. Spangler. *America's First Spaceman.* New York: Doubleday, 1962.

Sobel, Lester A., ed. *Space: From Sputnik to Gemini.* New York: Facts on File, 1965.

Sorensen, Theodore C. *Counselor: A Life at the Edge of History.* New York: HarperCollins, 2008.

———. *Decision-Making in the White House: The Olive Branch or the Arrows.* New York: Columbia University Press, 1963.

———. *Kennedy.* New York: Harper and Row, 1965.

Soviet Man in Space. Moscow: Foreign Languages Publishing House, 1961.

Spires, David N. *Beyond Horizons: A Half Century of Air Force Space Leadership*, rev. ed. Colorado Springs, CO: Air University Press, 1998.

Spudis, Paul D. *The Value of the Moon: How to Explore, Live, and Prosper in Space Using the Moon's Resources.* Washington, DC: Smithsonian Books, 2016.

Steadman, Bernice, and Jody M. Clark. *Tethered Mercury.* Traverse City, MI: Aviation Press, 2001.

Stern, Sheldon M. *The Week the World Stood Still: Inside the Secret Cuban Missile Crisis.* Stanford, CA: Stanford University Press, 2005.

Stone, Tanya Lee. *Almost Astronauts: 13 Women Who Dared to Dream.* Somerville, MA: Candlewick, 2009.

Strober, Deborah Hart, and Gerald S. Strober. *The Kennedy Presidency: An Oral History of an Era.* Washington, DC: Brassey's, 2003.

Suvorov, Vladimir, and Aleksandr Sabelnikov. *The First Manned Spaceflight: Russia's Quest for Space.* New York: Nova Science Publishing, 1999.

Swanson, Glen E., ed. *"Before This Decade Is Out": Personal Reflections on the Apollo Program.* Gainesville: University Press of Florida, 1999.

Swenson, Loyd, Jr., James M. Grimwood, and Charles C. Alexander. *This New Ocean: A History of Project Mercury.* NASA SP-4201. Washington, DC: Government Printing Office, 1966.

Syme, Anthony. *The Astronauts.* Sydney: Horwitz, 1965.

Tereshkova, Valentina. *Stars Are Calling.* Moscow: publisher TK, 1963.

Thompson, Neal. *Light This Candle: The Life and Times of Alan Shepard, America's First Spaceman.* New York: Crown, 2004.

Thompson, Wayne, and Bernard C. Nalty. *Within*

Limits: The U.S. Air Force and the Korean War. Forest Grove, OR: Pacific University Press, 2005.

Titov, Gherman. "An Amazing Art." In *The Soviet Circus: A Collection of Articles.* Moscow: Progress Publishers, 1967.

———. *700,000 Kilometers Through Space: Notes by Soviet Cosmonaut No. 2.* Moscow: Foreign Languages Publishing House, n.d.

Titov, Gherman, with Pavel Barashev and Yuri Dokuchayev. *Gherman Titov: First Man to Spend a Day in Space.* New York: Crosscurrents, 1962.

Titov, Gherman, and Martin Caidin. *I am Eagle!* New York: Bobbs-Merrill, 1962.

Trux, John. *Space Race: From Sputnik to Shuttle.* London: New English Library, 1986.

Tsiolkovsky, Konstantin. *Outside the Earth.* Fairfield, CT: Athena Books, 2006.

Tsymbal, Nikolai, ed. *First Man in Space: The Life and Achievements of Yuri Gagarin.* Moscow: Progress Publishing, 1984.

Tye, Larry. *Bobby Kennedy: The Making of a Liberal Icon.* New York: Random House, 2016.

Tyson, Neil deGrasse. *Space Chronicles: Facing the Ultimate Frontier.* New York: W. W. Norton, 2012.

Tyson, Neil deGrasse, and Avis Lang. *Accessory to War: The Unspoken Alliance Between Astrophysics and the Military.* New York: W. W. Norton, 2018.

U.S. House of Representatives, Committee on Science and Astronautics. *The Endless Frontier: History of the Committee on Science and Technology, 1959–1979.* Washington, DC: Government Printing Office, 1980.

Van Dyke, Vernon. *Pride and Power: The Rationale of the Space Program.* Urbana: University of Illinois Press, 1964.

Vladimirov, Leonid. *The Russian Space Bluff: The Inside Story of the Soviet Drive to the Moon.* London: Tom Stacey, 1971.

Walk into Space. Moscow: Novosti Press Agency Publishing House, 1965.

Ward, Bob. *Dr. Space: The Life of Wernher von Braun.* Annapolis, MD: Naval Institute Press, 2005.

Webb, James E. *Space Age Management: The Large-Scale Approach.* New York: McGraw-Hill, 1969.

Weiss, Otis L., ed. *The United States Astronauts and Their Families.* New York: World Encyclopedia Science Service, 1965.

Weitekamp, Margaret. *Right Stuff, Wrong Sex:*

America's First Women in Space Program. Baltimore: Johns Hopkins University Press, 2004.

Wendt, Guenter, and Russell Still. *The Unbroken Chain.* Burlington, ON: Apogee, 2001.

White, Frank. *The Overview Effect: Space Exploration and Human Evolution.* Boston: Houghton Mifflin, 1987.

Wilford, John Noble. *We Reach the Moon.* New York: Bantam, 1969.

Wolfe, Tom. *The Right Stuff.* New York: Farrar, Straus and Giroux, 1979.

Woods, Randall B. *LBJ: Architect of American Ambition.* New York: Simon and Schuster, 2006.

Zaloga, Steven J. *German V-Weapon Sites, 1943–1945.* London: Osprey Publishing, 2008.

Image Credits

Interior Image Credits

Frontispiece: Buyenlarge/Archive Photos/Getty Images

xviii: I. C. Rapoport/Archive Photos/Getty Images

xxiii: Papers of John F. Kennedy, Presidential Papers, President's Office Files, Courtesy of the John F. Kennedy Presidential Library and Museum, Boston

2: Everett Collection Historical/Alamy Stock Photo

32: Courtesy of the John F. Kennedy Presidential Library and Museum, Boston

52: Hulton Archive/Archive Photos/Getty Images

66: Frank Turgeon Jr., Courtesy of the John F. Kennedy Presidential Library and Museum, Boston

77: MPI/Archive Photos/Getty Images
80: INTERFOTO/Alamy Stock Photo
98: DPA/Getty Images
106: INTERFOTO/Alamy Stock Photo
147: Yale Joel/The LIFE Images Collection/Getty Images
150: Yale Joel/The LIFE Picture Collection/Getty Images
158: Schenectady Museum; Hall of Electrical History Foundation/CORBIS/Getty Images
203: Donald Uhrbrock/The LIFE Images Collection/Getty Images
204: ullstein bild Dtl./ullstein bild/Getty Images
214: QAI Publishing/UIG/Getty Images
254: Courtesy NASA
297: Ralph Morse/The LIFE Picture Collection/Getty Images
313: Ralph Morse/The LIFE Picture Collection/Getty Images
314: Paul Schutzer/The LIFE Picture Collection/Getty Images
339: CBS Photo Archive/Getty Images
342: Obtained from multiple sources
359: Bettmann/Getty Images
364: Popper Ltd./ullstein bild/Getty Images
393: Cecil Stoughton/White House Photographs/

The John F. Kennedy Presidential Library and Museum, Boston

403: Paul Slade/Paris Match/Getty Images

404: Courtesy NASA

434: Bob Gomel/The LIFE Images Collection/Getty Images

461: ©Cornell Capa © International Center of Photography/Magnum Photos

462: Bettmann/Getty Images

501: NASA Archive/Alamy Stock Photo

502: AFP/Getty Images

537: Ralph Morse/The LIFE Picture Collection/Getty Images

545: Ralph Morse/The LIFE Picture Collection/Getty Images

546: Bob Gomel/The LIFE Images Collection/Getty Images

568: Art Rickerby/The LIFE Picture Collection/Getty Images

583: Bob Gomel/The LIFE Images Collection/Getty Images

586: Robert Knudsen. White House Photographs. The John F. Kennedy Presidential Library and Museum, Boston

599: Bob Gomel/The LIFE Images Collection/Getty Images

606: Cecil Stoughton. White House Photographs. The John F. Kennedy Presidential Library and Museum, Boston
610: Ralph Morse/The LIFE Picture Collection/Getty Images
632: Ralph Crane/The LIFE Picture Collection/Getty Images
646: Bob Gomel/The LIFE Images Collection/Getty Images
675: Keystone-France/Gamma-Keystone/Getty Images
676: Bettmann/Getty Images
733: Science & Society Picture Library/SSPL/Getty Images
734: CBS Photo Archive/CBS/Getty Images
737: Consolidated News Pictures/Getty Images
748: Bettmann/Getty Images
753: Ralph Morse/The LIFE Picture Collection/Getty Images

Insert Image Credits

1: Heritage Images/Hulton Archive/Getty Images
2: From *Time*, January 6, 1958 © Time Inc. Used under license. TIME and Time Inc. are not affiliated with, and do not endorse products or services of, Licensee.

2–3: John Bryson/The LIFE Images Collection/Getty Images

4–5: Ralph Morse/The LIFE Picture Collection/Getty Images

6: Ralph Morse/The LIFE Picture Collection/Getty Images

6–7: Bettmann/Getty Images

8: National Archives and Records Administration, Lyndon Baines Johnson Library and Museum, Austin, Texas

8–9: Universal Images Group/Getty Images

10 (top, left): Courtesy of NASA

10 (bottom, left): Ralph Morse/The LIFE Picture Collection/Getty Images

11: Chris Howes/Wild Places Photography/Alamy Stock Photo

12–13: Courtesy of NASA

14: Interim Archives/Archives Photo/Getty Images

15: Ralph Morse/The LIFE Picture Collection/Getty Images

16 (top): NASA/Hulton Archive/Getty Images

16 (bottom): 8383/Gamma-Rapho/Getty Images

About the Author

Douglas Brinkley is the Katherine Tsanoff Brown Chair in Humanities and Professor of History at Rice University, the CNN Presidential Historian, and a contributing editor at *Vanity Fair.* He works in many capacities in the world of public history, including on boards, at museums, at colleges, and for historical societies. The *Chicago Tribune* dubbed him "America's New Past Master." The New-York Historical Society has chosen Brinkley as its official U.S. Presidential Historian. His recent book *Cronkite* won the Sperber Prize, while *The Great Deluge: Hurricane Katrina, New Orleans, and the Mississippi Gulf Coast* received the Robert F. Kennedy Book Award. He has received a Grammy Award for *Presidential Suite* and seven honorary doctorates in American studies. His two-volume, annotated *Nixon Tapes* recently won the Arthur S. Link–Warren F. Kuehl Prize for Documentary Editing. He is a member of the Century Association, the Council on Foreign Relations, and the James Madison Council of the Library of Congress. He lives in Austin, Texas, with his wife and three children.